Collins

Stude

CAMBRIDGE IGCSE™ COMBINED SCIENCE

Malcolm Bradley, Jackie Clegg, Susan Gardner, Sam Goodman, Sarah Jinks, Sue Kearsey, Gareth Price, Mike Smith and Chris Sunley

William Collins' dream of knowledge for all began with the publication of his firstbookin1819. A self-educated mill worker, he not only enriched millions of lives, but also founded a flourishingpublishing house. Today, staying true to this spirit, Collins books are packed with inspiration, innovation and practical expertise. They place you at the centre of a world of possibility and give you exactly what you need to explore it.

Collins. Freedom to teach.

An imprint of HarperCollins*Publishers*
The News Building
1 London Bridge Street
London
SE1 9GF

Browse the complete Collins catalogue at www.collins.co.uk

©HarperCollins*Publishers* Limited 2017

10 9 8 7 6 5 4

ISBN 978-0-00-819154-2

British Library Cataloguing in Publication Data
A catalogue record for this publication is available from the British Library.

Original material authored by **Malcolm Bradley, Jackie Clegg, Susan Gardner, Sam Goodman, Sue Kearsey, Gareth Price, Mike Smith and Chris Sunley**

New material authored by **Malcolm Bradley, Sarah Jinks and Chris Sunley**

Commissioned by **Joanna Ramsay**
Project editor **Rebecca Evans**
Project managed by **Catharine Tucker**
Developed by **Tim Jackson**
Proofread by **Sarah Ryan**
Cover design by **Angela English**
Cover artwork by **richcarey/iStock**
Internal design by **Jouve India Private Limited**
Typesetting by **QBS Media Services Private Limited**
Illustrations by **Jouve India Private Limited and QBS Media Services Private Limited**
Production by **Lauren Crisp**
Printed and bound by **Grafica Veneta S. P. A.**

All exam-style questions and sample answers have been written by the author. In examinations, marks may be given differently.

MIX
Paper from
responsible sources
FSC www.fsc.org **FSC C007454**

Acknowledgements

Cover & p1 richcarey/iStock, pp 10–11 Caroline Vancoillie /Shutterstock, p 12 STEVE GSCHMEISSNER/Alamy, pp 16–17 Jubal Harshaw/Shutterstock, p 18 TinyDevil/Shutterstock, p 19 Ed Reschke/Getty, p 20 Dimarion/Shutterstock, p 21 Melba Photo Agency/Alamy, p 24 Dr. Richard Kessel & Dr. Gene Shih/Getty Images, p 26 Dr. Stanley Flegler, Visuals Unlimited/Science Photo Library, p 27 Andrew Lambert Photography/Science Photo Library, p 29 Picsfive/Shutterstock, p 31 David Cook/BlueShiftStudios/Alamy, p 33 PHOTOTAKE Inc./Alamy, pp 36–37 LAGUNA DESIGN/Science Photo Library, p 40 Martin Shields/Alamy , p 41 ANDREW LAMBERT PHOTOGRAPHY/Science Photo Library, p 41 MARTYN F. CHILLMAID/Science Photo Library, p 42 ANDREW LAMBERT PHOTOGRAPHY/Science Photo Library, pp 46–47 dinsor/Shutterstock, p 48 mimaginephotography/Shutterstock, p 54 Martyn F. Chillmaid/Science Photo Library, pp 58–59 Anest/Shutterstock, p 60 MarcelClemens/Shutterstock, p 63 SciencePhotos/Alamy, p 66 Triff/Shutterstock, p 67 Dr Keith Wheeler/Science Photo Library, p 69 Nigel Cattlin /Alamy, p 69 Nigel Cattlin/Science Photo Library, pp 74–75 Angel Andrews/Shutterstock, p 76 FRANS LANTING, MINT IMAGES/Science Photo Library, p 78 HLPhoto/Shutterstock, p 80 Cate Turton/Department for International Development, pp 94–95 Artens/Shutterstock, p 96 Stocktrek Images, Inc./Alamy, p 98 D. Kucharski K. Kucharska/Shutterstock, p 98 Zastolskiy Victor/Shutterstock, p 98 Biophoto Associates/Science Photo Library, p 98 Dr Keith Wheeler/Science Photo Library, p 99 Nigel Cattlin/Alamy, p 100 Adam Hart-Davis/Science Photo Library, p 100 Nigel Cattlin, Visuals Unlimited/Science Photo Library, p 104 Alain Pol, ISM/Science Photo Library, p 106 Yiargo/Shutterstock, p 107 LeventeGyori/Shutterstock, p 107 Beerkoff/Shutterstock, p 115 National Cancer Institute/Science Photo Library, p 119 You Touch Pix of EuToch/Shutterstock, pp 122–123 Liya Graphics/Shutterstock, p 124 Sebastian Kaulitzki/Shutterstock, p 126 Science Photo Library / Alamy, p 132 Blend Images/ER Productions Ltd/Getty Images, p 134 Nickolay Vinokurov/Shutterstock, p 138 Fdimeo/Shutterstock, pp 140–141 R. BICK, B. POINDEXTER, UT MEDICAL SCHOOL/Science Photo Library, p 142 MARK CLARKE/Science Photo Library, p 144 Anest/Shutterstock, pp 150–151 Volodymyr Martyniuk/Shutterstock, p 152 Phototake Inc./Alamy, p 153 Nemeziya/Shutterstock, p 153 Oksix/Shutterstock, p 156 Glyn/Shutterstock, p 156 Dr Jeremy Burgess/Science Photo Library, p 157 Piyato/Shutterstock, p 158 Wildlife GMBH/Alamy, p 159 Phototake Inc./Alamy, p 159 Medical-on-Line/Alamy, p 159 Tim Gainey/Alamy, p 160 D. Virtser/Shutterstock, p 161 Pi-Lens/Shutterstock, p 162 Perry Mastrovito/Getty Images, p 166 Fracis Leroy, Biocosmos/Science Photo Library, p 167 Nic Cleave Photography/Alamy, p 170 Galyna Andrushko/Shutterstock, p 172–173 vovan/Shutterstock, p 174 Colin Pickett/Alamy, p 176 Anan Kaewkhammul/Shutterstock, p 179 David Hancock/Shutterstock, p 180 Frans Lanting Studio/Alamy, p 185 Peter Gudella/Shutterstock, p 185 Master135/Shutterstock, pp 188–189 Ivan_Sabo/Shutterstock, p 190 Jacques Jangoux/Science Photo Library, p 191 Jame McIlroy/Shutterstock, p 192 JPL/NASA, p 195 Earth Observations Laboratory, Johnson Space Center/NASA, p 197 Courtesy NASA/JPL-Caltech, pp 202–203 Denis Vrublevski/Shutterstock, p 204 MIKE HOLLINGSHEAD/Science Photo Library, p 204 Achim Baque/Shutterstock, p 208 cobalt88/Shutterstock, p 209 Lightspring/Shutterstock, p 213 Charles D. Winters/Science Photo Library, p 214 haveseen/Shutterstock, p 215 Andrew Lambert Photography/Science Photo Library, p 216 Martyn F. Chillmaid/Science Photo Library, p 217 Eldar nurkovic/Shutterstock, p 222 FikMik/Shutterstock, p 226 MrJafari/Shutterstock, p 228 Ho Philip/Shutterstock, p 229 David Parker & Julian Baum/Science Photo Library, p 234 Smit/Shutterstock, p 237 Blaz Kure/Shutterstock, p 238 jordache/Shutterstock, p 239 travis manley/Shutterstock, p 239 Marc Dietrich/Shutterstock, p 242 Voronin76/Shutterstock, p 250 Martyn F. Chillmaid/Science Photo Library, p 255 ggw/Shutterstock, p 257 Andrew Lambert Photography/Science Photo Library, p 260 PHOTOTAKE Inc./Alamy, pp 264–265 Galyna Andrushko/Shutterstock, p 266 Maximilian Stock Ltd/Science Photo Library, p 275 badahos/Shutterstock, p 283 Iakov Kalinin/Shutterstock, p 283 john t. fowler/Alamy, p 286 Anna Baburkina/Shutterstock, p 296 dgmata/Shutterstock, p 296 Ken Brown/iStockphoto, p 300 Cyril Hou/Shutterstock, p 303 Martyn F. Chillmaid/Science Photo Library, p 307 Blaz Kure/Shutterstock, p 307 Andrew Lambert Photography/Science Photo Library, p 309 Charles D. Winters/Science Photo Library, p 310 jcwait/Shutterstock, p 315 Johann Helgason/Shutterstock, p 316 Andrew Lambert Photography/Science Photo Library, p 319 Andrew Lambert Photography/Science Photo Library, pp 324–325 Piotr Zajc/Shutterstock, p 326 Shebeko/Shutterstock, p 333 Andrew Lambert Photography/Science Photo Library, p 334 Charles D. Winters/Science Photo Library, p 336 Andrew Lambert Photography/Science Photo Library, p 337 design56/Shutterstock, p 340 Andrew Lambert Photography/Science Photo Library, p 348 Slaven/Shutterstock, p 353 Centrill Media/Shutterstock, p 355 Ehrman Photographic/Shutterstock, p 356 Feraru Nicolae/Shutterstock, p 358 Lawrence Migdale/Science Photo Library, p 358 Richard treptow/Science Photo Library, p 358 Julia Reschke/Shutterstock, p 362 kilukilu/Shutterstock, p 363 Fokin Oleg/Shutterstock, p 363 Parmna/Shutterstock, p 363 Holly Kuchera/Shutterstock, p 363 Danicek/Shutterstock, p 365 David_Monniaux/Wikimedia Commons, p 369 Romas_Photo/Shutterstock, p 370 Martyn F. Chillmaid/Science Photo Library, p 371 Jackiso/Shutterstock, p 373 Nando Machado/Shutterstock, p 375 muzsy/Shutterstock, p 376 Prixel Creative/Shutterstock, pp 382–383 Photobank gallery/Shutterstock, p 384 Paul Rapson/Science Photo Library, p 384 BESTWEB/Shutterstock, p 390 Dawid Zagorski/Shutterstock, p 394 Yvan/Shutterstock, p 395 speedpix/Alamy, p 396 ggw/Shutterstock, p 397 Joe Gough/Shutterstock, p 400 Gwoeii/Shutterstock, p 403 JoLin/Shutterstock, pp 406–407 Camellia/Shutterstock, p 408 Rob kemp/Shutterstock, p 409 Hung Chung Chih/Shutterstock, p 413 wavebreakmedia ltd/Shutterstock, p 426 Dorling Kindersley/Getty, p 429 Andrea Danti/Shutterstock, p 430 Michal Vitek/Shutterstock, p 433 Jerritt Clark/Springer/Getty Images, p 434 Charles D. Winters/Science Photo Library, p 436 EcoPrint/Shutterstock, p 438 National Geographic Creative / Alamy Stock Photo, p 439 Kirk Geisler/Shutterstock, p 441 Michael Wesemann/Shutterstock, p 442 Martyn F. Chillmaid/Science Photo Library, p 451 Marcel Jancovic/Shutterstock, p 451 DenisNata/Shutterstock, pp 456–457 Hung Chung Chih/Shutterstock, p 458 ZoranOrcik/Shutterstock, p 460 3Dsculptor/Shutterstock, p 462 clearlens/Shutterstock, p 462 Adrian Hughes/Shutterstock , p 463 chatchai/Shutterstock, p 463 sizov/Shutterstock, p 473 Petr Malyshev, p 477 SkyLynx, p 479 Zoia Kostina/Shutterstock, p 480 Rido/Shutterstock, p 481 Peteri/Shutterstock, p 482 Ocean Power Delivery/Science Photo Library, pp 486–487 Katrina Leigh/Shutterstock, p 488 jele/Shutterstock, p 489 Alan Freed/Shutterstock, p 489 Geoffrey Kuchera/Shutterstock , p 491 Mike Blanchard/Shutterstock, p 497 Natursports/Shutterstock, p 497 jan kranendonk/Shutterstock, p 498 Dmitry Berkut/Shutterstock, p 504 Alexandra Lande/Shutterstock, p 507 Ulrich Mueller/Shutterstock, p 514 Vixit/Shutterstock, pp 518–519 Willyam Bradberry/Shutterstock, p 520 Jacob Wackerhausen, p 528 Aija Lehtonen/Shutterstock, p 529 Payless Images/Shutterstock, p 534 Nagy-Bagoly Arpad/Shutterstock, p 538 All-stock-photos/Shutterstock, p 539 Fedorov Oleksiy/Shutterstock, p 539 Darren Pullman/Shutterstock, p 541 Dario Sabljak/Shutterstock, p 547 Igor Klimov/Shutterstock, p 548 Losevsky Photo and Video/Shutterstock, p 549 NYTECH Corp, 03/WikiMedia Commons, pp 556–557 littlesam/Shutterstock, p 558 Tomasz Szymanski/Shutterstock, p 561 Nir Levy/Shutterstock, p 562 Andrew Howe/istockphoto, p 562 Jhaz Photography/Shutterstock, p 564 Flegere/Shutterstock, p 569 ra2studio/Shutterstock, p 374 Richard Wareham Fotographie/Alamy, pp 578–579 Yevhen Tamavski/Shutterstock, p 580 Bart Coenders/istockphoto, p 584 GIPhotoStock/Science Photo Library, p 585 Joshua Haviv/Shutterstock, p 592 Championfoto/Shutterstock, p 597 Alex Kuzovlev/Shutterstock, p 598 Photoseeker/Shutterstock, p 609 Jose Luis Calvo/Shutterstock, p 657 Jim Lopes/Shutterstock, p 666 Ed Phillips/Shutterstock

Contents

Physics

Getting the best from the book

Welcome to Collins *Cambridge IGCSE Combined Science*.

This textbook has been designed to help you understand all of the requirements of the Cambridge IGCSE Combined Science course.

SAFETY IN THE SCIENCE LESSON

This book is a textbook, not a laboratory or practical manual. As such, you should not interpret any information in this book that relates to practical work as including comprehensive safety instructions. Your teachers will provide full guidance for practical work and cover rules that are specific to your school.

A brief introduction to the section to give context to the science covered.

The section contents shows the separate topics to be studied matching the syllabus order.

The world's tallest known living tree, which has been named Hyperion, is a coast redwood tree growing in Northern California in the US. When it was measured in 2006 it was found to be 115.61 m tall (379.3 ft). Like all plants, from the smallest seedling to the tallest redwood giant, Hyperion makes its own food from three simple ingredients, sunlight, water and carbon dioxide. This process is known as photosynthesis and is one of the features of plants that distinguish them from animals.

SECTION CONTENTS
Plant nutrition

5
Plant nutrition

△ Saplings growing toward a controlled light source.

Knowledge check shows the ideas you should have already encountered in previous work before starting the topic.

Learning objectives cover what you need to learn in this topic.

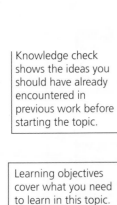

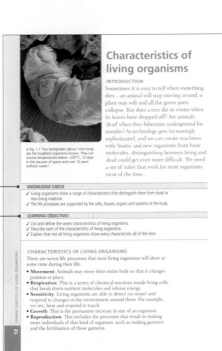

Characteristics of living organisms

INTRODUCTION

Sometimes it is easy to tell when something dies – an animal will stop moving around, a plant may wilt and all the green parts collapse. But does a tree die in winter when its leaves have dropped off? Are animals 'dead' when they hibernate underground for months? As technology gets increasingly sophisticated, and we can create machines with 'brains' and new organisms from basic molecules, distinguishing between living and dead could get even more difficult. We need a set of 'rules' that work for most organisms, most of the time.

△ Fig. 1.1 Tiny tardigrades (about 1 mm long) are the toughest organisms known. They can survive temperatures below −200°C, 10 days in the vacuum of space and over 10 years without water!

KNOWLEDGE CHECK
✓ Living organisms show a range of characteristics that distinguish them from dead or non-living material.
✓ The life processes are supported by the cells, tissues, organs and systems of the body.

LEARNING OBJECTIVES
✓ List and define the seven characteristics of living organisms.
✓ Describe each of the characteristics of living organisms.
✓ Explain that not all living organisms show every characteristic all of the time.

CHARACTERISTICS OF LIVING ORGANISMS

There are seven life processes that most living organisms will show at some time during their life.
- **Movement**: Animals may move their entire body so that it changes position or place.
- **Respiration**: This is a series of chemical reactions inside living cells that break down nutrient molecules and release energy.
- **Sensitivity**: Living organisms are able to detect (or sense) and respond to changes in the environment around them. For example, we see, hear and respond to touch.
- **Growth**: This is the permanent increase in size of an organism.
- **Reproduction**: This includes the processes that result in making more individuals of that kind of organism, such as making gametes and the fertilisation of those gametes.

- **Excretion**: This is the removal from the body of substances that are toxic (poisonous) and may damage cells if they stay in the body. Organisms also excrete substances that are in **excess**, where there is more in the body than is needed.
- **Nutrition**: This is the absorption of nutrients into the body. The nutrients are the raw materials needed by the cells to release energy and to make more cells for growth, development and repair.

All these characteristics will be described in greater detail in later Topics in this book.

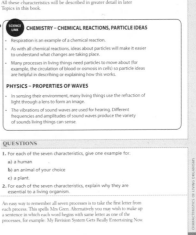

SCIENCE LINK — CHEMISTRY – CHEMICAL REACTIONS, PARTICLE IDEAS
- Respiration is an example of a chemical reaction.
- As with all chemical reactions, ideas about particles will make it easier to understand what changes are taking place.
- Many processes in living things need particles to move about (for example, the circulation of blood or osmosis in cells) so particle ideas are helpful in describing or explaining how this works.

PHYSICS – PROPERTIES OF WAVES
- In sensing their environment, many living things use the refraction of light through a lens to form an image.
- The vibrations of sound waves are used for hearing. Different frequencies and amplitudes of sound waves produce the variety of sounds living things can sense.

QUESTIONS
1. For each of the seven characteristics, give one example for:
   a) a human
   b) an animal of your choice
   c) a plant.
2. For each of the seven characteristics, explain why they are essential to a living organism.

An easy way to remember all seven processes is to take the first letter from each process. This spells Mrs Gren. Alternatively you may wish to make up a sentence in which each word begins with same letter as one of the processes, for example: My Revision System Gets Really Entertaining Now.

Science Link boxes help you to deepen your understanding of the connections between the different sciences. It is not necessary for you to learn the content of these boxes as they do not form part of the syllabus. However, they will help you to spot and understand the links between the different sciences and develop your scientific thinking skills.

magnesium + copper(II) oxide → magnesium oxide + copper
Mg(s) + CuO(s) → MgO(s) + Cu(s)

This reaction is an example of a redox reaction. The magnesium has been oxidised to magnesium oxide and the copper(II) oxide has been reduced to copper. Because the magnesium is responsible for the reduction of the copper(II) oxide, it is acting as a reducing agent. Similarly, the copper(II) oxide is responsible for the **oxidation** of the magnesium, so it is acting as an oxidising agent. In a redox reaction, the reducing agent is always oxidised and the oxidising agent is always reduced.

What will happen if copper is heated with magnesium oxide? Nothing happens, because copper is lower in the reactivity series than magnesium.

Using displacement reactions to establish a reactivity series
Displacement reactions of metals and their compounds in aqueous solution can be used to work out the order in the reactivity series.

In the same way that a more reactive element can push a less reactive element out of a compound, a more reactive metal ion in aqueous solution can displace a less reactive one.

For example, if you add zinc to copper(II) sulfate solution, the zinc displaces the copper because zinc is more reactive than copper. When the experiment is carried out, the blue colour of the copper(II) ion will fade as copper is produced and zinc ions are made:

zinc + copper(II) sulfate solution → zinc sulfate solution + copper
Zn(s) + Cu²⁺(aq) + SO₄²⁻(aq) → Zn²⁺(aq) + SO₄²⁻(aq) + Cu(s)

To build up a whole reactivity series, a set of reactions can be tried to see if metals can displace other metal ions. By following the general rule that a more reactive metal can displace a less reactive metal it is possible to establish the reactivity series.

For example, you may have seen the reaction of copper wire with silver nitrate solution. As the reaction proceeds, a shiny grey precipitate appears (this is silver) and the solution begins to turn blue as Cu(II) ions are produced from the copper.

copper + silver nitrate → copper(II) nitrate + silver
Cu(s) + 2AgNO₃(aq) → Cu(NO₃)₂(aq) + 2Ag(s)

This shows that silver can be displaced by copper, and so silver is below copper in the reactivity series.

END OF EXTENDED

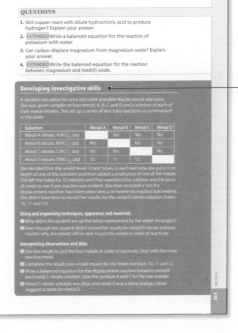

QUESTIONS
1. Will copper react with dilute hydrochloric acid to produce hydrogen? Explain your answer.
2. EXTENDED Write a balanced equation for the reaction of potassium with water.
3. Can carbon displace magnesium from magnesium oxide? Explain your answer.
4. EXTENDED Write the balanced equation for the reaction between magnesium and lead(II) oxide.

Developing investigative skills

A student was asked to carry out some possible displacement reactions. She was given samples of four metals A, B, C and D and a solution of each of their metal nitrates. She set up a series of test tube reactions as summarised in the table:

Solution	Metal A	Metal B	Metal C	Metal D
Metal A nitrate, A(NO₃)₃ (aq)		Yes	Yes	No
Metal B nitrate, B(NO₃)₂ (aq)	No		No	No
Metal C nitrate, C(NO₃)₂ (aq)	No	Yes		No
Metal D nitrate, D(NO₃)₂ (aq)	10	11	12	

She decided that she would need 12 test tubes. In each test tube she put a 1 cm depth of one of the solutions and then added a small piece of one of the metals. She left the tubes for 10 minutes and then examined the solution and the piece of metal to see if any reaction was evident. She then recorded a 'yes' if a displacement reaction had taken place and a 'no' where no reaction was evident. She didn't have time to record her results for the metal D nitrate solution (tubes 10, 11 and 12).

Using and organising techniques, apparatus and materials
❶ Why didn't the student set up the tubes represented by the white rectangles?
❷ Even though the student didn't record her results for metal D nitrate solution, explain why she would still be able to put the metals in order of reactivity.

Interpreting observations and data
❸ Use the results to put the four metals in order of reactivity. Start with the most reactive metal.
❹ Complete the results you would expect for the three reactions 10, 11 and 12.
❺ Write a balanced equation for the displacement reaction between metal B and metal C nitrate solution. (Use the symbols B and C for the two metals.)
❻ Metal D nitrate solution was blue and metal D was a shiny orange colour. Suggest a name for metal D.

Examples of investigations are included with questions matched to the investigative skills you will need to learn.

12

13

INORGANIC CHEMISTRY

METALS

360

361

GETTING THE BEST FROM THE BOOK

7

Getting the best from the book *continued*

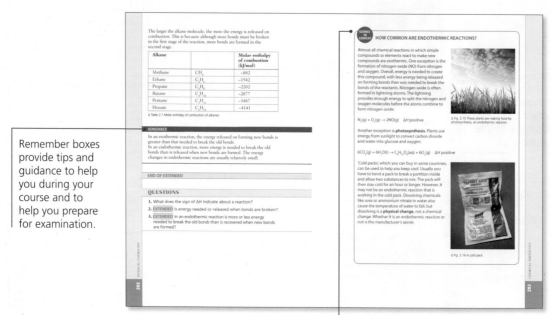

Remember boxes provide tips and guidance to help you during your course and to help you prepare for examination.

Science in context boxes put the ideas you are learning into real-life context. It is not necessary for you to learn the content of these boxes as they do not form part of the syllabus. However, they do provide interesting examples of scientific application that are designed to enhance your understanding.

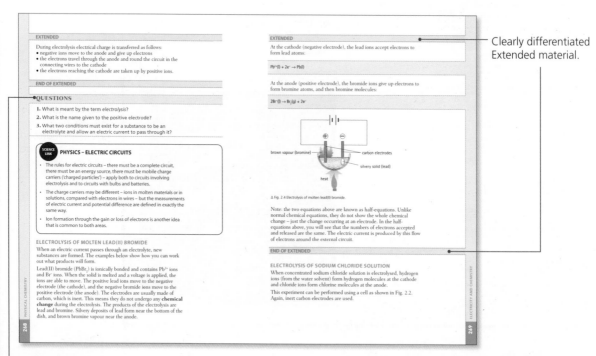

Clearly differentiated Extended material.

Questions to check your understanding.

A full checklist of all the information you need to cover the complete syllabus requirements for each topic.

End of topic questions allow you to apply the knowledge and understanding you have learned in the topic to answer the questions.

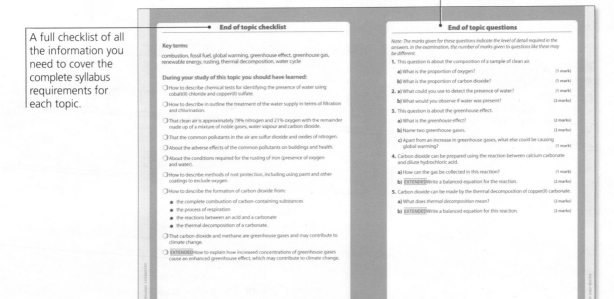

Exam-style questions to help you prepare for your exam in a focused way and help you get the best results.

Student sample with teacher's comments to show best practice.

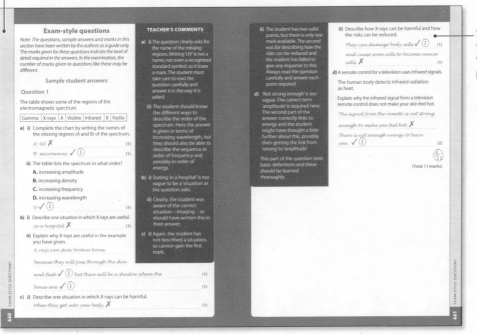

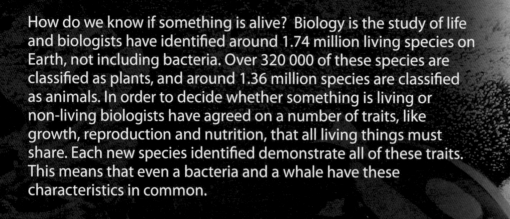

How do we know if something is alive? Biology is the study of life and biologists have identified around 1.74 million living species on Earth, not including bacteria. Over 320 000 of these species are classified as plants, and around 1.36 million species are classified as animals. In order to decide whether something is living or non-living biologists have agreed on a number of traits, like growth, reproduction and nutrition, that all living things must share. Each new species identified demonstrate all of these traits. This means that even a bacteria and a whale have these characteristics in common.

SECTION CONTENTS

1
Characteristics of living organisms

△ Many species of different kinds of organisms live on a coral reef.

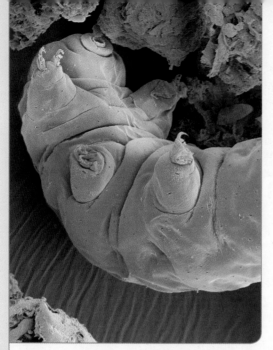

△ Fig. 1.1 Tiny tardigrades (about 1mm long) are the toughest organisms known. They can survive temperatures below −200°C, 10 days in the vacuum of space and over 10 years without water!

Characteristics of living organisms

INTRODUCTION

Sometimes it is easy to tell when something dies – an animal will stop moving around, a plant may wilt and all the green parts collapse. But does a tree die in winter when its leaves have dropped off? Are animals 'dead' when they hibernate underground for months? As technology gets increasingly sophisticated, and we can create machines with 'brains' and new organisms from basic molecules, distinguishing between living and dead could get even more difficult. We need a set of 'rules' that work for most organisms, most of the time.

KNOWLEDGE CHECK

✓ Living organisms show a range of characteristics that distinguish them from dead or non-living material.
✓ The life processes are supported by the cells, tissues, organs and systems of the body.

LEARNING OBJECTIVES

✓ List and define the seven characteristics of living organisms.
✓ Describe each of the characteristics of living organisms.
✓ Explain that not all living organisms show every characteristic all of the time.

CHARACTERISTICS OF LIVING ORGANISMS

There are seven life processes that most living organisms will show at some time during their life.

- **Movement**: Animals may move their entire body so that it changes position or place.
- **Respiration**: This is a series of chemical reactions inside living cells that break down nutrient molecules and release energy.
- **Sensitivity**: Living organisms are able to detect (or sense) and respond to changes in the environment around them. For example, we see, hear and respond to touch.
- **Growth**: This is the permanent increase in size of an organism.
- **Reproduction**: This includes the processes that result in making more individuals of that kind of organism, such as making gametes and the fertilisation of those gametes.

- **Excretion**: This is the removal from the body of substances that are toxic (poisonous) and may damage cells if they stay in the body. Organisms also excrete substances that are in **excess**, where there is more in the body than is needed.
- **Nutrition**: This is the absorption of nutrients into the body. The nutrients are the raw materials needed by the cells to release energy and to make more cells for growth, development and repair.

All these characteristics will be described in greater detail in later Topics in this book.

SCIENCE LINK

CHEMISTRY – CHEMICAL REACTIONS, PARTICLE IDEAS

- Respiration is an example of a chemical reaction.

- As with all chemical reactions, ideas about particles will make it easier to understand what changes are taking place.

- Many processes in living things need particles to move about (for example, the circulation of blood or osmosis in cells) so particle ideas are helpful in describing or explaining how this works.

PHYSICS – PROPERTIES OF WAVES

- In sensing their environment, many living things use the refraction of light through a lens to form an image.

- The vibrations of sound waves are used for hearing. Different frequencies and amplitudes of sound waves produce the variety of sounds living things can sense.

QUESTIONS

1. For each of the seven characteristics, give one example for:

   a) a human

   b) an animal of your choice

   c) a plant.

2. For each of the seven characteristics, explain why they are essential to a living organism.

An easy way to remember all seven processes is to take the first letter from each process. This spells Mrs Gren. Alternatively you may wish to make up a sentence in which each word begins with same letter as one of the processes, for example: My Revision System Gets Really Entertaining Now.

End of topic checklist

Key terms

excess, growth, movement, nutrition, reproduction, respiration, sensitivity

During your study of this topic you should have learned:

◯ How to describe the seven characteristics of life: movement, respiration, sensitivity, growth, reproduction, excretion and nutrition.

End of topic questions

Note: The marks given for these questions indicate the level of detail required in the answers. In the examination, the number of marks given to questions like these may be different.

1. Name and describe the seven processes of life. **(7 marks)**

2. Name two life processes necessary for an organism to release energy. **(2 marks)**

3. Plants cannot move about, as animals can. Does that mean animals are more alive than plants? Explain your answer. **(2 marks)**

4. During winter, an oak tree in the UK will lose its leaves and not grow. Is the tree still living during this time? Explain your answer using all the characteristics of life. **(4 marks)**

Multicellular organisms are made up of different cell types that each have a specific job to do. The human body is made up of about 200 different cell types, ranging from muscle and fat cells, to blood, skin and nerve cells.

All 'complex' cells (those that contain a nucleus) in all animals, plants and protoctista on Earth have the same basic structure. Scientists say that this is because we have all evolved from a single complex cell. This first complex cell evolved from a simple bacteria-like cell (without a nucleus) more than 1600 million years ago. This is the origin of all the millions of different species of plants, animals and protoctista that live on Earth today.

All cells need to exchange molecules such as water, oxygen and carbon dioxide with their environment. Cells need to transport molecules into the cell for use in metabolic processes and need to transport out the waste molecules they make.

All living organisms need to be able to transport water, oxygen, carbon dioxide and other molecules around their bodies.
For simple organisms such as bacteria, the distances travelled are very small, but more complex animals and plants have evolved highly specialised transport mechanisms to get vital substances from one part of the organism to another. If all the blood vessels in the human body – including arteries, veins and capillaries – were laid end to end, they would stretch for about 60 000 miles. That's nearly 100 000 km!

SECTION CONTENTS

a) Cell structure

b) Movement in and out of cells

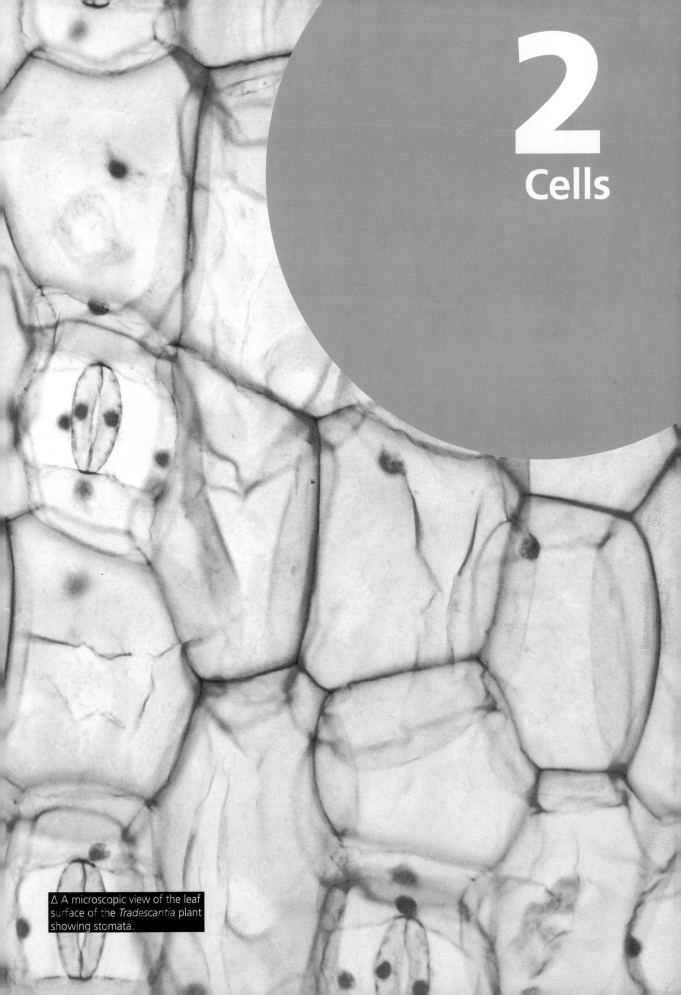

2
Cells

△ A microscopic view of the leaf surface of the *Tradescantia* plant showing stomata.

Cells

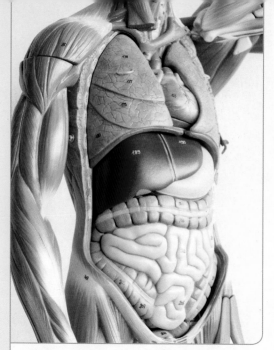

△ Fig. 2.1 The human body is made up of several systems of grouped organs, including the digestive system, the nervous system, the muscle/skeletal system and the respiratory system.

INTRODUCTION

Bringing together similar activities that have the same purpose can make things much more efficient. For example, bringing teachers and students together in a school helps more students to learn more quickly than if each teacher travelled to each student's home for lessons. The same is true in the body. Having groups of similar cells in the same place as a tissue, and grouping tissues into organs, helps the body carry out all the life processes much more efficiently and so stay alive.

KNOWLEDGE CHECK

✓ State that organisms are formed from many cells.
✓ Describe how cells may be specialised in different ways to carry out different functions.
✓ Define the terms *tissue*, *organ* and *organ system*.
✓ Describe how the organisation of the body systems contributes to the seven life processes.
✓ State that a microscope can be used to magnify specimens so we can see more detail.
✓ Cells need oxygen and glucose for respiration.
✓ Cells need to get rid of waste substances, such as carbon dioxide from respiration.

LEARNING OBJECTIVES

✓ Describe and compare the structures in plant and animal cells as seen under a light microscope.
✓ Describe the function of each type of cell structure seen under the light microscope.
✓ **EXTENDED** Relate structure to function in a range of specialised cells.
✓ Describe how to calculate the magnification of biological specimens seen under a microscope.
✓ Calculate the magnification and size of biological specimens using millimetres as units.
✓ Define the term *diffusion*.
✓ **EXTENDED** Define the term *osmosis*.
✓ Investigate and describe the effects on plant tissues of immersing them in solutions of different concentrations.

CELL STRUCTURE

The diagrams below show a typical animal cell and typical plant cells. These cells all have a nucleus and cytoplasm.

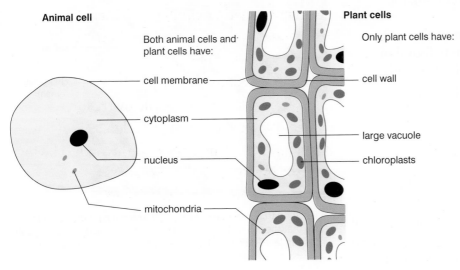

△ Fig. 2.2 The basic structures of an animal cell (for example, liver cell) and plant cells (for example, palisade mesophyll cells).

All living organisms are made of cells. Some, such as bacteria, protoctista and some fungi, are formed from a single cell; others, such as the majority of plants and animals, are **multicellular**, with a body made of many cells. All animal and plant cells have certain features in common:

- a **cell membrane** surrounds the cell
- **cytoplasm** inside the cell, in which all the other structures are found
- a large **nucleus**.

A typical animal cell is a human liver cell.

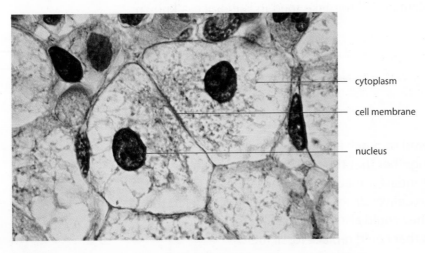

△ Fig. 2.3 Structures in animal cells seen using a light microscope. Note these cells have been stained to make some structures easier to see.

Plant cells also have features that are not found in animal cells, such as:

- a **cell wall** surrounding the cell membrane
- a large central **vacuole**
- green **chloroplasts** found in some, but not all, plant cells.

A typical plant cell is a palisade cell in the upper part of a leaf.

Functions of cell structures

Each structure in a cell has a particular role.

- The cell membrane holds the cell together and controls substances entering and leaving the cell.
- The cytoplasm supports many small cell structures and is where many different chemical processes happen. It contains water, and many solutes are dissolved in it.
- The nucleus contains genetic material in the **chromosomes**. These control how a cell grows and works. The nucleus also controls cell division.
- The plant cell wall is made of cellulose, which gives the cell extra support and defines its shape.
- The plant vacuole contains cell sap. The vacuole is used for storage of some materials, and to support the shape of the cell. If there is not enough cell sap in the vacuole, the whole plant may wilt.
- Chloroplasts contain the green pigment chlorophyll, which absorbs the light energy that plants need to make food in the process known as **photosynthesis**.

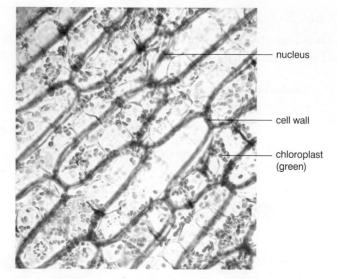

nucleus

cell wall

chloroplast (green)

△ Fig. 2.4 Structures in plant cells seen using a light microscope. Note that the cell membrane and vacuole are difficult to distinguish in this image. The chloroplasts are supported by the cytoplasm.

SCIENCE IN CONTEXT **ARTIFICIAL CELLS**

Scientists have discovered so much about how the structures of cells are formed and work together that they are starting to create artificial cells. This has great potential for medicine, because these cells could be used, for example, to deliver drugs inside the body directly to the cells that need them. They could also be used in biotechnology, for example, to make fuels that could replace fossil fuels.

Developing investigative skills

The photograph shows the view of some cells seen through a light microscope.

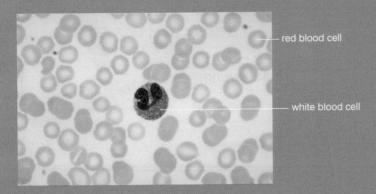
red blood cell

white blood cell

△ Fig. 2.5 A photograph of blood cells taken with a light microscope.

Demonstrate and describe techniques

❶ a) Describe how to set up a slide on a microscope so that the image is clearly focused.

b) Describe and explain what precautions should be taken when viewing a slide at high magnification.

c) Describe what precaution should be taken if using natural light to illuminate the slide, and explain why this is important.

Make observations and measurements

❷ a) Draw and label a diagram of the white blood cell shown in the light micrograph above.

b) If the ×10 eyepiece was used, and the ×40 objective, calculate the magnification of the image compared with the specimen on the slide. (Use the information on page 25 to help you with this.)

Analyse and interpret data

❸ Are the cells shown plant cells or animal cells? Explain your answer.

QUESTIONS

1. a) Using the photograph in Fig. 2.3, make a careful drawing of one of the cells using a sharpened pencil to make clear lines.

b) Label your drawing to show the three key structures of an animal cell.

2. List three cell structures that are found in plant cells but not in animal cells.

3. Name the part of a plant cell that does the following:

a) carries out photosynthesis

b) contains cell sap

c) stops the cell swelling if it takes in a lot of water.

Cell specialisation for function

Different types of cells carry out different jobs. Cells have special features that allow them to carry out their job. This is called **specialisation**. Good examples of specialised cells are:

- ciliated cells, red blood cells and sperm and egg cells in humans
- root hair cells and palisade mesophyll cells in flowering plants.

SCIENCE IN CONTEXT **STEM CELLS**

Every tissue in the human body contains a small number of undifferentiated cells. These are called stem cells, and their role is to divide and produce new differentiated cells within the tissue for growth and repair. Scientists are investigating how stem cells could be given to people to mend tissue that the body cannot mend, such as the spinal cord after an accident in which it is cut. This would make it possible for a person who is paralysed following an accident to move their whole body again.

Ciliated cells

Cilia are tiny hair-like projections that cover the surfaces of certain types of cells. Cilia can move and the cell can coordinate this movement to produce waves that pass over the cell. These waves of moving cilia can move liquid in particular directions.

Ciliated epithelial cells in the lining of the respiratory tract move a liquid called mucus. Tiny particles of dust or bacteria that are trapped in the mucus are carried along in this flow and pass up the tubes. They are then emptied, along with the mucus, into the oesophagus, where they are swallowed and pass into the stomach. In this way, the ciliated epithelium keeps the lungs clean. Smoking reduces the effectiveness of these cilia, which explains why smokers often have a cough – they cannot clear away the dirty mucus that collects in their lungs.

Δ Fig. 2.6 Secreting cells produce mucus that traps particles in the lungs. Cilia sweep the mucus out of the lungs and into the throat, where it is swallowed.

Palisade mesophyll cells

Palisade mesophyll cells are plant cells found in the upper part of a leaf. They have all the features of a plant cell (see Fig. 2.4) but contain a large number of chloroplasts. This is because most photosynthesis carried out by a plant happens in these cells (see more in Topic 5).

Red blood cells

Red blood cells in mammals are unusual in that they do not have a nucleus and so cannot divide. The whole of the cell is filled with a chemical called haemoglobin, which can pick up oxygen in the lungs and release it near the cells that need it deep inside the body. The shape of the cell means that the innermost part of the red blood cell is never far away from the outside, so diffusion of oxygen in and out happens very rapidly. Red blood cells are made in bone marrow and last only 120 days before they are destroyed in the spleen and liver.

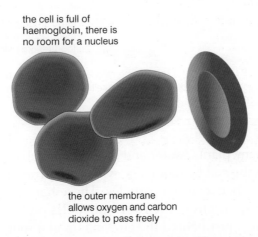

the cell is full of haemoglobin, there is no room for a nucleus

the outer membrane allows oxygen and carbon dioxide to pass freely

Δ Fig. 2.7 Red blood cells are specialised for carrying oxygen.

Human sex cells

The human sex cells (gametes – Topic 10) are the **sperm cell** and the **egg cell**. Sperm cells and egg cells have particular forms that are adapted to their roles in reproduction.

Sperm cells are relatively small compared with an egg cell. They have very little cytoplasm surrounding the nucleus, because they carry out few functions other than travelling to the egg cell for fertilisation. There is a small vesicle of enzymes, called the **acrosome**, at the front tip of the cell. The enzymes in the acrosome digest a hole in the egg cell membrane. This allows the nucleus of the sperm cell to enter the egg cell and fuse with its nucleus. The mid piece of the sperm cell contains many mitochondria. The mitochondria provide energy to move the tail, which moves the sperm towards the egg cell.

The human egg cell is a very large cell and is almost visible without a microscope. It cannot move on its own. The large amount of cytoplasm around the nucleus provides nutrients for when the cell is fertilised and starts to divide.

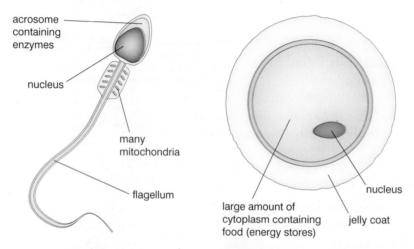

△ Fig. 2.8 Diagrams of a human sperm cell (left) and human egg cell (right). Note these are not drawn to scale. The volume of an egg cell is hundreds of times bigger than the volume of a sperm cell.

Root hair cells

In many plants, water and minerals are absorbed from the soil by root hairs, which penetrate the spaces between soil particles. These hairs are very fine extensions of the **root hair cells** on the root surface, just behind the growing tip of a root. The elongated shape of the cells increases the surface area available for absorption of water and dissolved mineral ions. As they age, root hairs develop a waterproof layer and become non-functional. New root hairs are constantly growing as the root pushes through the soil.

△ Fig. 2.9 The root hair cells greatly increase the surface area for absorption near the tips of roots.

QUESTIONS

1. Where would you find the following cells and what do they do?

   a) ciliated cells

   b) red blood cells

   c) root hair cells

2. Describe how the structures of sperm and egg cells are adapted to their functions in reproduction.

END OF EXTENDED

Size of specimens

Many of the structures that we study in biology are too small to be seen just using our eyes. We can use magnifying glasses and microscopes to examine details of plant and animal cells, and to take pictures and draw the diagrams you see in this book. But often we want to know the actual size of the specimen we are looking at. If we know the magnification we are using to look at a specimen then we can work out the size of a structure.

When using a microscope, the **magnification** of a specimen is calculated from the eyepiece and the objective used to view it.

- The magnification of an eyepiece for a light microscope may be ×4, ×5 or ×10.
- The magnification of an objective for a light microscope may be ×5, ×10, ×20 or ×40.

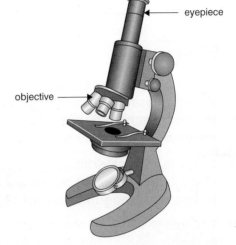

△ Fig. 2.10 A light microscope.

The magnification of the specimen is the magnification of the eyepiece multiplied by the magnification of the objective.

SCIENCE LINK **PHYSICS – PROPERTIES OF WAVES**

- Light microscopes make use of lenses and mirrors, so ideas about reflection and refraction are important.

- The magnification of a microscope is related to the refracting power of the lenses, which in turn depends on their shape and the refractive index of the glass.

1. If the microscope is set up with the ×5 eyepiece and ×20 objective, the magnification of a specimen viewed will be:

$$5 \times 20 = 100$$

We can work out the *size of a structure* from the image size seen under the microscope and the magnification used to view it.

$$\text{actual size} = \frac{\text{observed size}}{\text{magnification}}$$

The observed size is measured using a scale, such as a **graticule**, viewed through the microscope.

2. If the diameter of a cell observed under a microscope is 6 mm, and the magnification is ×400, the actual diameter of the cell is:

$$\frac{6}{400} = 0.015 \text{ mm.}$$

QUESTIONS

1. You are looking at an object that measures 0.5 mm and the image you see is 10 mm long. Your friend is looking at an object that is 0.1 mm long using the same magnification. What size of image does your friend see?

2. Imagine you are examining a specimen of blood under a microscope to look at red blood cells. Why might it be important to know the magnification of the lens you are using?

3. The image you are looking at is 2.5 mm long and you are using a magnification of 100. Write down the calculation you would use to work out the actual size of the object.

MOVEMENT IN AND OUT OF CELLS

If you put a red blood cell into pure water, it will eventually burst open. If you place the red blood cell into a salty solution instead, it will shrink. Surrounding every cell is the cell membrane. Imagine the cell membrane as a leaky layer that is strong enough to hold all the contents in the cell together, but that allows small particles to move through it. The cell membrane also has special 'gates' that allow certain, important particles through. Different cells have different kinds of 'gate' in them. So cell membranes play an essential role in controlling what goes in and out of cells, and therefore control the way that the cell functions.

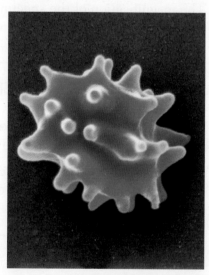

Δ Fig. 2.11 A red blood cell that has been placed in a salty solution loses water and shrinks.

SCIENCE LINK — CHEMISTRY – PARTICLE NATURE OF MATTER

- Particle ideas, particularly the properties of fluids, are important in describing and explaining how substances are able to move into and out of cells.

- The movement of water molecules is particularly important and particle ideas help explain how they can move through a semi-permeable membrane while other molecules do not.

Diffusion

Substances such as water, oxygen, carbon dioxide and food are made of particles (atoms, ions and molecules).

In liquids and gases the particles are constantly moving around. This means that they eventually spread out evenly. For example, if you dissolve sugar in a cup of water, even if you do not stir it, the sugar molecules eventually spread throughout the liquid. This is because all the molecules are moving around, colliding with and bouncing off other particles.

● water molecule

● sugar molecule

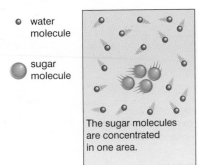

The sugar molecules are concentrated in one area.

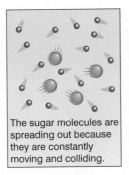

The sugar molecules are spreading out because they are constantly moving and colliding.

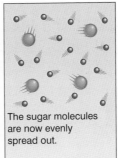

The sugar molecules are now evenly spread out.

△ Fig. 2.12 Diffusion of sugar molecules in a solution.

The sugar molecules have spread out from an area of high concentration, when they were added to the water, to an area of low concentration. Eventually, although all the particles are still moving, the sugar molecules are evenly spread out and there is no longer a **concentration gradient**.

Only while there is **net movement** (where there are more particles moving in one direction than another) from an area of high concentration to an area of lower concentration is there **diffusion**.

Diffusion is the net movement of molecules from a region of their higher concentration to a region of their lower concentration.

△ Fig. 2.13 Potassium permanganate(II) diffuses through a beaker of water as the solid crystal dissolves.

Diffusion can only occur when there is a difference in concentration between two areas. Particles are said to move *down* their concentration gradient. This happens because of the random movement of particles.

Diffusion in cells

Cells are surrounded by membranes. These membranes are leaky – they let tiny particles pass through them. Large particles can't get through, so cell membranes are said to be **partially permeable**.

Movement of particles across a cell membrane may happen more in one direction than the other if there is a difference in concentration on either side of the membrane (a concentration gradient). For example, in the blood vessels in the lungs there is a *low* oxygen concentration inside the red blood cells (because they have given up their oxygen to cells in other parts of the body) and a *high* oxygen concentration in the alveoli of the lungs. Therefore, oxygen diffuses from the alveoli into the red blood cells.

Other examples of diffusion include:

• carbon dioxide entering leaf cells
• digested food substances from the small intestine entering the blood.

Diffusion of some types of molecules and ions across the cell membrane is a **passive** process. It needs no input of energy from the cell.

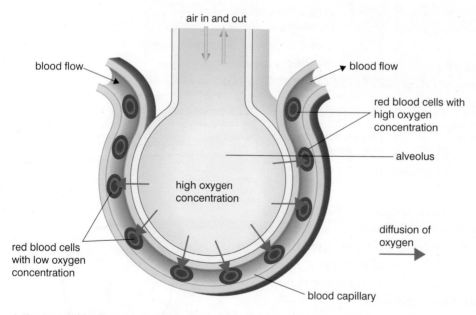

△ Fig. 2.14 In blood vessels in the lungs, oxygen diffuses down its concentration gradient from the air in the lungs into red blood cells.

KIDNEY FAILURE AND HAEMODIALYSIS

The kidneys are organs that depend on filtration and diffusion to produce urine and keep the concentration of many substances in the blood at a fairly constant level. People who suffer from kidney failure are unable to do this, and are very quickly at risk from the build-up of waste products, such as urea, in the body as a result of cell processes. In high concentrations these waste products can damage body cells and lead to death.

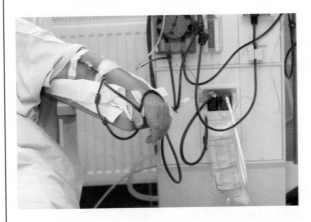

◁ Fig. 2.15 A patient undergoing kidney dialysis.

Haemodialysis is an artificial way of cleaning the blood by which substances diffuse out of the blood into dialysis fluid in a machine called a dialyser. The concentration of substances in the dialysis fluid has to be correct, so that all the waste products are removed and other substances are returned to the body at the right concentration.

QUESTIONS

1. In your own words, define the terms *net movement* and *diffusion*.

2. Explain why some particles can diffuse through cell membranes but not others.

Osmosis

Water molecules are small enough to diffuse through partially permeable membranes, such as cell membranes. However, because water molecules are so important to cells, and may be diffusing in a different direction to other molecules, this kind of diffusion has a special name – **osmosis**. Like diffusion, osmosis is a passive process and is a result of the random movement of particles.

Water molecules diffuse from a place where there is a high concentration of water molecules (such as a dilute sucrose sugar solution) to where there is a low concentration of water molecules (such as a concentrated sucrose sugar solution).

Osmosis is the diffusion (net movement) of water molecules from a region of their higher concentration to a region of their lower concentration through a partially permeable membrane.

Concentrations in solutions

Many people confuse the concentration of the solution with the concentration of the water. Remember, in osmosis it is the *water molecules* that we are considering, so you must think of the concentration of water molecules in the solution instead of the concentration of solutes dissolved in it.

- A low concentration of dissolved solutes means a high concentration of water molecules.
- A high concentration of dissolved solutes means a low concentration of water molecules.

So, the water molecules are moving from a *high concentration (of water molecules)* to a *low concentration (of water molecules)*, even though this is often described as water moving from a low-concentration solution to a high-concentration solution.

Water potential

The ability of a cell to draw water into itself is called its **water potential**. Pure water has a water potential of zero. As solutes are added to the water its water potential falls – it becomes more negative. So, a concentrated sugar solution has a lower water potential (more negative) than pure water.

When two regions of different water potential are separated by a partially permeable membrane, water moves from the region of higher water potential to lower water potential. Water molecules move *down* the **water potential gradient**.

Osmosis in plant cells

If a cell is placed in a solution that has a higher concentration of solute (and so a lower concentration of water molecules) than the cytoplasm inside the cell, water will leave the cell by osmosis and the cytoplasm will shrink.

If a cell is placed in a solution that has a lower concentration of solute (and so a higher concentration of water molecules) than the cytoplasm inside the cell, osmosis will result in water entering the cell.

Plant cells are surrounded by cell walls that are completely permeable. This means water and solute molecules pass easily through them.

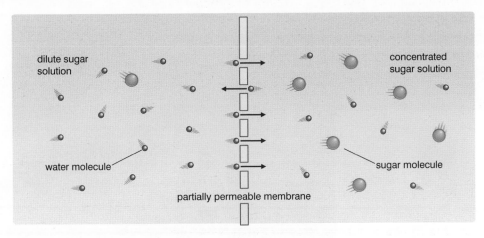

△ Fig. 2.16 Water molecules diffuse from an area of higher concentration (of water molecules) into an area of lower concentration (of water molecules). This kind of diffusion is known as osmosis.

In a solution of high concentration of solute (low concentration of water), water will leave a plant cell by osmosis. This can be seen in plant cells as the cytoplasm shrinks inside the cell. However, the whole cell doesn't shrink, because the plant cell wall controls the structure of the cell. The plant as a whole will show wilting.

In a solution of low concentration of solute (high concentration of water), water will enter the plant cell by osmosis. However, the plant cell does not eventually burst. The strong cell wall provides strength when the cytoplasm is full of water, preventing the cell from expanding any further and bursting. The pressure of water in the cytoplasm against the cell wall also provides strength, making the plant stand upright with its leaves held out to catch the sunlight.

△ Fig. 2.17 When the cells of the plant are not full of water, the cell walls are not strong enough to support the plant, and the plant collapses (wilts). When the cells are full of water, the plant stands upright.

Developing investigative skills

Strips of dandelion stem about 5 cm long and 3 mm wide were placed in sodium chloride solutions of different concentrations. After 10 minutes, the strips looked as shown in the diagram. (Note that the outer layer of a dandelion stalk is 'waterproofed' with a waxy layer to protect it from water loss to the environment.)

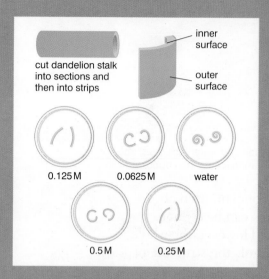

⊲ Fig. 2.18 Investigating osmosis.

Devise and plan investigations

❶ Write a plan for an experiment to carry out this investigation. Your plan should include:

a) instructions on how to prepare the stem samples

b) instructions on how to keep the stem samples until the experiment starts.

Make observations and measurements

❷ Using the diagram, describe the results of this investigation.

Analyse and interpret data

❸ Explain as fully as you can the results of this investigation.

❹ Use the results to suggest the normal concentration of cell cytoplasm. Explain your answer.

SCIENCE IN CONTEXT: STOMATA

Stomata (single: stoma) are the holes in the surface of a leaf (usually the undersurface) that allow air to move into and out of the leaf. This provides the oxygen for respiring cells and carbon dioxide for photosynthesising cells, and allows water vapour that has evaporated from cell surfaces inside the leaf to diffuse out into the atmosphere.

Each stoma is surrounded by two guard cells. These control the opening and closing of the stoma. Usually stomata are open during the day and close at night. The stoma opens and closes as the guard cells change shape. During the day the guard cells gain water from surrounding cells as a result of osmosis. This makes the cells turgid and, because the inner edge of the guard cell does not stretch, the cells curve and create a space between them – that is the stoma. During the night, the guard cells lose water by osmosis. The cells lose their turgidity and collapse a little, closing the stoma between them.

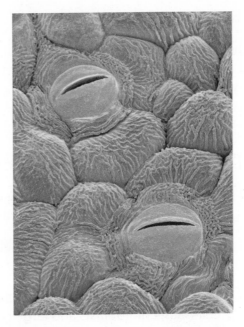

△ Fig. 2.19 Each stoma is surrounded by two guard cells.

REMEMBER

You will need to explain diffusion and osmosis in terms of particles and their concentration gradients. Be clear that, even when diffusion and osmosis stop because there is no concentration gradient, the particles in the solution continue to move – there is just no longer any net movement.

QUESTIONS

1. **EXTENDED** In your own words, define the term osmosis.

2. Explain how osmosis is:

   a) similar to diffusion

   b) different from diffusion.

3. Describe the role of the plant cell wall in supporting a plant that has been well watered.

End of topic checklist

Key terms

acrosome, cell membrane, cell wall, chloroplast, chromosome, cilia, ciliated cell, concentration gradient, cytoplasm, diffusion, egg cell, magnification, multicellular, net movement, nucleus, osmosis, partially permeable, passive, photosynthesis, root hair cell, specialisation, sperm cell, vacuole, water potential gradient

During your study of this topic you should have learned:

◯ How to describe structures inside cells including the nucleus, cytoplasm, cell membrane, cell wall, chloroplast and vacuole.

◯ EXTENDED That the cytoplasm, cell membrane, cell wall, chloroplast and vacuole have specific roles in cells.

◯ That plant and animal cells have some structures in common, but plants also have cell walls, chloroplasts and large central vacuoles that animal cells do not have.

◯ How to calculate the magnification and size of biological specimens.

  ● The magnification of a specimen seen under a microscope is the magnification of the eyepiece multiplied by the magnification of the objective used.

  ● The size of a structure seen under a microscope is the measured size divided by the magnification.

◯ To define diffusion as the net movement of particles from a region of their higher concentration to a region of their lower concentration, and as a passive process.

◯ That water diffuses through partially permeable membranes by osmosis.

◯ EXTENDED To define osmosis as the net movement of water molecules across a partially permeable membrane from their higher concentration (a dilute solution) to their lower concentration (a more concentrated solution), and is a passive process.

◯ How to investigate and describe osmosis in plant tissues.

End of topic questions

Note: The marks given for these questions indicate the level of detail required in the answers. In the examination, the number of marks given to questions like these may be different.

1. Describe the role of the following cell structures:

    a) nucleus   **b)** cell membrane   **c)** cytoplasm.   **(3 marks)**

2. Draw a table to compare the structures found in plant and animal cells.   **(12 marks)**

3. Here are some examples of statements written by students. Each statement contains an error. Identify the error and rewrite the statement so that it is correct.

    a) Animal cells are surrounded by a cell wall that controls what enters and leaves the cell.   **(1 mark)**

    b) All plant cells contain chloroplasts.   **(1 mark)**

    c) Both animal cells and plant cells contain a large central vacuole in the middle of the cell.   **(1 mark)**

4. Red blood cells are unusual because they contain no nucleus. When they are damaged, they have to be replaced with new cells from the bone marrow. Explain how this is different from other cells.   **(2 marks)**

5. An old-fashioned way of killing slugs in the garden is to sprinkle salt on them. This kills the slugs by drying them out. Explain why this works.   **(2 marks)**

6. Copy and complete the table to compare diffusion and osmosis.

	Diffusion	Osmosis
Active or passive?		
Which molecules move?		

    (4 marks)

7. Which of the following are examples of diffusion, osmosis or neither?

    a) Carbon dioxide entering a leaf when it is photosynthesising.   **(1 mark)**

    b) Food entering your stomach when you swallow.   **(1 mark)**

    c) A dried-out piece of celery swelling up when placed in a bowl of water.   **(1 mark)**

8. There are many membranes within a cell, separating off organelles that produce substances such as hormones and enzymes, or where cell processes such as photosynthesis and respiration occur. Explain fully the importance of these membranes and why it is an advantage to the cell to have them.   **(3 marks)**

There are around 90,000 different proteins in the human body, and thousands more in other organisms. What is unique about proteins is that each one may be a different shape. The different shapes make it possible for proteins to carry out different roles in the body.

All these many thousands of proteins are made up of chains of smaller molecules, called amino acids, which join together in long chains. All proteins are made up from different combinations of just 20 amino acids. It is the order of these amino acids in the chain, and the way that the chain folds up into its 3D shape, that gives each protein its unique function.

SECTION CONTENTS
Biological molecules

3

Biological molecules

△ Molecular model of the enzyme
alpha-amylase from the human pancreas.

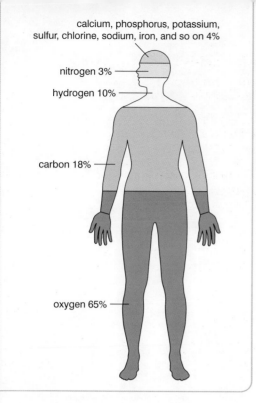

calcium, phosphorus, potassium, sulfur, chlorine, sodium, iron, and so on 4%

nitrogen 3%

hydrogen 10%

carbon 18%

oxygen 65%

△ Fig. 3.1 The proportions of elements in the human body.

Biological molecules

INTRODUCTION

Around 65% of your body mass is oxygen, another 18% is carbon and 10% is hydrogen. The remainder of your mass is made up of a large range of other elements, including nitrogen, sulfur, calcium and iron. These elements are combined in different ways to form all the compounds in your body.

KNOWLEDGE CHECK

✓ Most of the foods that we eat can be grouped into carbohydrates, proteins and fats.
✓ Carbohydrates, proteins and fats are formed from smaller molecules.

LEARNING OBJECTIVES

✓ Name the elements in carbohydrates, fats and proteins.
✓ Name the basic units from which carbohydrates, fats and proteins are made.
✓ State that water is important as a solvent.
✓ Describe tests for starch, reducing sugars, protein, fats and oils.

BIOLOGICAL MOLECULES

Carbohydrates, proteins and lipids

Most of the molecules found in living organisms fall into three main groups: carbohydrates, proteins and lipids. (Lipids are commonly called fats and oils.) All of these molecules contain carbon, hydrogen and oxygen. In addition, all proteins contain nitrogen and some also contain sulfur.

Carbohydrate molecules are made up of small basic units called **simple sugars**. These are formed from carbon, hydrogen and oxygen atoms, sometimes arranged in a ring-shaped molecule. One example of a simple sugar is glucose.

Simple sugar molecules can link together to form larger molecules. They can join in pairs, for example, **sucrose** (the 'sugar' we use in our food). They can also form much larger molecules called polysaccharides, for example, starch and glycogen, which are long chains of glucose molecules.

Protein molecules are made up of long chains of **amino acids** linked together. There are 20 different kinds of amino acid, in plant and animal cells, and they can join in any order, in long chains, to make all the different proteins within the plant or animal body. Examples include the structural proteins in muscle, as well as enzymes that help to control cell reactions.

A **lipid** is what we commonly call a fat or oil. At room temperature **fats** are solid and **oils** are liquid, but they have a similar structure. Both fats and oils are made from basic units called **fatty acids** and **glycerol**. There are three fatty acids in each lipid, and the fatty acids vary in different lipids. Lipids are important in forming cell membranes, and many other molecules in the body, such as fats in storage cells.

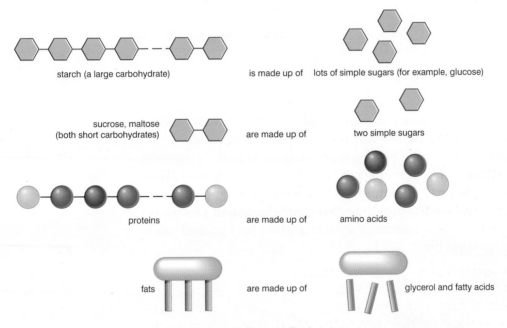

△ Fig. 3.2 Large biological molecules are formed from small sub units.

QUESTIONS

1. What are the basic units of:

a) lipids?

b) carbohydrates?

c) proteins?

2. Using the diagram of food molecules in Fig. 3.2, give two differences between the structure of a protein and a carbohydrate.

Water

Water is a very good solvent, which makes it essential for living organisms, as molecules such as glucose and small proteins can dissolve in it. Some of the blood and the cytoplasm of cells is made of water, meaning molecules that dissolve in water can be transported around the body and reach cells.

QUESTION

1. Why is water essential for living organisms?

Tests for food molecules

We can use simple tests to indicate whether or not a food contains particular food molecules, such as starch, glucose, proteins or lipids.

Test for starch

Starch is the storage molecule of plants and is found in many foods that are made from plant tissue. When iodine/potassium iodide solution is mixed with a solution of food containing starch, or dropped onto food containing starch, it changes from brown to dark blue. This happens when even small amounts of starch are present and can be used as a simple test for the presence of starch. The colour change is easiest to see if the test is examined against a white background, such as on a white spotting tile.

◁ Fig. 3.3 The blue–black colour shows there is starch in the biscuit.

Test for glucose

Glucose is a 'reducing sugar' that is important in respiration and photosynthesis. So it is commonly found in plant and animal tissues, and therefore in our food. Its presence can be detected using **Benedict's reagent**. The pale blue Benedict's solution is added to a prepared sample that contains glucose and is heated to 95 °C. If it changes colour or forms a precipitate, this indicates the presence of reducing sugars. A green colour means there is only a small amount of glucose in the solution. A medium amount of glucose produces a yellow colour. A significant amount of glucose produces a precipitate that is an orange–red colour.

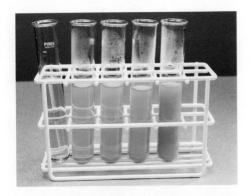

◁ Fig. 3.4 Benedict's reagent with a range of concentrations of reducing sugars (very low in the tube on the left, getting more concentrated towards the right).

Test for protein

The **biuret test** is used to check for the presence of protein. A small sample of the food under test is placed in a test tube.

◁ Fig. 3.5 A positive biuret test for protein.

An equal volume of biuret solution is carefully poured down the side of the tube. If the sample contains protein, a blue ring forms at the surface. If the sample is then shaken, the blue ring disappears and the solution turns a light purple.

Test for fat

This test depends upon the fact that fats and oils do not dissolve in water but do dissolve in ethanol. The test sample is mixed with ethanol. If fat is present, it will be dissolved in the ethanol to form a cloudy solution. The liquid formed is poured into a test tube of water, leaving behind any solid that has not dissolved. If there is any fat dissolved in the ethanol, it will form a cloudy white precipitate when mixed with the water.

△ Fig. 3.6 Fats have dissolved in the top layer of ethanol, making it appear cloudy.

QUESTIONS

1. Describe what you would see if you tested samples of the following with **i)** Benedict's solution and then **ii)** iodine solution:

   a) glucose syrup

   b) a cake made with wheat flour, table sugar (sucrose), fat and eggs.

   Explain your answers.

2. Explain how you would test the seed from a walnut tree to see if it contained stores of:

   a) fat

   b) protein.

SCIENCE LINK **CHEMISTRY – ATOMS, ELEMENTS AND COMPOUNDS**

- Biological molecules can be very complicated, but they are still made from the same simple atoms as all other materials.

- Ideas about conservation of mass mean that only the atoms available at the start of any change can be present at the end (but they must all be there!)

- Carbohydrates, composed of the elements carbon, hydrogen and oxygen, are particularly important in this topic.

- Some of the larger molecules (such as starch) are made by joining smaller, simpler molecules together (glucose in the case of starch).

- Chemical tests, such as adding iodine, are used to identify a range of different substances.

End of topic checklist

Key terms

amino acid, Benedict's reagent, biuret test, carbohydrate, fat, fatty acid, glycerol, lipid, oil, protein, simple sugar, starch, sucrose

During your study of this topic you should have learned:

○ That carbohydrates, proteins, fats and oils all contain the elements carbon, hydrogen and oxygen.

○ That proteins also contain the element nitrogen and some may contain sulfur.

○ That large carbohydrates, such as starch and glycogen, are made up of smaller carbohydrates (reducing sugars) such as glucose.

○ Proteins are made of smaller molecules called amino acids.

○ To state that fats and oils are made of smaller molecules called fatty acids and glycerol.

○ To state that water is an essential solvent for living organisms, and forms the basis of cell cytoplasm in which many metabolic reactions occur.

○ To describe the use of the iodine test to identify the presence of starch.

○ To describe the use of Benedict's reagent to test for the presence of simple reducing sugars such as glucose.

○ To describe the use of the biuret test to identify the presence of proteins.

○ To describe the use of the ethanol emulsion to identify the presence of fats and oils.

End of topic questions

Note: The marks given for these questions indicate the level of detail required in the answers. In the examination, the number of marks given to questions like these may be different.

1. **a)** Explain why carbon, hydrogen and oxygen are the most common elements found in the human body.

   (1 mark)

   b) Why does the body need other elements, in addition to those mentioned in part **a)**?

   (1 mark)

2. A sample of bread was ground up. Some of the breadcrumbs were tested with Benedict's reagent and some with iodine solution. The rest of the crumbs were mixed with Substance A. After 20 minutes, some of the mixture was tested with Benedict's reagent and some with iodine solution. The results of the tests are shown in the table.

	Test with Benedict's solution	Test with iodine solution
Before adding Substance A	no precipitate	change to blue–black colour
After 20 min with Substance A	orange–red precipitate	no colour change

   a) Describe what the results show.

   (2 marks)

   b) What was Substance A? Explain your answer.

   (4 marks)

Many useful enzymes come from bacteria and other microorganisms, and these organisms are harnessed both in industry and in the home for uses as wide-ranging as washing detergents, to baking leavened bread. Yeast is a single-celled organism that produces enzymes that break down the sugars in flour, and in the process tiny bubbles of carbon dioxide gas are released, which cause the bread to rise.

The first enzyme used commercially in washing products was introduced in the 1960s. It was a protease that broke down protein-based stains such as blood, and it was extracted from a bacterium. Since then, a much wider range of enzymes has been added to washing products, to digest fats, starches and other molecules.

SECTION CONTENTS

4
Enzymes

△ Enzymes released into the gut, and attached to the gut surface (shown here), digest food so that nutrients can be absorbed.

Enzymes

INTRODUCTION

Many of our staple foods, such as rice, potato, pasta and bread, contain large quantities of starch. Take a mouthful of one of these, without anything else, and you won't taste a lot to start with. But continue chewing on it for a few minutes, to mix it with saliva and reduce it to a slush, and you will find it starts to taste sweeter. This is because there are enzymes in saliva that start to break down the starch into smaller sugar molecules that taste sweet. Enzymes are essential in digestion, to break down the large molecules in our food into molecules that are small enough for diffusion through the cells of the gut wall and into our bodies.

△ Fig. 4.1 Enzymes in the mouth, stomach and small intestine will break down this food into much smaller molecules.

KNOWLEDGE CHECK

✓ Food is digested in the gut into smaller molecules.

LEARNING OBJECTIVES

✓ Define the term *catalyst*.
✓ Describe enzymes as proteins that are biological catalysts.
✓ EXTENDED Describe enzyme action.
✓ Describe the effect of temperature and pH on the rate of an enzyme-controlled reaction.
✓ Investigate the effect of temperature and pH on enzyme activity.
✓ EXTENDED State that enzymes catalyse reactions in which substrates are converted to products.
✓ EXTENDED Describe the importance of the shape of the active site.
✓ EXTENDED Explain the effect of temperature and pH on enzyme activity.
✓ EXTENDED Explain the specificity of enzymes.

ENZYMES

A **catalyst** is a substance that changes the speed of a reaction, often speeding it up. Catalysts are used in many industrial processes, for example, making ammonia. Living cells also use catalysts to change the rate of reactions that happen inside them. These are known as *metabolic reactions* because they are the reactions of the **metabolism** (all the processes that keep a living organism alive). This makes enzymes very important to all living organisms.

Catalysts that control metabolic reactions are **enzymes**, and, because they work in living cells, they are called **biological catalysts**. Enzymes are proteins. They help cells carry out all the life processes quickly. Without them, most metabolic reactions would happen too slowly for life to carry on.

Some enzymes help two or more small molecules join together, for example, when the polysaccharides starch and glycogen are built from glucose. Other enzymes help large molecules break down into smaller ones, for example, when proteins are broken down into separate amino acids.

A molecule that an enzyme joins with at the start of a reaction is called a **substrate**, and the molecule that is formed by the end of the reaction is called a **product**. So, during a reaction, substrate molecules are changed to product molecules.

QUESTIONS

1. Define the term *catalyst*.

2. Explain what is meant by *biological catalyst*.

3. Explain why cells need enzymes.

4. Define the terms *substrate* and *product*.

EXTENDED

Enzyme action

Enzymes are proteins and, like all proteins, they have a three-dimensional (3D) shape produced by the way the molecule folds up. Enzymes have a space in the molecule with a particular 3D shape. This space matches the shape of the substrate molecule. We say the shapes are **complementary**, because the substrate fits neatly into the space in the enzyme, like fitting two jigsaw pieces together.

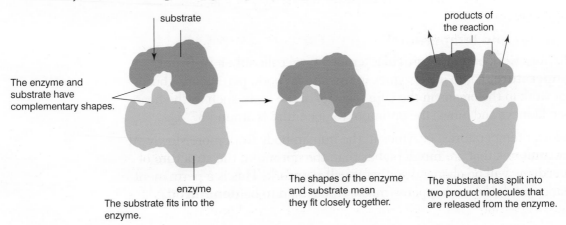

substrate

products of the reaction

The enzyme and substrate have complementary shapes.

enzyme
The substrate fits into the enzyme.

The shapes of the enzyme and substrate mean they fit closely together.

The substrate has split into two product molecules that are released from the enzyme.

△ Fig. 4.2 In this reaction, the enzyme helps a substrate molecule split into two product molecules.

The space in the enzyme shape into which the substrate fits is called the enzyme's **active site**. The substrate fits tightly into the active site, forming an enzyme–substrate complex. This makes it easier for the bonds inside the substrate to be rearranged to form the products. Once the products are formed, they no longer fit the active site, so they are released, leaving the active site free and the enzyme unchanged. This means the enzyme molecule is able to bind with another substrate molecule.

Explaining enzyme action

Enzymes are **specific**, which means that each enzyme only works with one substrate or a group of similar-shaped substrates. For example:

- amylase is a type of carbohydrase enzyme produced in the mouth, which starts the digestion of starch in food into simple sugars
- proteases are digestive enzymes that break down proteins into smaller units
- lipases are digestive enzymes that break down lipids in foods.

The complementary shapes of the enzyme and substrate helps to explain the fact that enzymes are specific, because only a substrate with the correct shape can fit into the active site and so be affected by the enzyme.

END OF EXTENDED

QUESTIONS

1. **EXTENDED** Describe how an enzyme causes a substrate molecule to change into product molecules.
2. **EXTENDED** Explain what is meant by the active site of an enzyme.
3. **EXTENDED** Explain how the shape of the active site is related to the specificity of an enzyme.

Enzymes and temperature

Enzymes work best at a particular temperature, called their **optimum temperature**. For many enzymes in the human body, particularly those that work in the organs in the core (centre) of the body, such as the heart, liver, kidneys and lungs, the optimum temperature is around 37 °C.

At lower temperatures, enzymes in the human body work more slowly. At temperatures that are much higher than the optimum, the structure of an enzyme will be changed so that it will not work. This is a permanent change, and when it happens the enzyme is said to be **denatured**.

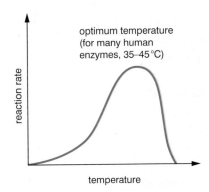

optimum temperature
(for many human
enzymes, 35–45°C)

reaction rate

temperature

△ Fig. 4.3 Many enzymes work best at an optimum temperature.

Remember the relationship between enzyme activity, temperature and pH, particularly when discussing excretion and homeostasis, as this helps to explain why maintaining particular conditions in the body is so important for health.

Investigating the effect of temperature on enzymes

The effect of temperature on an enzyme can be tested by measuring the rate of reaction of the enzyme at different temperatures. One method is shown in the Developing investigative skills box on page 52. Alternatively, you could use the following method to investigate the optimum temperature of amylase.

Starch is broken down to glucose by the enzyme amylase. Starch reacts with iodine solution by turning it blue–black. Glucose does not react with iodine solution, leaving it bright orange. If you mix starch solution with amylase solution and place different tubes of the mixture in water baths of different temperature and take a sample for testing with iodine solution every minute or so, you can see at which temperature the amylase works fastest. The sample that is the first to stop reacting with iodine solution comes from the tube kept at the optimum temperature for amylase.

EXTENDED

The effect of temperature

Kinetic energy is the energy of moving particles. Particles that have a greater kinetic energy move more or move faster. The kinetic energy of molecules that are free to move will cause them to bump into surrounding molecules. The kinetic energy of atoms held within larger molecules by bonds will cause them to vibrate.

An enzyme molecule and substrate molecule can only form an enzyme–substrate complex when they bump into each other with sufficient energy, and the substrate fits into the active site.

- At a low temperature the enzyme and substrate molecules move slowly, so they may take a long time to bump into each other with enough force to join and start the reaction.

- As the temperature increases, the molecules gain more energy and move faster, so the chance of them bumping into each other and joining together increases. The rate of reaction increases up to the optimum temperature.
- Beyond the optimum temperature, the atoms in the enzyme molecule are vibrating so much that they start to change the shape of the active site. This means the substrate doesn't fit as well, so the chances of an enzyme–substrate molecule forming decreases. The rate of reaction decreases.
- If the temperature increases too much, the bonds between atoms in the enzyme molecule start to break, changing the shape of the active site permanently and denaturing the enzyme.

END OF EXTENDED

Developing investigative skills

Developed black-and-white negative film consists of a celluloid backing covered with a layer of gelatin. Where the film has been exposed the gelatin layer contains tiny particles of silver, which make that area black. Gelatin is a protein and is easily digested by proteases.

Strips of exposed film were soaked in protease solution at different temperatures. When the gelatin had been digested, the silver grains fell away from the celluloid backing, leaving transparent film. The table shows the results.

Tube	Temperature in °C	Time to clear
1	10	6 min 34 s
2	20	3 min 15 s
3	30	2 min 43 s
4	40	3 min 55 s
5	50	8 min 33 s

Devise and plan investigations

❶ Describe how you would set up this investigation to get results like those shown in the table.

Analyse and interpret data

❷ Draw a graph using the data in the table.

❸ Describe the shape of the graph.

❹ Explain the shape of the graph.

Evaluate data and methods

❺ How could you modify this experiment to get a more accurate estimate of the optimum temperature for this enzyme?

ENZYMES

Enzymes and pH

Enzymes also often work best at a particular pH, called their **optimum pH**. Extremes of (very high or very low) pH can slow down the rate of reaction of the enzyme and even denature it.

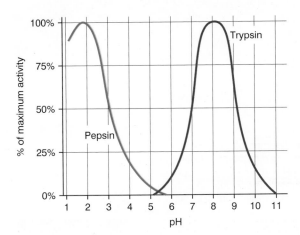

△ Fig. 4.4 Pepsin (an enzyme found in the stomach) and trypsin (an enzyme released into the small intestine) have different optimum pHs.

Different enzymes have different optimum pHs, depending on where they are normally found in the body. Pepsin digests proteins in the stomach, which is a highly acidic environment. Trypsin digests proteins in the small intestine, where conditions are more alkaline.

Investigating the effect of pH on enzymes

You can investigate the effect of pH on the enzyme amylase using a similar method to the one above for temperature.

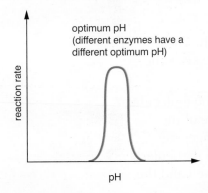

△ Fig. 4.5 Many enzymes work best at an optimum pH.

Set up one tube for each pH to be investigated and add buffer solution, which will keep the contents at a particular pH. Add starch solution to each tube, and then amylase solution. Take a sample from each tube every minute or so and test for starch using iodine solution. The sample that is the first to stop turning iodine solution blue–black comes from the tube where digestion of starch to glucose was fastest, and therefore from the tube kept at the optimum pH for that enzyme.

△ Fig. 4.6 The tubes show the results of an experiment on the digestion of meat. Pepsin is a protease enzyme that is released in the stomach where it starts the digestion of proteins in food. Acid is also secreted into the stomach contents, reducing the pH and providing the optimum pH for pepsin. The left tube shows that acid has no effect on the meat. The middle tube shows that pepsin on its own digests the meat slowly. Only when the pepsin is mixed with acid can the enzyme work quickly to digest the protein.

QUESTIONS

1. Describe the effect of temperature on the rate of an enzyme-controlled reaction.

2. Compare the optimum pHs for pepsin and trypsin, shown in the graph in Fig. 4.4, and explain the differences.

The effect of pH

Proteins are made of amino acids, joined together in a chain. The amino acids then interact with nearby amino acids, which causes the chain to fold up into the 3D shape of the enzyme.

Some of the interactions between amino acids in the enzyme molecule depend on the pH of the surrounding solvent. So, the shape of the enzyme will depend on the surrounding pH. If the pH changes too much from the optimum pH, the shape of the enzyme, and particularly its active site, will change. So, the substrate will not fit as well and the rate of reaction will decrease.

QUESTIONS

1. Explain the effect of temperature on enzyme activity:

   a) at temperatures below the optimum

   b) at temperatures above the optimum.

2. Explain the effect of pH on pepsin (see graph in Fig. 4.4) in terms of the active site of the enzyme.

END OF EXTENDED

SCIENCE LINK **CHEMISTRY – CHEMICAL REACTIONS**

- Catalysts are present in a range of chemical reactions that happen in a biological context. Remember that they help the chemical reaction to happen without being 'used up'.

- As with any chemical reactions, the reactions involving enzymes are affected by temperature, so, ideas about how the motion and energy of particles vary at different temperatures are important.

- Enzymes are affected by the acidity of their environment, which links to ideas of pH, acids, bases and alkalis.

End of topic checklist

Key terms

active site, biological catalyst, catalyst, complementary, denature, enzyme, kinetic energy, metabolism, optimum pH, optimum temperature, product, specific, substrate

During your study of this topic you should have learned:

○ How to define the term catalyst.

○ How to define enzymes as proteins that are biological catalysts, which control the rate of metabolic reactions.

○ How to describe that enzymes and substrates have complementary shapes so that they fit closely together.

○ How to investigate the optimum temperature of enzymes at which the rate of reaction occurs most rapidly; the rate is slower at lower temperatures, and also at higher temperatures when the enzyme molecule starts to denature.

○ That enzymes may have an optimum pH at which the rate of reaction happens most rapidly.

○ **EXTENDED** How to explain the active site as the space in the enzyme into which the substrate fits neatly during a reaction.

○ **EXTENDED** How to explain that a temperature lower than the optimum causes a slower rate of an enzyme-controlled reaction because the molecules move around more slowly and so don't come into contact with each other as often.

○ **EXTENDED** How to explain that a temperature higher than the optimum causes a slower rate of an enzyme-controlled reaction because the vibration of atoms in the enzyme slightly changes the shape of the active site so that the substrate does not fit as easily into it.

○ **EXTENDED** How to explain that a very high temperature denatures the enzyme as interactions between amino acids break and change the shape of the active site completely.

○ **EXTENDED** How to explain that pH affects the interactions between amino acids in the enzyme molecule and so the ability of the substrate to fit into the active site.

○ **EXTENDED** That enzymes catalyse reactions in which substrates are converted to products.

○ **EXTENDED** How to describe enzyme action with reference to the active site, substrate and product.

End of topic questions

Note: The marks given for these questions indicate the level of detail required in the answers. In the examination, the number of marks given to questions like these may be different.

1. Describe how you would investigate the optimum temperature for a particular enzyme. **(4 marks)**

2. Sketch a graph to show the effect of temperature on the rate of reaction for an enzyme from humans. Label the value of the optimum temperature on your graph. **(2 marks)**

3. The body has many mechanisms for keeping internal conditions within limits. One of the internal conditions that is controlled is the concentration of carbon dioxide in the blood. Carbon dioxide gas is acidic and highly soluble.

   a) Which process in cells produces carbon dioxide? **(1 mark)**

   b) How is this gas removed from the body? **(1 mark)**

   c) What would you expect to happen to the amount of carbon dioxide in the body during exercise? Explain your answer. **(2 marks)**

   d) What effect would this have on conditions inside cells if the carbon dioxide was not removed? **(1 mark)**

   e) What problem would this cause for enzymes and the cell processes that they control? **(2 marks)**

4. EXTENDED Explain fully the shape of the graph you drew for Question 2. **(5 marks)**

The world's tallest known living tree, which has been named Hyperion, is a coast redwood tree growing in Northern California in the US. When it was measured in 2006 it was found to be 115.61 m tall (379.3 ft). Like all plants, from the smallest seedling to the tallest redwood giant, Hyperion makes its own food from three simple ingredients, sunlight, water and carbon dioxide. This process is known as photosynthesis and is one of the features of plants that distinguish them from animals.

5
Plant nutrition

Δ Saplings growing toward a controlled light source.

Plant nutrition

INTRODUCTION

From space, we can see where plants do or do not grow. We can distinguish different environments by looking at where the land is green, brown or white. The green areas are a result of chlorophyll in photosynthesising plants. We can also see where land use is changing, by looking at how the green areas of rainforests are slowly becoming brown as a result of deforestation.

△ Fig. 5.1 The green on this satellite image shows plant growth on Earth.

KNOWLEDGE CHECK

✓ Plants make their own food in their leaves using photosynthesis.
✓ Plant structures, such as the leaf and root cells, are adapted for their functions in nutrition.

LEARNING OBJECTIVES

✓ State the word equation for photosynthesis.
✓ Define the term *photosynthesis*.
✓ Investigate the need for chlorophyll, light and carbon dioxide for photosynthesis.

✓ EXTENDED Write the balanced symbol equation for photosynthesis.

✓ EXTENDED Explain the importance of photosynthesis for plant nutrition.

✓ EXTENDED Investigate the effect of varying light intensity and temperature on the rate of photosynthesis.

✓ Identify structures in a leaf.

✓ EXTENDED Explain how a leaf is adapted for photosynthesis.

✓ Describe the importance of some mineral ions in plant growth.

✓ EXTENDED Explain the effect of the deficiency of some minerals on a plant.

PLANT NUTRITION

Photosynthesis

Plant tissue contains the same types of chemical molecules (carbohydrates, proteins and lipids) as animal tissue. However, whereas animals eat other organisms to get the nutrients they need to make these molecules, plants make these molecules from basic building blocks, beginning with the process of **photosynthesis**.

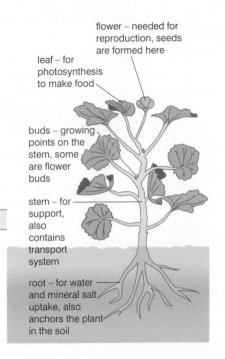

flower – needed for reproduction, seeds are formed here

leaf – for photosynthesis to make food

buds – growing points on the stem, some are flower buds

stem – for support, also contains transport system

root – for water and mineral salt uptake, also anchors the plant in the soil

△ Fig. 5.2 Anatomy of a plant.

EXTENDED

In photosynthesis, plants combine the raw materials carbon dioxide (from the air) and water (from the soil) to form glucose, a simple sugar and also a carbohydrate. This process transfers energy from light (usually from sunlight) into chemical energy in the bonds of the glucose. The light is absorbed by the green pigment **chlorophyll** in plants.

END OF EXTENDED

Photosynthesis is fundamental to almost all life on Earth, because most organisms, other than plants, get their energy from the chemical energy in the food that they eat, whether that is herbivores getting energy directly from plants or carnivores consuming the herbivores.

Oxygen is also produced in photosynthesis. Although some is used inside the plant for respiration (releasing energy from food), most is not needed and is given out as a **waste product**.

The process of photosynthesis can be summarised in a word equation:

$$\text{carbon dioxide} + \text{water} \xrightarrow[\text{light energy}]{\text{chlorophyll}} \text{glucose} + \text{oxygen}$$

EXTENDED

Photosynthesis can also be summarised as a balanced symbol equation:

$$6CO_2 + 6H_2O \xrightarrow[\text{light energy}]{\text{chlorophyll}} C_6H_{12}O_6 + 6O_2$$

END OF EXTENDED

REMEMBER

To write a really good answer you will need to know, and be able to balance, the chemical equation for photosynthesis.

Much of the glucose formed by photosynthesis is converted into other substances, including **starch**. Starch molecules are large carbohydrates made of lots of glucose molecules joined together. Starch is insoluble and so can be stored in cells without affecting water movement into and out of the cells by **osmosis**. Some plants, for example, potato and rice plants, store large amounts of starch in particular parts of the plant (tubers or seeds). We use these parts as sources of starch in our food.

Some glucose is converted to **sucrose** (a type of sugar formed from two glucose molecules joined together). This is still soluble, but not as reactive as glucose, so can easily be carried around the plant in solution.

The energy needed to join simple sugars to make larger carbohydrates comes from respiration.

END OF EXTENDED

QUESTIONS

1. Write the word equation for photosynthesis.

2. Explain the importance of light in photosynthesis.

3. a) EXTENDED Write the balanced symbol equation for photosynthesis.

  b) EXTENDED Annotate your equation to show where each of the reactants come from, and each of the products go to.

4. EXTENDED Explain why the transfer of energy from light to chemical energy in plant cells is essential for life on Earth.

SCIENCE LINK

CHEMISTRY – STOICHIOMETRY

- Photosynthesis is a chemical reaction, so this is an opportunity to link in ideas of word equations and balanced symbol equations.

- Remember that the formula for each compound is fixed, so to keep in line with the law of conservation of mass we need to think about the complete molecules when balancing an equation.

- Photosynthesis is an example of an endothermic reaction.

PHYSICS – PROPERTIES OF WAVES

- Photosynthesis requires the energy from light to work.

- The colours in the light link to different wavelengths in the electromagnetic spectrum – some wavelengths are more useful in photosynthesis than others.

Investigating photosynthesis

We can use the iodine test to show that photosynthesising parts of a plant produce starch. Before carrying out this test, though, we must start by leaving the plant in a dark place for 24 hours. This will make sure that the plant uses up its stores of starch (this is known as destarching) and means that any starch identified by the test is the result of photosynthesis during the investigation.

- The production of starch after photosynthesis can be shown simply by placing a destarched plant in light for an hour. Remove one leaf and place it in boiling water for a few minutes to soften it. Then place the leaf in boiling ethanol heated in a beaker of boiling water, not over a Bunsen because ethanol fumes are flammable. This removes the chlorophyll in the leaf. When the leaf has lost its green, wash it in cold water before placing it in a dish and adding a few drops of iodine solution. The leaf should turn blue–black, indicating the presence of starch.

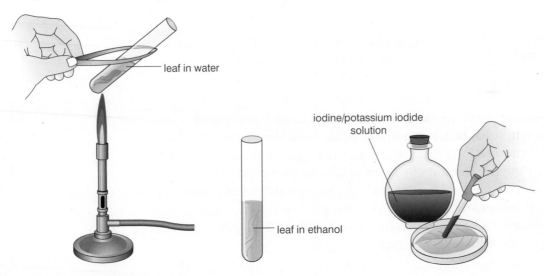

leaf in water

iodine/potassium iodide solution

leaf in ethanol

△ Fig. 5.3 Preparing and testing a leaf for starch.

- The investigation above can be adjusted to show the need for light by covering part of the leaf before the destarched plant is brought into the light. Only the part of the leaf that received light should test positive for the presence of starch, showing that photosynthesis is linked to the production of starch.

- This investigation can also be adjusted to show the need for chlorophyll by using variegated leaves. Variegated leaves are partly green (where the cells contain chlorophyll) and partly white (where there is no chlorophyll). A variegated leaf after this

△ Fig. 5.4 Light was excluded from all of the lower leaf except an L-shaped window. After exposure to light, only the L shape tests positive for starch.

investigation will show the presence of starch where there was chlorophyll but not in the parts of the leaf that had no chlorophyll.

- A simple test to show the need for carbon dioxide can be carried out by setting up two bell jars on glass sheets. Sodium or potassium hydroxide reacts with carbon dioxide, removing it from the air. So a dish of one of the hydroxides is placed in one bell jar. Carbon dioxide is added to the other bell jar by burning a candle in it, which also removes some of the oxygen. Similar destarched plants are placed in each bell jar, and the base of the jar sealed to the glass sheet, for example, with petroleum jelly. After a few hours in the light, a leaf from each plant is tested for starch, which should show that the plant with the least carbon dioxide produces little starch.

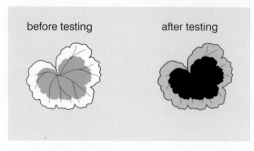

△ Fig. 5.5 Only the green parts of a variegated leaf can photosynthesise, as shown by the leaf on the right, which has been tested for starch.

QUESTIONS

1. Describe a test that would show the need for chlorophyll in photosynthesis.

2. What precautions should be taken when boiling ethanol to remove chlorophyll in a leaf? Explain your answer.

Investigating the rate of photosynthesis

Measuring starch production is an indirect measurement of photosynthesis, because starch is made from the glucose produced in photosynthesis. You can investigate photosynthesis more directly by measuring the amount of oxygen produced by a plant. The oxygen is usually collected over water, and these investigations are most simply done using aquatic plants (plants that grow in water), such as pondweed, using the apparatus shown in the Developing investigative skills box on the following page.

- To prove that photosynthesis produces oxygen, simply use the glowing splint test on the gas collected. The splint should reignite, showing that the gas is oxygen.
- The investigation can be adjusted to test for the effect of light intensity on the rate of photosynthesis as described in the Developing investigative skills box.
- The investigation can be adjusted to test for the effect of temperature by placing the beaker of pondweed in water baths of different temperatures and measuring the rate at which bubbles of oxygen are produced.

In each of these investigations, all other factors that may affect the rate of photosynthesis must be controlled and kept constant as far as possible.

Developing investigative skills

You can investigate the effect of light on photosynthesis by shining a light on a water plant and measuring how quickly bubbles are given off, as shown in Fig. 5.6.

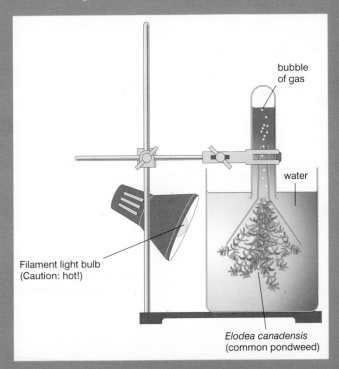

bubble
of gas

water

Filament light bulb
(Caution: hot!)

Elodea canadensis
(common pondweed)

△ Fig. 5.6 Apparatus for the investigation into the effect of light on photosynthesis.

The results below were gathered using this apparatus.

	Distance to lamp/cm				
	5	10	15	20	25
Number of gas bubbles given off in 5 minutes	67	57	40	20	4

Devise and plan investigations

❶ a) Explain why the rate of bubble production can be used as a measure of the rate of photosynthesis.

b) Explain how you would identify the gas produced by the plant.

Analyse and interpret data

❷ a) Use the data in the table to draw a suitable graph.

b) Describe and explain the shape of the graph.

Evaluate data and methods

❸ Light is not the only factor that can affect the rate of photosynthesis.

a) Which other factor might have had an effect on these measurements.

b) Suggest how the method could be changed to avoid this problem.

QUESTIONS

1. Describe how each of the following factors affects the rate of photosynthesis:

   a) light intensity

   b) temperature.

2. Explain why the factors have the effects you described in Question 1.

END OF EXTENDED

Leaf structure

Photosynthesis takes place mainly in the leaves, although it can occur in any cells that contain green chlorophyll. Leaves are adapted to make them very efficient as sites for photosynthesis, gas exchange, transport and support.

◁ Fig. 5.7 The leaves of trees are often arranged so that they do not overlap each other, which makes it possible for the tree to capture as much light energy as possible.

Figure 5.8 shows the arrangement of cells and tissues inside a leaf.

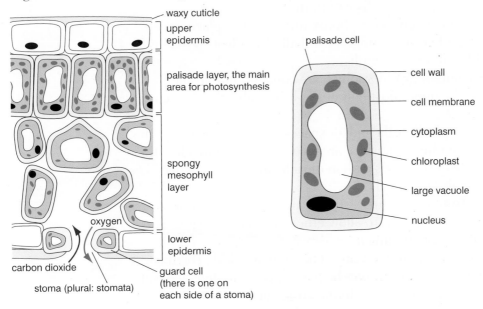

△ Fig. 5.8 Cells in a section of a leaf (left), and a palisade cell (right), which contains many chloroplasts, for photosynthesis.

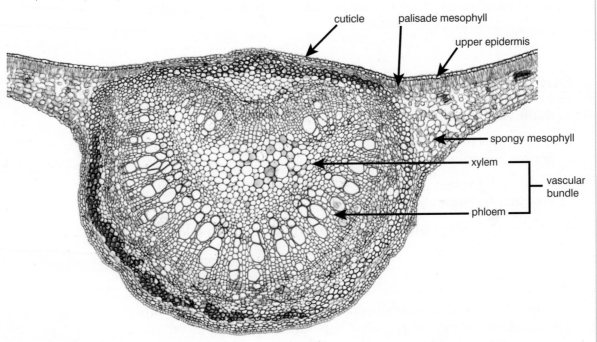

△ Fig. 5.9 Photomicrograph of a section of a dicotyledonous plant leaf. Stomata are not easily seen in this section.

Adaptations for photosynthesis

Many structures in a leaf are adapted so that photosynthesis can be carried out as efficiently as possible.

- The waxy **cuticle** that covers the leaf, particularly the upper surface, prevents the loss of water from epidermal cells and helps to stop the plant from drying out too quickly.

- The transparent upper **epidermis** allows as much light as possible to reach the photosynthesising cells within the leaf.
- The **palisade cells**, where most photosynthesis takes place, are tightly packed together in the uppermost half of the leaf so that as many as possible can receive sunlight.
- **Chloroplasts**, containing chlorophyll, are concentrated in the palisade cells in the uppermost half of the leaf to absorb as much sunlight as possible.
- The **spongy mesophyll cells** and air spaces in the lower part of the leaf provide a large internal surface area to volume ratio to allow the efficient exchange of the gases carbon dioxide and oxygen between the cells and the air in the leaf.
- Many pores or **stomata** (singular: stoma) allow the movement of gases into and out of the leaf, to allow efficient gas exchange between the leaf and the air surrounding it.
- The **vascular bundles** form the veins in the stem and leaf. The thick cell walls of the tissue in the bundles help to support the stem and leaf.
- **Phloem** tissue transports sucrose, formed from glucose in photosynthesising cells, away from the leaf. **Xylem** tissue transports water and minerals to the leaf from the roots. The leaf is broad and thin, this is to maximise surface area for palisade cells to receive sunlight.

END OF EXTENDED

QUESTIONS

1. Name four tissues in a leaf.

2. EXTENDED List as many adaptations of a plant leaf for photosynthesis as you can.

3. EXTENDED Explain why a large surface area inside the leaf is essential for photosynthesis.

4. EXTENDED Explain why a transparent epidermis is an adaptation for photosynthesis.

Mineral requirements of plants

Photosynthesis produces carbohydrates, but plants contain many other types of elements and compounds. Carbohydrates contain just the elements carbon, hydrogen and oxygen, but the amino acids that make up proteins also contain nitrogen. So plants need a source of nitrogen. Other substances in plants contain different elements; for example, chlorophyll molecules contain magnesium and nitrogen. Without a source of magnesium and nitrogen, a plant cannot produce chlorophyll and so cannot photosynthesise.

These additional elements are dissolved in water in the soil as **mineral ions**. Plants absorb mineral ions through their roots, using active transport because the concentration of the ions in the soil is lower than in the plant cells.

Mineral deficiencies

Plants that are not absorbing enough mineral ions show particular symptoms of deficiency. For example:

- a plant with a nitrate ion deficiency has stunted growth
- a plant with magnesium ion deficiency has leaves that are yellow between the veins, particularly in older leaves as the magnesium is transported in the plant to the new leaves.

△ Fig. 5.10 A plant showing symptoms of nitrate ion deficiency.

△ Fig. 5.11 A plant showing symptoms of magnesium ion deficiency.

QUESTIONS

1. Explain why plants need a supply of mineral ions.

2. Explain what plants use the following mineral ions for:

   a) nitrogen ions

   b) magnesium ions.

3. EXTENDED Describe and explain the deficiency symptoms in a plant for the following mineral ions:

   a) nitrate ions

   b) magnesium ions.

End of topic checklist

Key terms

chlorophyll, chloroplast, cuticle, epidermis, mineral ion, palisade cell, osmosis, phloem, photosynthesis, spongy mesophyll cell, starch, stomata, sucrose, vascular bundle, waste product, xylem

During your study of this topic you should have learned:

○ How to define photosynthesis as the process by which plants make carbohydrates (glucose) from raw materials using light.

○ How to investigate the necessity of chlorophyll, light and carbon dioxide for photosynthesis, using appropriate controls.

○ EXTENDED That photosynthesis takes place in chloroplasts in plant cells, using chlorophyll, and converts light energy into chemical energy.

○ The word equation for photosynthesis:

$$\text{carbon dioxide} + \text{water} \xrightarrow[\text{light energy}]{\text{chloroplast}} \text{glucose} + \text{oxygen}$$

○ EXTENDED The balanced symbol equation for photosynthesis:

$$6CO_2 + 6H_2O \xrightarrow[\text{light energy}]{\text{chloroplast}} C_6H_{12}O_6 + 6O_2$$

○ EXTENDED How to outline the use and storage of the carbohydrates made in photosynthesis.

○ EXTENDED How to investigate the increase in rate of photosynthesis as light intensity increases and as temperature increases up to an optimum temperature, after which it decreases as enzymes denature.

○ That chloroplasts are found mainly in cells in the upper part of a leaf.

○ To identify the tissues that can be seen in leaves, including: cuticle, upper epidermis, lower epidermis including stomata and guard cells, palisade mesophyll, spongy mesophyll, phloem and xylem.

○ **EXTENDED** To explain how the leaf is adapted to maximise the rate of photosynthesis by:

- being broad and thin
- the cells where most photosynthesis takes place are in the palisade mesophyll near the upper surface of the leaf
- the transparent upper epidermis lets lots of light through
- the spongy mesophyll maximises the internal surface area of the leaf for diffusion
- stomata allow gases to diffuse into and out of the leaf and are opened and closed by guard cells
- xylem transports water to the leaf
- phloem transports sugars away from the leaf.

○ To describe how plants use mineral ions to convert the sugars from photosynthesis into other essential substances, such as chlorophyll, which contains magnesium, and amino acids, which contain nitrogen.

○ **EXTENDED** The deficiency of nitrate ions causes stunted growth with yellowing leaves because the plant cannot make proteins for growth and chlorophyll.

○ **EXTENDED** The deficiency of magnesium ions causes yellowing leaves as the plant cannot make chlorophyll.

End of topic questions

Note: The marks given for these questions indicate the level of detail required in the answers. In the examination, the number of marks given to questions like these may be different.

1. **EXTENDED** Explain why gardeners may add a liquid feed containing nitrogen and magnesium ions to the water for the plants that they are growing. **(2 marks)**

2. **EXTENDED** Using what you have learnt about the effect of concentration gradient and surface area to volume ratio, explain the adaptations of a leaf for photosynthesis. **(4 marks)**

3. It is commonly stated that 'Plants produce oxygen during the day and carbon dioxide at night.' Explain fully the limits of this statement. **(5 marks)**

4. Photosynthesis is how plants make sugar. State the word equation for photosynthesis. **(3 marks)**

Nutrition is one of the seven characteristics of living organisms. For humans and other animals, nutrition is the taking in of nutrients (including organic substances and minerals) that contain the raw materials needed by the body to make essential molecules, such as proteins, which are used as the building blocks to maintain healthy growth and tissue repair.

We think of malnutrition as not having enough food, but many people may be malnourished because they don't eat enough of the kinds of foods that contain all the nutrients that their bodies need, or they eat too much of the types of foods that will supply more energy than the body needs, so the excess is stored as fat.

SECTION CONTENTS

a) Diet

b) Alimentary canal

c) Digestion

6
Animal nutrition

Δ Green beans contain vital nutrients essential to maintaining growth and tissue repair.

Animal nutrition

△ Fig. 6.1 Different animals have different nutritional requirements, which must be satisfied by diet.

INTRODUCTION

Human taste buds have evolved to give us useful information about what we are putting into our mouths. Sour or bitter tastes can indicate food that is decaying or poisonous, and so is dangerous to eat. Sweet (presence of sugars), salty and savoury (presence of proteins) tastes indicate nutrients that are essential for healthy growth. These were particularly important in our hunter-gatherer past, when it could be difficult to find foods containing these nutrients. They are not so useful to us now because, for many of us, foods containing large quantities of these are easily available. The urge to eat foods containing high levels of sugars and salt has led to problems of obesity and disease, particularly heart disease in people who have increasingly sedentary lifestyles.

KNOWLEDGE CHECK

✓ Animals eat other organisms to get the food they need for their life processes.
✓ The organs, tissues and cells of the digestive system are adapted to digest and absorb nutrients from food.
✓ Different groups of people need different diets.

LEARNING OBJECTIVES

✓ Describe what is meant by a *balanced diet*.
✓ **EXTENDED** Explain how a balanced diet varies in different groups of people.
✓ Describe the sources and functions of nutrients in human nutrition.
✓ **EXTENDED** Describe the deficiency symptoms for some nutrients in the human diet.
✓ **EXTENDED** Describe some effects of malnutrition in humans.
✓ Describe the structures and functions of organs in the human alimentary canal and related organs of the digestive system.
✓ Describe the processes involved in human nutrition, including *ingestion*, *digestion*, *absorption* and *egestion*.
✓ Identify and describe the functions of the main regions of the alimentary canal.
✓ **EXTENDED** State the functions of enzymes and hydrochloric acid in chemical digestion.
✓ **EXTENDED** Outline the role of enzymes in digestion.
✓ **EXTENDED** Describe the functions of hydrochloric acid in digestion.
✓ **EXTENDED** Describe the role of protease, lipase and amylase in the alimentary canal.

DIET

Essential nutrients

To keep healthy, humans need a diet that includes all the nutrients that our cells and tissues use, such as:

- **proteins** – these are broken down to make amino acids. The amino acids are used to form other proteins needed by cells, including enzymes. Protein sources include eggs, milk and milk products (cheese, yoghurt, and so on), meat, fish, legumes (peas and beans), nuts and seeds.
- **carbohydrates** – which are broken down to simple sugars for use in respiration. This releases energy in our cells and enables all the life processes to take place. Good sources of carbohydrate include rice, bread, potatoes, pasta and yams.
- **fats** – these are deposited in many parts of the body, including just below the skin. Some fat helps to maintain body temperature. Fat is also a store of energy to supply molecules for respiration if the diet does not contain enough energy for daily needs. Fat is present in meat and can also come in the form of oils, milk products (butter, cheese), nuts, avocados and oily fish.
- **vitamins** and **minerals** – these substances are needed in tiny amounts for the correct functioning of the body. Vitamins and minerals cannot be produced by the body, and cooking food destroys some vitamins. For example, vitamin C is best supplied by eating raw fruit and vegetables.

Essential vitamins and minerals	Job	Good food source	EXTENDED Deficiency disease
vitamin D	for strong bones and teeth	fish, eggs, liver, cheese and milk	rickets (softening of the bones)
iron	needed to make haemoglobin in red blood cells	red meats, liver and kidneys, leafy green vegetables, for example, spinach	anaemia (reduction in number of red blood cells, person soon becomes tired and short of breath)

△ Table 6.1 Vitamins and minerals, their roles, sources and effects of deficiency.

- **fibre (roughage)** – which is made up of the cell walls of plants. Good sources are leafy vegetables, such as cabbage, and unrefined grains, such as brown rice and wholegrain wheat. It adds bulk to food so that it can be easily moved along the digestive system by peristalsis. This is important in preventing constipation. Fibre is thought to help prevent bowel cancer.

- water – which is the major constituent of the body of living organisms and is necessary for all life processes. Water is continually being lost through excretion and sweating, and must be replaced regularly through food and drink in order to maintain health. Most foods contain some water, but most fruit and vegetables contain a lot of water.

△ Fig. 6.2 A healthy meal contains a good balance of the foods your body needs and nothing in too large an amount.

QUESTIONS

1. Which three groups of food molecules do we need most of in a healthy diet?
2. Give examples of foods that are good sources of each group of food molecules.
3. Which other substances are needed in our diet?
4. Explain the role of each of these substances in our diet.

The right balance

A **balanced diet** contains all of these nutrients in the right proportions to stay healthy because we need more of some nutrients than of others. As most foods contain more than one kind of nutrient, trying to work out what a balanced diet looks like can be difficult. Governments use images like the ones in Figs. 6.3 and 6.4, of food on a plate, to guide people on what proportion of different foods to eat.

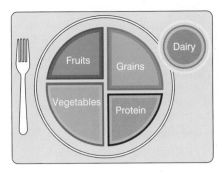

△ Fig. 6.3 Guidance from the USDA (United States Department of Agriculture) on the proportions of different nutrients in a balanced diet.

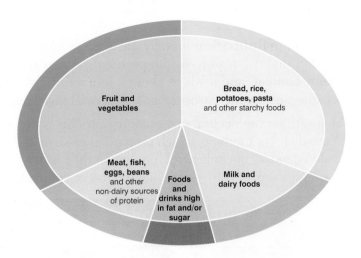

△ Fig. 6.4 Guidance from the UK Government on the proportions of different nutrients in a balanced diet.

EXTENDED

Different groups of people may have different needs for nutrients at different times in their lives, so this balance can change. For example, children need a higher proportion of protein than adults because they are still growing rapidly. Also, some groups of people have a greater need for a specific nutrient. During pregnancy, for example, a woman needs more iron than usual, to supply what the growing baby needs for making blood cells.

Even with the right proportions of nutrients in our foods, we can still be eating an unhealthy diet. This is because many of our foods, particularly carbohydrates but also fats and proteins, can contribute to the energy our bodies need. If we eat food that supplies more energy than we use, the extra will be deposited as energy stores of fat. This can lead to **obesity**, which is related to many health problems, such as heart disease and diabetes. Controlling the portion size at each meal, keeping between-meal snacks to a minimum and increasing levels of exercise can help to reduce the risk of becoming overweight.

Energy requirements depend on body size, stage of development and level of exercise, as shown in Table 6.2.

	Energy used in a day (kJ)	
	Male	Female
6-year-old child	7500	7500
12 to 15-year-old teenager	12 500	9700
adult manual worker	15 000	12 000
adult office worker	11 000	9800
pregnant woman		10 000
breastfeeding woman		11 000

△ Table 6.2 Daily energy requirements for different people.

Malnutrition

The term **malnutrition** literally means 'bad nutrition' and applies to any diet that will lead to health problems. A diet that is too high in energy content, and leads to obesity, is one form of malnutrition, because obesity increases the risk of several diseases.

Malnutrition can occur if one or more nutrients is in too high a proportion in the diet. For example, a high proportion of saturated fats in the diet can lead to deposits of cholesterol forming on the inside of arteries, increasing blood pressure and also increasing the risk of coronary heart disease.

Δ Fig. 6.5 Starvation is most commonly seen in places where crops have failed due to drought (**famine**) or when people have been displaced as a result of war. However, it can also happen in people who choose to starve themselves, for example, by crash dieting or as a result of conditions such as anorexia.

Malnutrition also occurs if any of the substances needed for a healthy body are in too low a proportion in the diet. For example, a lack of a vitamin or a mineral can cause deficiency diseases, as shown in Table 6.1. Too little fibre in the diet can lead to **constipation**, in which food moves too slowly through the **alimentary canal**, increasing the risk of diseases such as diverticulitis and bowel cancer.

Developing investigative skills

Combustion (burning) of food releases heat energy. The word equation for combustion is:

food + oxygen → carbon dioxide + water (+ heat energy)

This reaction is similar to respiration inside cells, so we can use a combustion experiment to model the energy that is released from foods during respiration.

A crisp/potato chip and a plant leaf were tested in an investigation to see which released the most energy by combustion. Here are the results.

	Sample	
	Crisp	**Leaf**
mass of sample in grams	22	12
temperature of water after burning in °C	27	16
temperature of water before burning in °C	15	15
temperature rise in °C	12	1
energy released by the sample in joules	1260	105

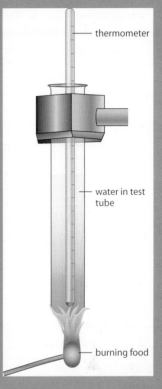

△ Fig. 6.6 Apparatus for burning food.

Demonstrate and describe techniques

❶ **a)** Look at the diagram and describe what happens during the experiment.

b) Identify any areas of safety that should have been considered and suggest how risks could be controlled.

Analyse and interpret data

❷ **a)** Use the results to calculate the energy released per gram of each sample.

b) Explain why you need to do this.

❸ Which part of the sample released the most energy per gram?

❹ A plant seed contains a similar amount of energy as a crisp. Suggest why some animals that eat the leaves of plants for most of the year change to eating seeds (nuts) when they are available.

Evaluate data and methods

❺ The apparatus shown does not give accurate results for the amount of energy in the burning material. Explain why, and suggest a method that would increase the accuracy of the results.

Starvation occurs when there is too little energy provided by the diet. In this state, the body will start to break down its energy stores. Initially this uses the fat stores but, when those have run out, the body will start to break down muscle tissue to produce substances that can be used in respiration. This can damage the muscle tissue of the heart, and also the immune system, increasing the risk of many diseases.

QUESTIONS

1. Explain why different groups of people need different amounts of nutrients. Give examples in your answer.

2. Explain why someone following a healthy diet needs to consider energy as well as nutrients.

3. Explain why the following are considered to be a result of *malnutrition*:

   a) obesity

   b) starvation

   c) constipation.

END OF EXTENDED

ALIMENTARY CANAL

Eating food involves several different processes:

- **ingestion** – taking food and drink into the body (through the mouth in humans)
- **digestion** – breaking down of large food molecules into smaller water-soluble molecules using both chemical and mechanical methods
- absorption of digested food molecules from the intestine into the blood and lymph vessels
- **egestion** – removal of substances through the anus that were ingested but not absorbed by the body (**faeces**).

All these different processes take place in different parts of the alimentary canal.

The alimentary canal is a continuous tube through the body, from the mouth where food is ingested, through the oesophagus, stomach, small intestine and large intestine, to the anus where faeces is egested. You could say that materials in the alimentary canal aren't truly in the body. Not until food molecules are absorbed, do they cross cell membranes into body tissue. Then **waste products** can be *excreted* through other organs.

The digestive system includes the alimentary canal and the other organs that contribute to digestion, for example, the liver, pancreas and gall bladder. Table 6.3 describes the functions of each of the organs in the digestive system.

Part of digestive system	What happens there
mouth	teeth and tongue break down food into smaller pieces
salivary glands	produce liquid saliva, which moistens food so it is easily swallowed and contains the enzyme amylase to begin breakdown of starch
oesophagus	each lump of swallowed and chewed food, called a bolus, is moved from the mouth to the stomach by waves of muscle contraction called peristalsis
stomach	acid and protease enzymes are secreted to start protein digestion; movements of the muscular wall churn up food into a liquid
liver	cells in the liver make bile; amino acids not used for making proteins are broken down to form urea, which passes to the kidneys for excretion; excess glucose is removed from the blood and stored as glycogen in liver cells
gall bladder	stores bile from the liver; the bile is passed along the bile duct into the small intestine, where it neutralises the stomach acid in the chyme
pancreas	secretes digestive enzymes in an alkaline fluid into the duodenum
small intestine (duodenum and ileum)	secretions from the gall bladder and pancreas enter the first part of the small intestine (duodenum) to complete the process of digestion; digested food molecules and water are absorbed in the ileum
large intestine (colon)	water is absorbed from the remaining material
rectum	the remaining, unabsorbed, material (**faeces**), plus dead cells from the lining of the alimentary canal and bacteria, are compacted and stored
anus	faeces is egested through a sphincter

△ Table 6.3 The functions of parts of the human digestive system.

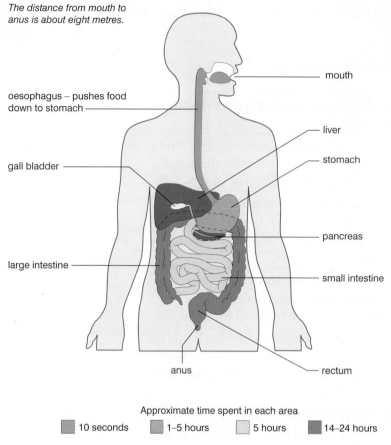

The distance from mouth to anus is about eight metres.

mouth

oesophagus – pushes food down to stomach

gall bladder

liver

stomach

pancreas

large intestine

small intestine

anus

rectum

Approximate time spent in each area

| 10 seconds | 1–5 hours | 5 hours | 14–24 hours |

△ Fig. 6.7 The human digestive system.

Food moves along the alimentary canal because of the contractions of the muscles in the walls of the alimentary canal. This is called **peristalsis**. Fibre in the food keeps the bolus bulky and soft, making peristalsis easier.

Peristalsis moves food along the digestive system.

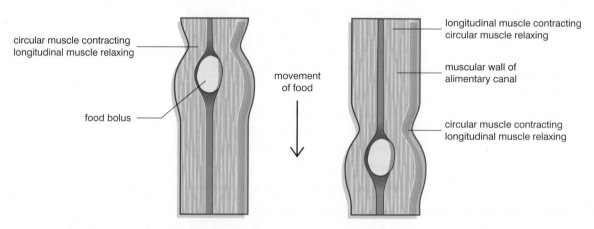

circular muscle contracting longitudinal muscle relaxing

food bolus

movement of food

longitudinal muscle contracting circular muscle relaxing

muscular wall of alimentary canal

circular muscle contracting longitudinal muscle relaxing

△ Fig. 6.8 Peristalsis moves food along the digestive system.

1. Sketch the diagram of the digestive system shown in Fig. 6.7. Label the organs, and add notes to each organ to explain its function in the system.

2. Explain the difference between egestion and excretion.

3. Explain how the muscles of the alimentary canal wall move food.

SCIENCE LINK

CHEMISTRY – PARTICLE MODEL OF MATTER, ACIDS

- Breaking down food into molecules small enough to be absorbed into the bloodstream involves a number of different chemical reactions, making use of enzymes and other chemicals, such as acid in the stomach.

PHYSICS – FORCES AND PRESSURE

- An early stage in breaking down food is the mechanical process of biting and chewing the food into smaller pieces.

- Different teeth have different shapes to help the process in particular ways – the different types of teeth vary the pressure on the food and help with different actions such as cutting, gripping and grinding.

DIGESTION

Different types of digestion

If food is to be of any use to us, the food molecules must enter the blood so that they can travel to every part of the body. Many of the foods we eat are made up of large, **insoluble** molecules that cannot cross the wall of the alimentary canal and the cell membranes of cells lining the blood vessels. This means the food molecules have to be broken down into small, **soluble** molecules that can easily cross cell membranes and enter the blood. Breaking down the molecules is called digestion, both mechanical and chemical methods are used.

EXTENDED

Mechanical and chemical digestion

- **Mechanical** and **physical digestion** occurs mainly in the mouth, where food is broken down physically into smaller pieces by the biting and chewing action of the teeth. It also happens in the small intestine, where bile helps to emulsify fats, which means breaking them into small droplets.

- **Chemical digestion** is the breakdown of large food molecules into smaller ones using chemicals such as enzymes.

Some molecules, for example, glucose, vitamins, minerals and water, are already small enough to pass through the alimentary canal wall and do not need to be digested.

QUESTION

1. Explain the difference between chemical and mechanical/physical digestion.

END OF EXTENDED

Chemical digestion

Chemical digestion in the alimentary canal is the result of enzymes. **Digestive enzymes** are a group of enzymes that are produced in the cells lining parts of the digestive system and are **secreted** (produced) into the alimentary canal to mix with the food.

The digestive enzymes include:

- carbohydrases that break down carbohydrates, one example of which is **amylase**
- **proteases**
- **lipases**.

REMEMBER

The -*ase* at the end of the name means that it is an enzyme, and the first part usually names the substrate that the enzyme works on.

Each of the food groups (carbohydrates, proteins and fats) contains many different molecules. As each enzyme is specific to its substrate, this means that in each group of digestive enzymes there are many different enzymes.

EXTENDED

Different enzymes are made in different parts of the digestive system, as shown in Table 6.4.

Enzyme	Where produced	Substrate	Final products*
amylase	salivary glands (mouth) pancreas	starch	glucose
protease (many types)	stomach wall pancreas	proteins	amino acids
lipase (many types)	pancreas	fats and oils (lipids)	fatty acids and glycerol

△ Table 6.4 Digestive enzymes.

*These are the soluble substances produced at the end of digestion. The substances in food and drink may go through many stages of digestion by different enzymes as they pass through the alimentary canal.

SCIENCE IN CONTEXT — LACTOSE

Lactose (from *lactis*, meaning milk) is the disaccharide sugar in milk, which is broken down in the alimentary canal by the enzyme lactase to the simple sugars glucose and galactose.

Like all young mammals, human babies produce lactase, which helps them to digest the lactose in breast milk. In most mammals the production of lactase decreases as the young mature, because the adult diet does not include milk. This also happens in adults from many human cultures in which adults generally do not drink milk, such as in South-East Asia. However, there are human cultures in Europe, India and parts of East Africa, where mammals such as sheep, goats or cattle are kept to supply meat and milk for food. In these human groups the adults continue to produce lactase and are able to digest the lactose in milk. Adults who cannot do this are *lactose intolerant*. In these people bacteria in the alimentary canal break down the lactose, producing gas, which causes great discomfort.

Details of digestion

Amylase only partly digests starch, to the disaccharide (two simple sugars) maltose. Digestion to the monosaccharide glucose is completed by the enzyme maltase, which is attached to the epithelial cell membranes of the small intestine.

Similarly, there are several proteases involved in the breakdown of proteins to amino acids. Different proteases are produced in different parts of the digestive system. Protein digestion starts in the stomach, when the first protease is produced and is continued in the small intestine where are other proteases are produced.

The right conditions

Remember that different enzymes work better in different conditions. Those enzymes that digest food in the stomach work best in acid conditions. Special cells in the lining of the stomach secrete hydrochloric acid into the stomach to create the right conditions for the enzymes. The acid is also helpful in killing microorganisms taken in with the food.

This happens because the low pH denatures enzymes in the microorganisms so that they cannot function properly.

QUESTIONS

1. Explain why enzymes are needed in the digestive system.

2. a) Which enzyme has starch as its substrate?

   b) Which product is formed by the digestion of starch by this enzyme?

3. Describe the role of stomach acid

END OF EXTENDED

End of topic checklist

Key terms

alimentary canal, amylase, balanced diet, carbohydrate, chemical digestion, constipation, digestion, digestive enzyme, egestion, famine, faeces, fat, fibre, ingestion, insoluble, lipase, malnutrition, mechanical digestion, mineral, obesity, peristalsis, physical digestion, protease, protein, secretion, soluble, starvation, vitamin, waste product

During your study of this topic you should have learned:

◯ How to describe the roles of the main components of a healthy human diet, these being: carbohydrates, proteins, lipids, vitamins (for example, A, C and D), minerals (for example, calcium and iron), water and dietary fibre.

◯ That a balanced diet includes all the components needed for health in the right proportions.

◯ EXTENDED How to describe the effects of malnutrition.

◯ EXTENDED How to explain the causes and effects of malnutrition.

◯ EXTENDED That diet provides energy as well as nutrients and different groups of people have different requirements.

◯ That the human alimentary canal is made up of the mouth, oesophagus, stomach, small intestine and large intestine.

◯ That ingestion takes place in the mouth, digestion is the breakdown of large food molecules into smaller ones by mechanical and chemical digestion, absorption is the taking of nutrients from the small intestine, and egestion is the removal of waste food from the body.

◯ That chemical digestion breaks down large food molecules into small molecules that can be absorbed.

◯ To state the functions of digestive enzymes including amylase (digest starch to simpler sugars), lipases (digest fats to fatty acids and glycerol) and proteases (digest proteins to amino acids).

◯ EXTENDED How to describe the digestion of starch, and the role of pepsin and trypsin in the alimentary canal.

◯ EXTENDED That amylase is secreted in the mouth, stomach and by the pancreas; lipases are secreted by the pancreas; proteases are secreted in the stomach and by the pancreas: enzymes secreted by the pancreas pass into the small intestine.

◯ EXTENDED How to explain that hydrochloric acid in the stomach denatures harmful microorganisms and provides the right pH for gastric enzymes.

◯ To state that digested food molecules are absorbed in the small intestine.

End of topic questions

Note: The marks given for these questions indicate the level of detail required in the answers. In the examination, the number of marks given to questions like these may be different.

1. Describe the importance of the following in a healthy diet:

    a) vitamins C and D   **(2 marks)**

    b) the minerals calcium and iron   **(2 marks)**

    c) water   **(1 mark)**

    d) dietary fibre.   **(2 marks)**

2. Identify the organs of the digestive system involved, and their role, in each of the following processes:

    a) ingestion   **(2 marks)**

    b) digestion   **(5 marks)**

    c) absorption.   **(2 marks)**

3. This is the diet schedule for a male Olympic athlete training for a competition, not including drinks during training.

breakfast	large bowl of cereal, such as porridge or muesli
	half pint semi-skimmed milk plus chopped banana
	1–2 thick slices wholegrain bread with olive oil or sunflower spread and honey or jam
	glass of fruit juice + 1 litre fruit squash
post-training 2nd breakfast	portion of scrambled eggs
	portion of baked beans
	1–2 pieces of grilled tofu
	portion of grilled mushrooms or tomatoes
	2 thick slices wholegrain bread with olive oil spread
	1 litre fruit squash
lunch	pasta with bolognese or chicken and mushroom sauce
	mixed side salad
	fruit
	1 litre fruit squash
post-training snack	4 slices toast with olive oil or sunflower spread and jam
	large glass of semi-skimmed milk
	fruit
	500 ml water
evening meal	grilled lean meat or fish
	6–7 boiled new potatoes, large sweet potato or boiled rice
	large portion of vegetables, for example, broccoli, carrots, corn or peas
	1 bagel
	1 low-fat yoghurt and 1 banana or other fruit
	750 ml water and squash
bedtime snack	low-fat hot chocolate with 1 cereal bar

a) Identify the foods that contribute to each of these food types:

   i) carbohydrates (5 marks)

   ii) proteins (3 marks)

   iii) lipids (2 marks)

   iv) vitamins and minerals (2 marks)

   v) dietary fibre. (1 mark)

b) Which food type is most represented in this diet? (1 mark)

c) Explain why this food type is so important in this diet. (1 mark)

d) Which food group would you expect to be more represented in an athlete's diet in the early stages of training? Explain your answer. (2 marks)

e) Explain why this diet is not suitable for everyone. (2 marks)

Transport is essential to plants and animals. Plants live in many different environments on Earth, from the hot, humid rainforests, to the high, cold mountains. Like animals, plants need to be well adapted to the conditions in which they live to enable them to survive. In deserts, for example, water is usually very limited, and the air temperature may range from above 40 °C during the day to below zero at night. Most plants would not be able to survive in these conditions, but those that do live in the desert have developed some very special adaptations that allow them to store water, reduce the loss of water from their leaves, and even help to insulate them against the very cold desert nights.

An average-sized human adult carries about 4.5–5.0 litres of blood. Blood is pumped around the bodies of mammals by the heart to carry oxygen, nutrients and water to every organ, tissue and cell, and transport waste products away from cells. The human heart is about the size of an adult fist, located just about in the centre of the chest. It beats over 100 000 times a day, and the characteristic double sound of a beat corresponds to the two sets of heart valves opening and closing in order.

SECTION CONTENTS

a) Transport in plants

b) Transport in mammals

7
Transport

Δ The roots of this tree have grown away from the trunk in order to access the highest quantities of nutrients, which are most plentiful in the top layers of soil.

Transport

INTRODUCTION

In one day, hundreds of litres of water will be absorbed from the soil and transpired through a fully grown tree in the Amazonian rainforest. This has a major impact on the environment in the rainforests. It reduces the amount of water in the soil. It also cools the air around the trees as the water evaporates into the air. The increase in water vapour in the air also affects where rainfall occurs. So the trees are effectively controlling the climate.

Δ Fig. 7.1 These fine clouds over the Amazon rainforest are formed from water transpired by trees earlier in the day.

KNOWLEDGE CHECK

✓ Cells in a plant leaf make glucose by photosynthesis, which is converted to sucrose and transported to other parts of the plant in phloem cells.
✓ Xylem vessels transport water and mineral ions from the roots of a plant through the stem to the leaves.
✓ The heart and blood vessels form the human circulatory system.
✓ Cells need a continuous supply of oxygen and glucose for respiration, which are supplied by the blood in a human body.

LEARNING OBJECTIVES

✓ State the functions of xylem tissue and identify xylem in sections of plant structures.
✓ State the functions of phloem tissue and identify phloem in sections of plant structures.
✓ Identify root hair cells and state their functions.
✓ Describe and investigate the pathway taken by water through a plant.
✓ **EXTENDED** Explain the importance of the large surface area of root hairs.
✓ Define the term *transpiration*.
✓ Describe and investigate the effect of temperature and humidity on transpiration rate.
✓ **EXTENDED** Explain the effects of variation of temperature and humidity on transpiration rate.
✓ Describe the circulatory system as a system of continuous blood vessels, with a pump and valves to ensure one-way flow of blood.
✓ **EXTENDED** Describe the double circulatory system of mammals and explain its advantages.
✓ Describe the structure of the heart and how its activity can be monitored.
✓ **EXTENDED** Explain how the structure of the heart is linked to its function.
✓ Investigate the effect of exercise on heart rate.
✓ **EXTENDED** Describe coronary heart disease and how it is linked to diet, stress, smoking, genetic factors, age and gender.
✓ Describe the structure of arteries, veins and capillaries and identify some of the main arteries and veins in the human body.

TRANSPORT IN PLANTS

In plants, water and dissolved substances are transported throughout the plant in a series of tubes or vessels. There are two types of transport vessel in plants, called **xylem** and **phloem**.

- Xylem tissue contains long, hollow xylem cells that form long tubes through the plant. The tubes are the hollow remains of dead cells. The thick strong cell walls help to support the plant. Xylem tubes are important for carrying water and dissolved mineral ions, which have entered the plant through the roots to all the parts of the plant that need them. They are particularly important for supplying the water that the leaf cells need for photosynthesis.

- Phloem cells are living cells that are linked together to form continuous phloem tissue. Dissolved food materials, particularly sucrose and amino acids that have been formed in the leaf, are transported all over the plant from the leaves. For example, sucrose will be carried to any cell that needs glucose for respiration. Sucrose is less reactive than glucose and, therefore, is easier to transport without causing problems for other cells. Sucrose may also be carried to parts of the plant where it will be stored, often as another carbohydrate, such as starch which is stored in seeds and root tubers. This transport of sucrose and other materials is called **translocation**.

In roots the xylem and phloem vessels are usually grouped separately, but in the stem and leaves they are found together as **vascular bundles** or **veins**.

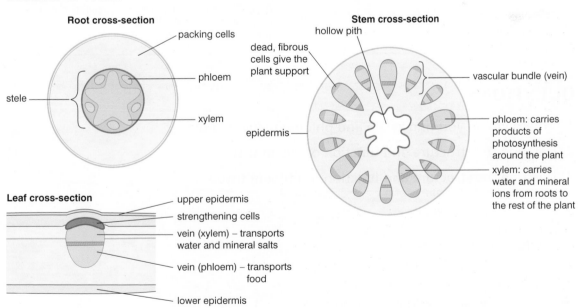

△ Fig. 7.2 The positions of xylem and phloem tissue in a root, stem and leaf.

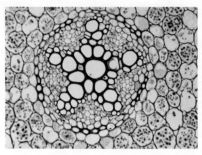 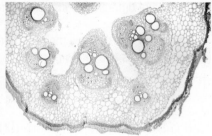 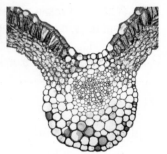

Δ Fig. 7.3 Photomicrographs of (left) a cross-section of the middle of a buttercup root, (middle) a cross-section of part of a pumpkin stem, (right) a cross-section of a vein in a meadow-beauty leaf.

TREE RINGS

The wood of a tree is mostly xylem tissue. Every year, new xylem cells are produced from a ring of cells just inside the bark of the tree. When the tree is growing rapidly, the new xylem cells are large. In temperate regions, such as the UK, the rate of growth and the size of new cells decrease as autumn approaches, and growth stops during winter. The difference in size of cells produced over one year gives the tree its 'rings' and makes it possible to estimate the age of the tree.

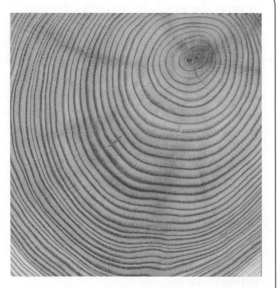

Δ Fig. 7.4 Growth rings occur in temperate climates when new xylem cells alternately grow (in spring and summer) and stop growing (in winter).

QUESTIONS

1. Where would you find xylem and phloem tissue in a plant?

2. Describe the structure and function of xylem tissue.

3. Describe the structure and function of phloem tissue.

Water uptake

Plants absorb water and dissolved mineral ions from the soil through **root hair cells**. Root hair cells are found in a short region just behind the growing tip of every root. They are very delicate, and easily damaged. As the root grows, the hairs of the cells are lost, and new root hair cells are produced near the tip of the root.

△ Fig. 7.5 The root of this germinating seed has many fine root hair cells that greatly increase its surface area.

EXTENDED

Root hair cells are specially adapted for absorption of substances, because they have a fine extension that sticks out into the soil. This greatly increases their surface area for absorption.

Soil water contains minute amounts of dissolved mineral ions. So dissolved mineral ions are usually in higher concentration inside root cells than in the soil water. This means that essential mineral ions cannot usually enter the root by **diffusion**, because that would be against their concentration gradient. Instead, the cell membranes of root hair cells are adapted to take in mineral ions such as nitrates and magnesium ions by **active transport**.

END OF EXTENDED

Water enters the root hair cells, then passes across the root from cortex cell to cortex cell by osmosis. It then enters the xylem tissue in the root and can move from there to all other parts of the plant, including the leaves.

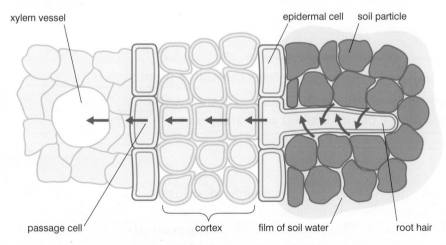

△ Fig. 7.6 The passage of water across a root.

In the leaves, water moves out of the xylem cells in the vascular bundle, into the cells of the spongy mesophyll by osmosis.

Investigating water movement through a plant

The movement of water through the above-ground parts of a plant can be investigated by adding food colouring to the water given to the plant. Food colouring is soluble and is carried through the plant with the water in the xylem. After a day or two in coloured water, the veins of the leaves and flowers of a plant will show the colour.

△ Fig. 7.7 A section across a celery stalk that has been standing in coloured water for a day will show colour mainly within the veins (vascular bundles) of the stalk.

△ Fig. 7.8 A carnation flower that has been standing in coloured water shows the colour in its petals.

QUESTIONS

1. Describe the route that water takes as it moves through a plant.

2. How could you investigate the above-ground route that water takes as it moves through a plant? Explain your answer.

Transpiration

Water is a small molecule that easily crosses cell membranes. Inside the leaf, water molecules cross the cell membranes of the spongy mesophyll cells into the air spaces. This process is called **evaporation** because the liquid water in the cells becomes water vapour in the air spaces. Whenever the **stomata** in a leaf are open, water molecules diffuse from the air spaces out into the air (where there are usually fewer water molecules). So, in addition to using water in the process of photosynthesis, plants lose water by evaporation from the leaf. This loss of water from the leaves is called **transpiration**.

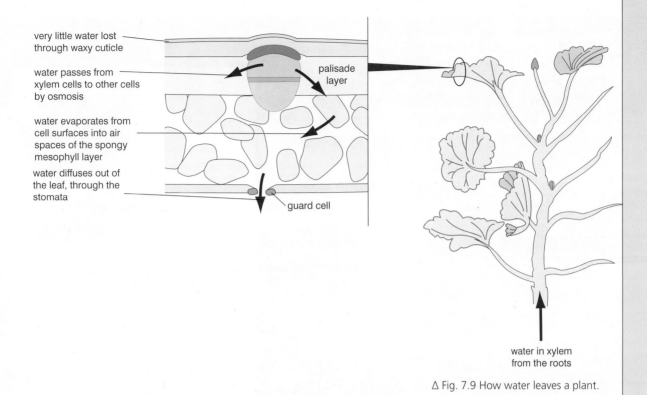

very little water lost through waxy cuticle

water passes from xylem cells to other cells by osmosis

water evaporates from cell surfaces into air spaces of the spongy mesophyll layer

water diffuses out of the leaf, through the stomata

palisade layer

guard cell

water in xylem from the roots

△ Fig. 7.9 How water leaves a plant.

Factors that affect the rate of transpiration

Several factors affect the rate of transpiration.

EXTENDED

The rate of transpiration from a leaf will be affected by anything that changes the concentration gradient of water molecules between the leaf and the air. The steeper the concentration gradient, the faster the rate of transpiration.

END OF EXTENDED

Transpiration is faster when:

- the temperature is higher
- the air is dry (low humidity).

- Temperature – Increased temperature means particles have more energy, which results in faster movement of the particles. The faster particles move, the easier it is for them to evaporate from cell surfaces into the air spaces, diffuse out of the leaf and move away. So, increased temperature increases the rate of transpiration.
- **Humidity** – This is a measure of the concentration of water vapour in the air. When the air is very **humid**, it feels damp because there is a high concentration of water vapour in the air. When the air feels dry, the humidity is low. The concentration of water molecules inside the air spaces in the leaf is high. The higher the humidity of the air, the lower the concentration gradient between the air outside and inside the leaf, and so the lower the rate of transpiration.

Developing investigative skills

The diagram shows apparatus called a potometer that can be used to investigate the effect of a range of factors on the rate of transpiration. As water evaporates from the leaf surface, the bubble of air in the potometer moves nearer to the leafy twig.

Devise and plan investigations

❶ Suggest how you could use a potometer to measure the effect of the following factors on transpiration: (a) temperature, (b) humidity.

Analyse and interpret data

The table below shows the results of an investigation using a potometer in five different sets of conditions.

Conditions	Time for water bubble to move 5 cm, in seconds
humid, warm air	135
dry, warm air	75
humid, cold air	257
dry, cold air	122

❷ For each of the following factors, identify which data in the table should be compared to show the effect of the factor, and explain why those are the right data to compare.

a) temperature

b) humidity

❸ Using the data you have identified, draw a conclusion about the effect of the following on the rate of transpiration.

a) temperature

b) humidity

Evaluate data and methods

❹ Explain why the time taken for the bubble to move 5 cm is a measure of transpiration.

❺ What else could make the bubble move?

❻ Explain how you could improve the reliability of the conclusions you drew in Question 3.

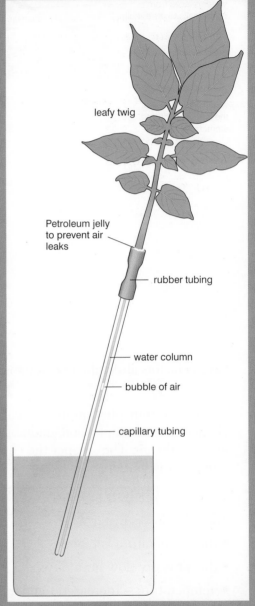

△ Fig. 7.10 A potometer.

(labels on figure: leafy twig; Petroleum jelly to prevent air leaks; rubber tubing; water column; bubble of air; capillary tubing)

QUESTIONS

1. Define the word *transpiration*.

2. Copy the diagram of the plant and leaf section in Fig. 7.9 and add your own annotations to explain how water moves through the plant. Include the following words in your labels: evaporation, osmosis, diffusion, transpiration.

3. Explain the advantage to plants of closing their stomata at night in terms of water loss.

4. Explain, in terms of the movement of water molecules, why transpiration rate is faster when:

   a) the temperature is higher

   b) the humidity of the air is lower.

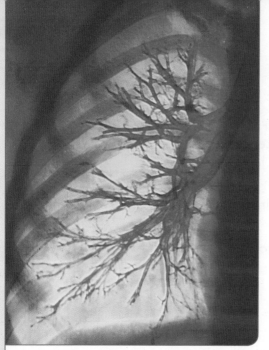

△ Fig. 7.11 This photograph shows the larger blood vessels that are found in a lung. The millions of capillaries are not visible.

TRANSPORT IN MAMMALS

Almost no cell in your body is more than 20 µm (0.02 mm) from a blood vessel. This is because the blood delivers a constant supply of oxygen and glucose, for respiration, without which the cells will rapidly die. So it's not surprising that, no matter where you cut yourself, you will bleed. Many of the blood vessels that penetrate the tissues are extremely narrow – about 5 to 10 µm wide, which is about the width of one red blood cell. It has been calculated that if you placed the blood vessels of an adult in a line it would wrap four times around the equator of the Earth.

Transport in mammals usually takes place inside a **circulatory system**. A circulatory system is formed from a system of continuous tubes (blood vessels) that carry blood around the body. The tubes are connected to a pump, the heart, which forces the blood through the circulation. Valves in the heart and in some of the blood vessels make sure that blood circulates in only one direction.

EXTENDED

Different circulatory systems

Fish have a single circulatory system, which means that blood passes all parts of the body in one circuit before returning to the heart. The mammalian circulatory system (as in humans) is more complex than that of a fish. It is described as a **double circulation** because the blood passes through the heart twice for each time that it passes through the body tissues. This is because when blood leaves the left side of the heart it passes through the tissues of the lungs before returning to the heart for pumping around the rest of the body.

Separating the circulation to the lungs from the circulation to the body means that the blood in the two circulations can be at different pressures.

- Blood leaving the left side of the heart is normally below 4 kPa. Blood does not travel far to the lung tissue, so there is little loss of pressure before it reaches the capillaries surrounding the alveoli. This lower pressure prevents damage to the delicate capillaries that pass through lung tissue.
- Blood leaving the right side of the heart has to travel all round the body and back to the heart. So it needs to start at a much higher pressure, at about 16 kPa as it leaves the heart. By the time it reaches the capillaries within body tissues, the pressure has dropped to below 3 kPa and so will not damage them.

END OF EXTENDED

1. What is the role of the heart in the circulatory system?

2. What prevents blood flowing the wrong way round through the circulatory system?

3. EXTENDED Explain what is meant by a *double circulatory system*.

Heart

The **heart** is a muscular organ that pumps blood by expanding in size as it fills with blood, and then contracting, forcing the blood on its way through the blood vessels. Blood is pumped away from the heart in arteries and returns to the heart through veins.

EXTENDED

The heart is two pumps in one. The right side and left side are separated by a layer of tissue called the septum. The right side pumps blood to the lungs to collect oxygen. The left side then pumps the **oxygenated** blood around the rest of the body. The **deoxygenated** (without oxygen) blood then returns to the right side, to be sent to the lungs again.

END OF EXTENDED

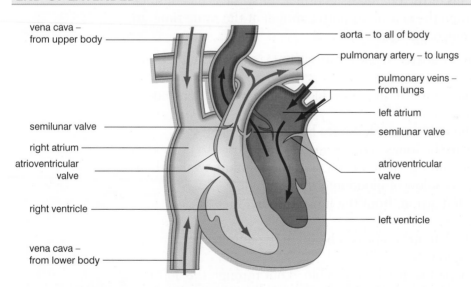

△ Fig. 7.12 Structure of the human heart. Oxygenated blood is shown in red and deoxygenated blood is shown blue. (Knowledge of the names of heart valves is not required.)

REMEMBER

We always draw diagrams of the circulatory system and heart as if looking in a mirror, or at another person. So in the diagram the 'left' side of the heart/circulation in a body is drawn on the right side of the diagram.

The heart consists of four chambers: two **atria** (single: **atrium**) and two **ventricles**. The walls of the chambers are formed from thick muscle. Blood that flows towards the heart passes through blood vessels called **veins**. Blood that leaves the heart passes through blood vessels called **arteries.** To make sure that blood only flows in one direction through the heart, there are **valves** at the points where blood vessels enter and leave the heart, and between the atria and ventricles. These close when the heart contracts, to prevent backflow of blood.

The heart has its own separate blood supply, to provide the muscle tissue with oxygen and glucose for respiration so that it can contract. These blood vessels are called the coronary arteries and coronary veins. You can see some of these on the outside of a whole heart.

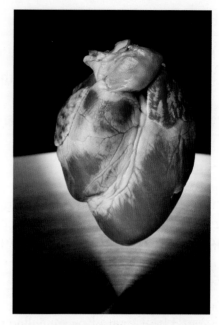

△ Fig. 7.13 Whole human heart.

Blood flow through the heart

Blood passes through the chambers of the heart in a particular sequence as the walls of the chambers contract. First the atria contract at the same time, then the ventricles both contract at the same time, to move the blood through the heart.

- Blood from the body arrives at the heart via the vena cava, and enters the right atrium.
- Contraction of the right atrium passes blood through the tricuspid valve to the right ventricle.
- Contraction of the right ventricle forces blood out through the pulmonary artery to the lungs. The tricuspid valve closes to prevent backflow of blood into the right atrium, and the semilunar valve then closes to prevent backflow of blood into the ventricle.
- Blood enters the left atrium from the lungs through the pulmonary vein.
- Contraction of the left atrium passes blood to the left ventricle through the bicuspid valve.
- Contraction of the left ventricle forces blood out through the aorta towards the rest of the body. The bicuspid valve closes to prevent backflow of blood into the left atrium, and then the semilunar valve closes to prevent backflow of blood into the ventricle.

END OF EXTENDED

QUESTIONS

1. Name the four chambers of the mammalian heart.

2. Distinguish between *arteries* and *veins*.

3. EXTENDED Starting in the vena cava, list the chambers and blood vessels in the order that blood passes through them until it reaches the aorta.

Heart rate

Heart rate is the measure of how frequently the heart beats, generally given as beats per minute. We can take measurements of heart rate by feeling for a pulse point, where the blood flows through an artery near to the skin, such as in the wrist or at the temple.

Taking the pulse rate is actually measuring the expansion and relaxation of the artery wall as the blood passes through it. However, as each pulse of blood is created by one contraction of the ventricles, we say that we are measuring *heart beats*.

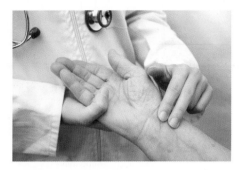

△ Fig. 7.14 Taking the pulse at the wrist.

Heart rate can also be measured by listening to the heart. The 'lub, dup' sounds of one complete contraction are the sounds of the valves inside the heart as they open and shut.

Resting heart rate is the rate at which the heart beats when the person is at rest. On average it is between 60 and 80 beats per minute for an adult human, but this range is very variable. Resting heart rate may vary as a result of:

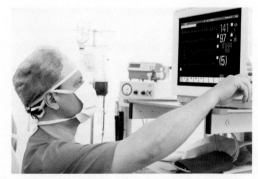

△ Fig. 7.15 During an operation, a doctor continually checks the patients ECG trace to make sure all is well.

- age – children usually have a faster average than adults
- fitness – a trained athlete may have a resting rate as low as 40 beats per minute because their heart contains more muscle and can pump out more blood with each contraction
- illness – infection can raise resting heart rate, but some diseases of the circulatory system can slow resting heart rate.

Changing heart rate

Heart rate increases during activity. The harder you exercise, the faster the heart beats.

The increase in heart rate increases the amount of blood that is pumped around the body, and the speed with which it reaches body tissues. This supplies oxygen and glucose more rapidly to respiring cells, particularly in the muscles, and removes waste products more rapidly.

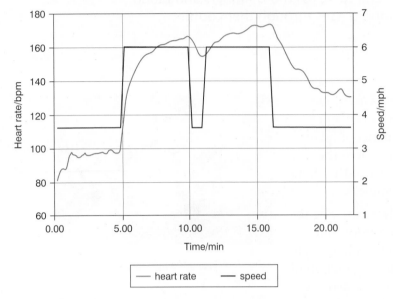

△ Fig. 7.16 How heart rate changes with exercise.

END OF EXTENDED

Developing investigative skills

The effect of exercise on heart rate can be measured by taking pulse measurements after different levels of exercise.

Devise and plan investigations

In an investigation of the effect of exercise on the heart rate, a student was asked to exercise at different levels for 2 minutes, at which point the pulse rate was measured. The student was then allowed to rest for 5 minutes and then continue to exercise at the next level of activity.

❶ a) Explain why the pulse rate was taken after 2 minutes of exercise and not sooner.

b) Explain why the student rested for 5 minutes before starting the next level of exercise.

Analyse and interpret data

Table 7.1 shows the results of an investigation into the effect of exercise on heart rate of one student.

	Resting	Walking	Jogging	Running
Heart rate beats per minute)	72	81	96	122

△ Table 7.1 Heart rate with activity.

❷ Describe the pattern shown in the data.

❸ Use the data to draw a conclusion for the investigation.

❹ Explain why heart rate responds like this to different levels of exercise.

Evaluate data and methods

❺ How reliable is this conclusion, and what could have been done during the investigation to improve the reliability?

QUESTIONS

1. EXTENDED Explain why resting heart rate in an adult is given as a range of values and not a single value.

2. Describe the effect of activity on heart rate.

3. EXTENDED Explain the effect of activity on heart rate.

EXTENDED

Coronary heart disease

The muscle of the heart has its own blood supply to provide the oxygen and sugars it needs for respiration. The heart cannot get these materials from the blood that flows through it, so there are coronary arteries and veins that pass through heart muscle. If the blood flow through these coronary blood vessels is reduced, it can reduce the amount of oxygen and sugars getting to the muscle cells, and so reduce the amount of energy that they can release through respiration.

Blockage of the coronary arteries can occur when layers of cholesterol are deposited on the inner lining of the blood vessel. This causes **coronary heart disease**. Even partial blockage can cause a health problem, such as angina (heart pains) or high blood pressure. A full blockage will cause a heart attack, which may result in death.

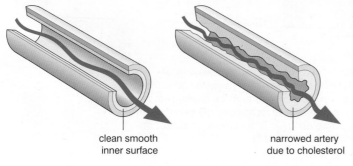

clean smooth
inner surface

narrowed artery
due to cholesterol

△ Fig. 7.17 Deposits of cholesterol inside arteries makes it more difficult for blood to flow through freely, increasing the risk of diseases of the circulatory system.

Some factors can increase the risk of a blockage of the coronary arteries:

- diet: high levels of saturated fats in the diet (particularly from red meats) may cause increased deposits of cholesterol
- smoking: chemicals in tobacco smoke that pass into the blood can damage the delicate lining of arteries, which increases the chance that deposits of cholesterol are laid down at these points.

Stress is not a direct cause of coronary heart disease, but response to stress, such as smoking, drinking alcohol or eating for comfort, particularly over a long time, can increase the risk of heart disease.

Genes can give some people a tendency towards developing coronary heart disease. As this tendency can be inherited, the disease may appear more commonly in some families than in others, and more frequently in some particular ethnic groups than others. For example, African-Americans are at greater risk of developing coronary heart disease than white Americans.

Preventing coronary heart disease

Medical advice for preventing coronary heart disease includes changing the diet and amount of exercise taken.

Many studies of the relationship between saturated fat and heart disease have concluded that reducing the amount of saturated fat in the diet should reduce the risk of heart disease. However, it is difficult to prove this relationship because people don't just change their diet when they are advised to live more healthily. For example, they may also change how much they exercise.

The amount of exercise taken each day does seem to affect the risk of heart disease. Someone who has a sedentary lifestyle (mainly sitting) can significantly reduce their risk of an early death by just a little exercise every day. This exercise seems to strengthen the heart muscle and make it able to cope with sudden increases in heart rate more easily, such as when you run.

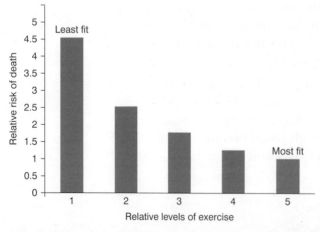

Δ Fig. 7.18 The relationship between risk of death and level of exercise. *'Relative level of exercise' is a comparative measure of fitness comparing how much the energy the body uses compared to a person at rest (group 1). Those in group 5 use 5 times as much energy being active as those in group 1. Relative risk of death is a comparative measure of the risk of death compared against the fittest group, who have a relative risk of 1. Those who are the least fit have a 4.5 times bigger risk of death than those who are the most fit.*

END OF EXTENDED

1. Explain why the heart needs its own blood supply.

2. Identify four possible risk factors for coronary heart disease.

3. **EXTENDED** Explain how a person can reduce their risk of coronary heart disease.

Blood vessels

Figure 7.19 shows a simplified layout of the human circulatory system, including the major blood vessels. The name of a major blood vessel is often related to the organ it supplies: *coronary* for heart (from the Latin *corona* for 'crown' because the blood vessels surround the top of the heart like a crown), *renal* for kidneys (from the Latin *renes* meaning 'the kidneys'), *pulmonary* for lungs (from the Latin *pulmonis* meaning 'lungs'). Learn the names of the blood vessels that are associated with the heart, the lungs and kidneys.

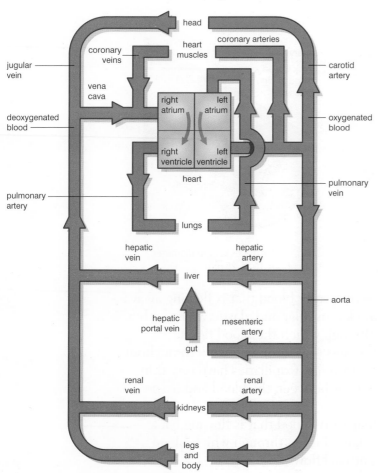

Δ Fig. 7.19 Plan of the human circulatory system.

The largest vein and the largest artery in the body have special names. The **vena cava** is the vein that carries blood to the heart, and the **aorta** is the artery that receives blood from the heart.

Remember, **a** for arteries that travel **a**way from the heart. **V**eins carry blood into the heart and contain **v**alves.

The blood vessels are grouped into three different types: arteries, capillaries and veins.

vein:
thin-walled, carrying blood at low pressure

artery:
thick-walled, carrying blood at high pressure

capillary:
very small; the walls may be just one cell thick

△ Fig. 7.20 Veins vary in diameter from about 5 to 15 mm. Capillaries are very small, with a diameter of around 0.01 mm.

△ Fig. 7.21 Arteries vary in diameter from about 10 to 25 mm.

- **Arteries** are large blood vessels that carry blood that is flowing away from the heart. Arteries have thick muscular and elastic walls, with a narrow central space (lumen) through which the blood flows.
- **Capillaries** are the tiny blood vessels that form a network throughout every tissue and connect arteries to veins. Capillaries have very thin walls. All the exchange of substances between the blood and tissues happens in the capillaries.
- **Veins** are large blood vessels that carry blood that is flowing back towards the heart. Veins have a large lumen through which blood flows. Valves in the veins prevent backflow.

Structure related to function

The structure of different blood vessels enables them to carry out their function most efficiently.

- Blood carried in the arteries is at higher pressure than in the other vessels. The highest pressure is in the aorta, the blood vessel that leaves the left ventricle. The thick walls of arteries help to protect them from bursting when the pressure increases as the pulse of blood enters them. The recoil of the elastic wall after the pulse of blood has passed through helps to maintain the blood pressure and even out the pulses. By the time the blood enters the fine capillaries, the change in pressure during and after a pulse has been greatly reduced.

Blood vessels	Blood pressure (kPa)
aorta	>13
arteries	13–5.3
capillaries	3.3–1.6
veins	1.3–0.7
vena cava	0.3

△ Table 7.2 Blood pressure in different blood vessels.

- The thin walls of capillaries helps to increase the rate of **diffusion** of substances by keeping the distance for diffusion between the blood and cell cytoplasm to a minimum.
- By the time blood leaves the capillaries and enters the veins, there is no pulse and the blood pressure is very low. The large lumen (centre) of the veins allows blood to flow easily back to the heart. The contraction of body muscles, such as in the legs, helps to push the blood back toward the heart against the force of gravity. The valves make sure that blood can flow only in the right direction, back towards the heart.

normal blood flow

veins have valves to stop the blood flowing backwards

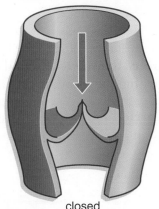

open

closed

△ Fig. 7.22 Valves in the veins make sure that blood can only move in one direction, toward the heart.

END OF EXTENDED

1. Name the following blood vessels:

a) the vessels that carry blood to the kidneys

b) the vessel that carries blood from the heart toward the body

c) the vessels that carry blood from the lungs back toward the heart.

2. Describe the differences in structure of arteries, capillaries and veins.

3. EXTENDED Explain how the structure of arteries helps to reduce and even out the blood pulses from the heart.

SCIENCE LINK

CHEMISTRY – PARTICLE NATURE OF MATTER

- In liquids, such as the blood, the particles are able to move over each other, which allows this flow to happen.

- Substances need to move into and out of the blood stream, so ideas about the relative size of the different particles become important.

PHYSICS – SIMPLE KINETIC THEORY, FORCES

- The pressure of the blood, being forced around the body by the heart, also links to the properties of fluids.

- As the blood is pumped along, the pressure in the fluid is transmitted in all directions, so the sides of the blood vessels have to be strong enough to cope with the forces involved.

Blood

The human circulatory system carries substances around the body. Table 7.3 shows some of the important substances transported around the human body. These substances are carried within the blood, in different forms.

Substance	Carried from	Carried to
molecules absorbed from digested food, for example, glucose, amino acids, fatty acids	small intestine	all parts of the body
water	intestines	all parts of the body
oxygen	lungs	all parts of the body
carbon dioxide	all parts of the body	lungs
urea (waste)	liver	kidneys
hormones	glands	all parts of the body (different hormones affect different parts)

△ Table 7.3 Substances carried by the blood around the body.

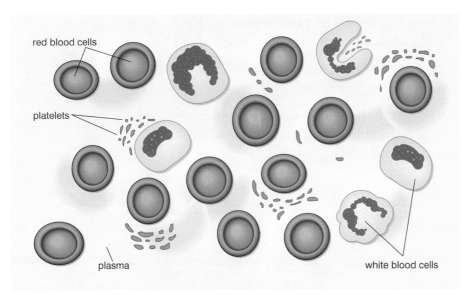

△ Fig. 7.23 Blood is mostly water, containing cells and many dissolved substances.

Blood contains plasma, red blood cells, white blood cells and platelets. Each of these has a particular function in the body.

Plasma

Plasma is the straw-coloured, liquid part of blood. It mainly consists of water, which makes it a good solvent for many substances. Digested food molecules, such as glucose and amino acids, easily dissolve in plasma. Urea, which is formed by the liver from excess amino acids, is also soluble in plasma. Many hormones are also soluble and are carried around the body dissolved in plasma. Carbon dioxide dissolves in water to form carbonic acid (H_2CO_3), and most carbon dioxide is carried in the blood in this form.

Red blood cells

Red blood cells are the most numerous cell type in blood. Their main function is to carry oxygen around the body. The oxygen is attached to molecules of **haemoglobin** inside the cells, which give red blood cells their colour.

White blood cells

There are several different types of white blood cell, but they all play an important role in defending the body against disease. They are part of the **immune system** that responds to infection by trying to kill the **pathogen** (the disease-causing organism). Some kinds of white blood cell kill pathogens by engulfing (flowing around the pathogen until it is

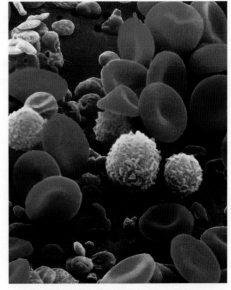

△ Fig. 7.24 Red blood cells (shown in red), white blood cells (yellow) and platelets (pink).

completely enclosed), which is known as **phagocytosis**. Other kinds of white blood cell produce chemicals called antibodies that attack pathogens.

Platelets

Platelets are small fragments of much larger cells that are also important in protecting us from infection by causing blood to clot where there is damage to a blood vessel.

SCIENCE IN CONTEXT / BLOOD TESTS

Doctors often send blood off to be tested when a patient is unwell. These tests not only check the concentrations of different substances in the blood, for example, glucose, they also count the number of different kinds of cells. Taking a blood sample can be quick and easy for a doctor to do, and the tests can help in diagnosing what is wrong with the patient.

- An abnormal concentration of glucose may indicate diabetes.
- Too few white blood cells may indicate liver or bone marrow disease.
- Too few red blood cells can make the blood look paler than normal. Low numbers of red blood cells cause anaemia, which usually results in the patient feeling more tired than usual.

Any abnormal results are followed up with other tests to confirm diagnosis.

QUESTIONS

1. Draw up a table to show the components of blood and the roles that they play in the body.

2. EXTENDED Using all of your knowledge, explain how the structure of a red blood cell is adapted to its function.

End of topic checklist

Key words

active transport, aorta, artery, atrium, capillary, circulatory system, coronary heart disease, deoxygenated, diffusion, double circulation, evaporation, haemoglobin, heart, heart rate, humidity, immune system, oxygenated, pathogen, phagocytosis, phloem, platelet, root hair cell, stomata, translocation, transpiration, valve, vascular bundle, vein, vena cava, ventricle, xylem

During your study of this topic you should have learned:

○ That xylem tissue transports water and mineral ions from plant roots to other parts of the plant.

○ That phloem tissue carries sucrose and amino acids from where they are made in the leaves to other parts of the plant.

○ That root hair cells are the site of absorption of water and mineral ions into a plant.

○ EXTENDED How to explain the importance of the large surface area of root hair cells.

○ To explain how water is absorbed from the soil through root hair cells and crosses the root cells to the xylem. It then passes up the stem in the xylem to the leaf and moves through the spongy mesophyll cells to the air spaces.

○ To explain that transpiration is the evaporation of water from the surfaces of a plant, mostly through the stomata.

○ To explain that the rate of transpiration increases with increased temperature, and decreased humidity.

○ To describe a circulatory system as consisting of tubes that carry blood around the body, where the blood is pushed by a pump called the heart.

○ EXTENDED To describe the double circulatory system of humans in which blood from the right side of the heart is pumped through the lungs, then back to the left side of the heart and to the rest of the body.

○ The heart pumps blood through arteries, then through capillaries, and finally through the veins back to the heart.

○ The heart is formed from four chambers, two atria and two ventricles, which have muscular walls to push blood through the heart. Valves prevent backflow of blood so that it only flows in one direction.

End of topic checklist continued

○ **EXTENDED** The heart pumps blood by the contraction of the atria followed by the ventricles, with one-way valves preventing the backflow of blood. The septum separates oxygenated from deoxygenated blood.

○ Heart rate increases during exercise.

○ **EXTENDED** Increase in heart rate during exercise supplies oxygen and glucose more rapidly to active muscle cells and removes carbon dioxide more rapidly.

○ **EXTENDED** To describe coronary heart disease as caused by the blockage of a coronary artery.

○ Diet, smoking, stress and genetic tendency are all risk factors for coronary heart disease.

○ To describe the structure of arteries including the thick muscular walls and thin central lumen.

○ **EXTENDED** The muscular walls of arteries resist the pressure of blood as it enters and even out the change in pressure as blood flows through.

○ How to describe the structure of veins and capillaries.

○ **EXTENDED** The thin capillary walls make it easier for substances such as carbon dioxide, oxygen and glucose to be exchanged with cells.

○ **EXTENDED** The large lumen in veins helps blood to flow easily through them, and the valves prevent the blood flowing in the wrong direction.

○ Human blood is formed from liquid plasma that carries red blood cells, white blood cells and platelets around the body.

○ Plasma is mostly water in which many substances dissolve, such as glucose, urea and hormones.

○ Red blood cells have a disc shape, contain large amounts of haemoglobin, and are small and flexible so that they can carry oxygen efficiently to all the cells in the body.

○ White blood cells protect aginst infection, by the phagocytosis of pathogens and by producing antibodies.

○ **EXTENDED** White blood cells include phagocytes that engulf pathogens, and lymphocytes which produce antibodies specific to a pathogen.

End of topic questions

Note: The marks given for these questions indicate the level of detail required in the answers. In the examination, the number of marks given to questions like these may be different.

1. Look at the photos in Fig. 7.3 of sections through a plant leaf, stem and root.

   a) Using a sharp pencil, make careful diagrams that show the positions of all the main tissues in the sections. (Only draw a few cells in each tissue, to show their form. It will take too long to draw all the cells.) **(3 marks)**

   b) Use the labelled diagrams in Fig. 7.2 to help you to clearly label your drawings to show the position of the xylem vessels. **(3 marks)**

   c) Clearly label the position of the phloem vessels in each of your drawings. **(3 marks)**

2. Flower sellers sometimes take white flowers and produce flowers of unusual colours for sale. Explain how they may do this, and why it is possible. **(4 marks)**

3. **EXTENDED** Cactus plants have many adaptations to help them survive in a dry desert. One of these adaptations is that they close their stomata during the day and open them at night. Explain fully the advantage of this adaptation. **(4 marks)**

4. People who suffer from anaemia often have a low red blood cell count (fewer blood cells per mm³ blood). One of the symptoms of anaemia is becoming tired more easily than usual. Explain why this symptom occurs. **(4 marks)**

5. Doctors look for risk factors in a person's lifestyle to help advise their patients on how to live more healthily.

    a) Describe four risk factors for coronary heart disease. **(4 marks)**

    b) EXTENDED For each of your answers to part **a)**, explain what the patient can or cannot do about them. **(4 marks)**

6. EXTENDED The blood pressure of blood leaving the right ventricle of the human heart is 3 kPa, and from the left ventricle is around 16 kPa. Explain how the heart can produce these different pressures, and why this difference is important for the body. **(4 marks)**

End of topic questions continued

7. **EXTENDED** The graph below shows the death rate from coronary heart disease (CHD) in some European countries compared with the proportion of the energy in the average diet in those countries that is provided by saturated fat.

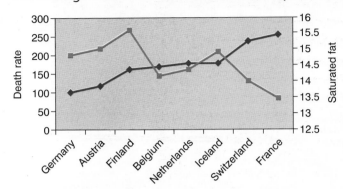

△ Fig. 7.20 Death rates and saturated fat intake for some European countries.

a) Saturated fat consumption is a risk factor for coronary heart disease. Explain what we mean by a *risk factor*. **(1 mark)**

b) Explain why saturated fat consumption is a risk factor for coronary heart disease. **(1 mark)**

c) Analyse the shape of the graph. **(3 marks)**

d) Some people think this graph indicates that there is no correlation between saturated fat consumption and CHD. Evaluate this idea. **(4 marks)**

Our lungs are the organs that allow the body to take in oxygen from the air and expel carbon dioxide that is produced in cells. We breathe in and out about 500 ml of air during every breath. Oxygen from this air passes into tiny air sacs in the lungs, which are called alveoli, and diffuses into the capillaries that lie just underneath them. From here, the oxygen-rich blood is passed to the heart, where it is pumped around the rest of the body, delivering oxygen and nutrients from the digestive system to cells. Cells use the nutrients and oxygen to carry out cell respiration. The blood is then passed back to the lungs to offload carbon dioxide and pick up a fresh supply of oxygen. When we talk about respiration we often think of the process of breathing, but cellular respiration is a series of chemical reactions that take place in cells to generate the energy that they need to carry out their specific functions.

SECTION CONTENTS

8

Gas exchange and respiration

△ Gas exchange takes place in the alveoli, located in the lungs.

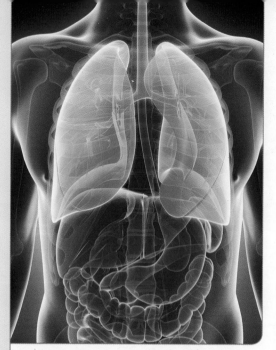

△ Fig. 8.1 The lungs are the site of gas exchange in humans.

Gas exchange and respiration

INTRODUCTION

Respiration uses substances that are gases in air and produces gases that need to be returned to the environment. These gases must get into and out of the body fast enough to support the rate at which body processes work. For single-celled organisms this isn't a problem. They have a large surface area to volume ratio, and diffusion across the cell membrane can supply and remove the gases at a fast enough rate.

Larger organisms cannot do this. Not only do they have a much smaller surface area to volume ratio, which slows the rate of diffusion, many of them also live on land, where the delicate surface required for gas exchange would dry out. Different groups of organisms have different solutions to these problems. Plants exchange gases inside the leaf; insects have internal tubes (a tracheal system) inside the body where they exchange gases; fish have gills; and many vertebrates, including humans, have lungs.

KNOWLEDGE CHECK

✓ Animals breathe in oxygen from the air and breathe out carbon dioxide.
✓ Humans use lungs for breathing.
✓ Organisms need energy for all the life processes that keep them alive.
✓ Plants get this energy from the sugars they make in photosynthesis.
✓ Animals get this energy from their food.
✓ Plants take in oxygen and give out carbon dioxide as a result of photosynthesis.

LEARNING OBJECTIVES

✓ **EXTENDED** Describe how the gas exchange surfaces of the lung are adapted for diffusion.
✓ Identify on a diagram the larynx, trachea, bronchi, bronchioles, alveoli and associated capillaries.
✓ Describe the differences in composition of gases in inspired air and expired air.
✓ **EXTENDED** Explain the differences in composition of gases in inspired air and expired air.
✓ Describe how you would investigate the effect of physical activity on the rate and depth of breathing.
✓ **EXTENDED** Explain why the rate and depth of breathing increase with increasing activity.
✓ **EXTENDED** Explain the role of mucus and cilia in the lungs in protecting against damage and infection.

✓ **EXTENDED** State that tobacco smoking can cause chronic obstructive pulmonary disease, lung cancer and coronary heart disease.

✓ **EXTENDED** Describe the effects of tobacco smoke on the gas exchange system.

✓ State that respiration releases energy in living organisms, and involves the action of enzymes in cells.

✓ Give examples of how energy released from respiration is used in the human body.

✓ **EXTENDED** Describe aerobic respiration as the release of energy from glucose using oxygen from the air.

✓ Give the word equation for aerobic respiration as:

glucose + oxygen → carbon dioxide + water

✓ **EXTENDED** Give the balanced symbol equation for aerobic respiration as:

$C_6H_{12}O_6 + 6O_2 \rightarrow 6CO_2 + 6H_2O$

✓ Investigate the uptake of oxygen by respiring organisms.

GAS EXCHANGE

Animals need to exchange gases with the environment, to supply oxygen for respiration in cells and to remove the waste product of respiration – carbon dioxide. These gases are exchanged at surfaces by diffusion. So, gas exchange surfaces, such as in the human lungs, need adaptations to maximise the rate at which diffusion occurs.

EXTENDED

An effective gas exchange surface has:

- a large surface area
- a short distance over which substances have to diffuse, so cells across which diffusion occurs are usually thin
- a good blood supply
- good ventilation to deliver more oxygen and remove carbon dioxide from the body rapidly.

END OF EXTENDED

The human respiratory system

Breathing is the way that oxygen is taken into our bodies and carbon dioxide is removed. When we breathe, air is moved into and out of our lungs. This involves different parts of the respiratory system within the thorax (the chest cavity).

When we breathe in, air enters though the nose and mouth. In the nose the air is moistened and warmed. The air passes over the **larynx**, where it may be used to make sounds, for example, when we talk. The air travels down the **trachea** (windpipe) to the **lungs**. The air enters the lungs through the **bronchi** (singular: bronchus), which branch and divide to form a network of **bronchioles**.

EXTENDED

Cells called goblet cells in the lining of the trachea, bronchi and bronchioles secrete **mucus**, which is a slimy liquid. This traps microorganisms, which might be pathogens, and dust particles that are breathed in. The lining of the trachea and bronchi are covered in tiny hairs called **cilia**, which are found on the surface of ciliated cells. The cilia sweep in a coordinated motion to move the mucus up from the lungs, up the trachea to the back of the mouth, where it can be swallowed. The combined action of mucus and cilia helps to prevent dirt and microorganisms entering the lungs and causing damage and infection.

cilia

goblet cells

△ Fig. 8.2 Section through the tracheal epithelium, showing goblet cells, which secrete mucus, and cilia, which sweep the mucus along the epithelial surface.

END OF EXTENDED

At the end of the bronchioles are air sacs. The bulges on an air sac are called **alveoli** (singular: alveolus). The alveoli are covered in tiny blood capillaries. This is where oxygen and carbon dioxide are exchanged between the blood and the air in the lungs. This is called **gas exchange**. The movement of air across the alveolar surface is called **ventilation**.

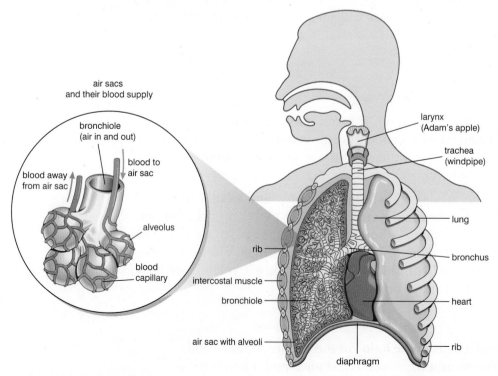

air sacs
and their blood supply

bronchiole
(air in and out)

blood to
air sac

blood away
from air sac

alveolus

blood
capillary

larynx
(Adam's apple)

trachea
(windpipe)

lung

rib

bronchus

intercostal muscle

bronchiole

heart

air sac with alveoli

rib

diaphragm

△ Fig. 8.3 The human respiratory system.

The alveoli are where oxygen and carbon dioxide diffuse into and out of the blood. For this reason, the alveoli are described as the *site of gas exchange*, or the *respiratory surface*.

EXTENDED

The alveoli are adapted for efficiency in exchanging gases by diffusion. They have:

- thin permeable walls, which keep the distance over which diffusion of gases takes place between the air and blood to a minimum
- a moist lining, in which the gases dissolve before they diffuse across the cell membranes
- a large surface area – there are hundreds of millions of alveoli in a human lung, giving a surface area of around 70 m^2 for diffusion
- high concentration gradients for the gases, because the blood is continually flowing past the air sacs, delivering excess carbon dioxide and taking on additional oxygen, and because of ventilation of the lungs, which refreshes the air in the air sacs.

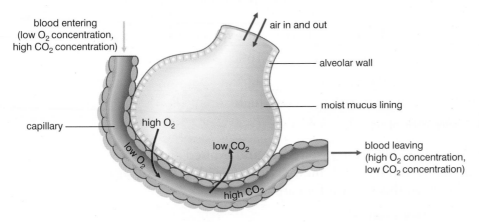

blood entering (low O$_2$ concentration, high CO$_2$ concentration)

air in and out

alveolar wall

moist mucus lining

capillary

high O$_2$

low O$_2$

low CO$_2$

high CO$_2$

blood leaving (high O$_2$ concentration, low CO$_2$ concentration)

△ Fig. 8.4 Gas exchange in an air-filled alveolus.

REMEMBER

To write a really good answer, be careful how you describe the process of gas exchange between the air in the lungs and the blood. Remember that diffusion is a passive process, so that it only occurs while there is a concentration gradient. Avoid answering in simple terms, which imply that the movement of oxygen is only from the air to the blood, and that the movement of carbon dioxide is only from the blood to the air.

END OF EXTENDED

QUESTIONS

1. **EXTENDED** Explain, as fully as you can, why the lungs show adaptations for a rapid rate of diffusion.

2. List the structures of the human respiratory system and, for each structure, explain its role in breathing.

3. **EXTENDED** Sketch a diagram of an alveolus and annotate it to show how it is adapted for efficient gas exchange. (Hint: remember to refer to diffusion.)

4. **EXTENDED** What is the role of the cilia and mucus in the human respiratory system?

Inspired air and expired air

The air we breathe in and out contains many gases. Oxygen is taken into the blood from the air we breathe in. Carbon dioxide and water vapour are added to the air we breathe out. The other gases in the air we breathe in are breathed out almost unchanged, except for being warmer.

	In inspired air	**In expired air**
oxygen	21%	16%
carbon dioxide	0.04%	4.5%
water	variable	high

△ Table 8.1 Differences in composition of inspired and expired air.

We can compare the carbon dioxide in inspired air and expired air using the apparatus shown in Fig. 8.5. Limewater reacts with carbon dioxide and turns cloudy, so this is a test for carbon dioxide.

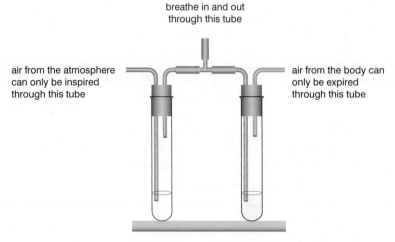

breathe in and out through this tube

air from the atmosphere can only be inspired through this tube

air from the body can only be expired through this tube

△ Fig. 8.5 Limewater in the tubes shows that expired air contains much more carbon dioxide than inspired air. Note: one-way valves must be used when setting up this apparatus.

The composition of inspired and expired air changes because:

- oxygen is removed from the blood by respiring cells and used for cellular respiration, so blood returning to the lungs has a lower concentration of oxygen than blood leaving the lungs
- carbon dioxide is produced by respiration and diffuses into the blood from respiring cells; the blood transports the carbon dioxide to the lungs, where it diffuses into the alveoli
- water vapour concentration increases because water evaporates from the moist linings of the alveoli into the expired air as a result of the warmth of the body.

Other gases remain unaffected because they are not used or produced by the body.

Investigating the effect of exercise on breathing

There are two aspects of breathing that can change during exercise – the rate of breathing and the volume of breath.

- Rate of breathing is usually counted as the number of breaths per minute.
- The volume of a breath can be measured in dm^3 using a spirometer. A simple spirometer can be made using a 2-litre plastic bottle that has been marked down the side with volumes of water. (This can be done by adding $500\,cm^3$ of water at a time, and marking the volume on the side of the bottle with a waterproof marker.) When the bottle is full of water, turn it upside-down into a water trough without allowing any air into the bottle. Insert a flexible plastic tube into the neck of the bottle and secure the bottle and tube in position. Clean the other end of the tubing with antiseptic solution. (Alternatively, add a mouthpiece to the end of the tubing that can easily be removed and sterilised after each test.) To measure the volume of a breath, ask the person to wear a noseclip and then to breathe out a normal breath into the tube. The scale on the bottle can be used to measure the volume of air breathed out.
(Safety note: this apparatus must only be used for measuring one breath. The bottle must be set up again before measuring another breath. This is because carbon dioxide build-up in the air in the bottle over several breaths can be dangerous.)

Devise and plan investigations

❶ Design an investigation into the effect of exercise on breathing. (Hint: think carefully about how many people to test, and how to test them, in order to get reliable results.)

Demonstrate and describe techniques

❷ This investigation could involve vigorous exercise. What risks will you need to prepare for, and how should they be minimised?

Analyse and interpret data

The data in the table are the results from an investigation into the effect of exercise on breathing in four people. They were first tested at rest and then after 2 minutes of running on a treadmill set at the same speed.

Person	A		B		C		D	
	Rate (breaths/ minute	Breath volume (dm³)	Rate (breaths/ minute	Breath volume (dm³)	Rate (breaths/ minute	Breath volume (dm³)	Rate (breaths/ minute	Breath volume (dm³)
at rest	13	0.5	15	0.4	12	1.2	18	0.6
after exercise	19	1.3	23	0.9	18	1.3	26	1.5

❸ Explain how these data should be adjusted before they can give a reliable answer to the question 'How does exercise affect breathing?'

The results of an investigation like the one in the Developing investigative skills box above, should show that both the rate of breathing and depth of breathing increase with the level of activity. However, a trained athlete will show a smaller change in rate and depth of breathing than an untrained person.

EXTENDED

The rate and depth of breathing increase with level of activity because as the muscles contract faster they respire faster and so make carbon dioxide more quickly. Carbon dioxide is an acidic gas that dissolves easily in water-based solutions, such as the cytoplasm of a cell and blood plasma. The more carbon dioxide there is in solution the more acidic the solution. A change in pH can affect the activity of many cell enzymes, so it is important that carbon dioxide is removed from the cells and the body as quickly as possible.

The increase in carbon dioxide concentration as a result of increased physical activity is detected as the blood flows past receptors in part of the brain. The receptors send impulses to the lungs, causing an increase in the rate and depth of breathing, which helps to remove the extra carbon dioxide as quickly as possible.

HOW BREATHING RATE IS CONTROLLED

Rate of breathing is controlled by the part of the brain that measures not the oxygen concentration of the blood but the carbon dioxide concentration. This is because a small increase in carbon dioxide concentration in body fluids could have a much more damaging effect on the body than a small decrease in oxygen concentration.

END OF EXTENDED

QUESTIONS

1. Describe the differences in composition between inspired air and expired air.

2. Describe the effects of exercise on the rate and depth of breathing.

3. EXTENDED Explain the differences in composition between inspired air and expired air.

4. EXTENDED Explain what would happen to cells if rate and depth of breathing did not change during exercise.

EXTENDED

The effects of smoking

When a person smokes tobacco, the chemicals in the smoke are taken into the lungs. Those chemicals that are small enough molecules can then diffuse into the blood and be carried around the body. Many of the chemicals in tobacco smoke have damaging effects, not only on the respiratory system, but also on other systems in the body.

Smoking tobacco can cause chronic pulmonary obstructive diseases (**COPD**), such as bronchitis and emphysema.

- The tar in tobacco smoke is a mixture of chemicals that form a black sticky substance in the lungs. This sticky layer can coat the tiny hair-like cilia lining the tubes of the lungs, making it more difficult for them to clear out dust and microorganisms. This can result in many lung infections and a thick cough as the smoker tries to clear sticky mucus from the lungs. The irritation and infection can cause a disease called **bronchitis**.

- Continued coughing, in order to clear tar and smoke particles from lungs, over a long time damages the alveoli, breaking down the divisions between them and so reducing their surface area. This causes a disease called **emphysema**, in which the patient has difficulty getting enough oxygen into their blood for any kind of activity. They may have to breathe pure oxygen to make sure their damaged lungs can absorb enough oxygen into their body.

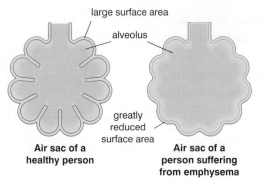

large surface area

alveolus

greatly reduced surface area

Air sac of a healthy person

Air sac of a person suffering from emphysema

◁ Fig. 8.6 Repeated coughing over a long period breaks down the surface of each alveolus, reducing the surface area for exchange of gases. This condition is called emphysema.

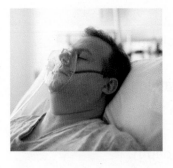

◁ Fig. 8.7 People with emphysema may have to breathe air containing a high concentration of oxygen, to make sure that their damaged lungs can absorb enough oxygen into their bodies. Breathing masks such as the one shown attached to oxygen tanks can provide this for patients.

Many other gases in tobacco smoke are **carcinogenic**, meaning they cause cells to become cancerous and take over tissue. Smoking is the greatest cause of lung cancer. Smoking is also linked to many other kinds of cancer in the body, and to heart disease. People who smoke are more likely to suffer a heart attack or heart pains (angina) than people who do not smoke.

Tobacco smoke also contains carbon monoxide, which is a toxic gas. It combines with haemoglobin in red blood cells and so prevents the cells from carrying oxygen. This reduces the amount of oxygen that gets to tissues, which in extreme cases can lead to cell death. In lower amounts carbon monoxide can result in breathlessness, when the body cannot get sufficient oxygen to cells for activity. During pregnancy, smoking passes through the placenta to the developing fetus. This can reduce the rate of growth of the fetus, resulting in a low birth weight, which can cause complications during the birth and health problems through life.

Nicotine in tobacco smoke alters people's moods – smokers often say they feel more relaxed but alert after smoking. Nicotine is also highly addictive, which makes it difficult for smokers to give up.

QUESTIONS

1. Give two examples of COPD, and describe the symptoms of each.

2. Describe the effects of carbon monoxide and nicotine on the body.

END OF EXTENDED

RESPIRATION

When we talk about respiration generally, we usually mean breathing (or ventilation), when gases are exchanged across a respiratory surface. Here we are focusing specifically on **cellular respiration**, which is the release of energy from the chemical bonds in food molecules such as glucose. This only takes place inside cells, and every living cell carries out cellular respiration.

REMEMBER

Be clear in your answers that you are using the term *respiration* to mean *cellular respiration*, and to use *ventilation* not *respiration* when talking about breathing.

Every cell in a living organism requires energy, and this energy comes from respiration, which is the breakdown of chemical bonds in food molecules such as glucose to release energy in a form that can be used in cells.

In human cells, this energy is used:

- to produce the contraction of muscle cells
- to produce new chemical bonds during the **synthesis** (formation) of new protein molecules
- to produce new chemicals needed for cell division and for the growth of cells
- for the active transport of molecules across cell membranes
- to produce the movement of nerve impulses along nerve cells
- for the maintenance of a constant core body temperature.

Note that we usually refer to glucose as the *nutrient molecule* or *food molecule* that is broken down in respiration. This is because it is the molecule most commonly used in this reaction in the body. If glucose is in short supply, then other molecules may be used instead from the breakdown of fats or proteins.

Respiration is a series of reactions and, like other reactions in cells, these are controlled by enzymes. So any change in a cell that affects enzymes (such as a change in temperature or pH) will affect the rate of respiration.

Aerobic respiration

Most plant and animal cells use oxygen during cellular respiration. Respiration that uses oxygen to release energy from glucose is called **aerobic respiration**. Water and carbon dioxide are produced as waste products. This is very similar to burning fuel except that in our bodies enzymes control the process.

Aerobic respiration can be summarised by a word equation:

glucose + oxygen → water + carbon dioxide (+ energy)

EXTENDED

It can also be written as a symbol equation:

$$C_6H_{12}O_6 + 6O_2 \rightarrow 6H_2O + 6CO_2 \text{ (+ energy)}$$

END OF EXTENDED

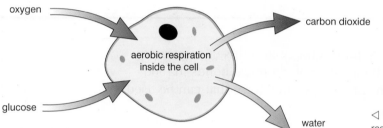

◁ Fig. 8.8 Aerobic respiration in a cell.

The oxygen needed for respiration comes from the air (except for a small proportion in photosynthesising plants, which comes from photosynthesis). The carbon dioxide from cellular respiration is released to the air, and the water is either used in the body or excreted through the kidneys.

SCIENCE IN CONTEXT   WATER FROM RESPIRATION

A camel can survive for many days without drinking liquid water, which means it survives well in desert conditions. The camel's hump is not a store of water, but a store of fat. Over a long period without food, the fat is broken down to release substances for aerobic respiration. As water is one of the products of aerobic respiration, this also helps the camel to survive longer without drinking water.

△ Fig. 8.9 Wild camels live in dry areas and so must go many days without drinking water.

A complete lack of food and drinking water are conditions that would kill a human in a few days, because we don't metabolise fat as well as the camel does, or retain water as well. So, before setting out into the desert for a long trip, make sure your camel has a large hump (and you have plenty of food and water).

During aerobic respiration, many of the chemical bonds in the glucose molecule are broken down. This releases a lot of energy: around 2900 kJ of energy is released for each mole of glucose molecules used in aerobic respiration.

SCIENCE LINK

CHEMISTRY – PHYSICAL AND CHEMICAL CHANGES, STOICHIOMETRY

- Respiration is a chemical change releasing energy. The energy changes occur as a result of bond-breaking and bond-making.

- Using word equations and balanced symbol equations to describe chemical changes helps us to understand the particle interactions that are taking place.

PHYSICS – ENERGY

- Energy is released during respiration and this links to several key ideas about energy.

- Ideas about conservation of energy indicate how much glucose is required to maintain a healthy organism and how that might change depending on age and lifestyle.

- Energy transfers are not always useful and this links to processes for maintaining body temperature as non-useful energy often transfers as heat.

QUESTIONS

1. **a)** Write out the word equation for aerobic respiration.

   b) Annotate your equation to show where the reactants come from.

   c) Annotate your equation to show what happens to the products of the reaction in a human.

   d) Describe how your answer to part **c)** might differ for a camel on a long journey without water, and explain your answer.

2. Where does respiration take place in the body?

3. Give three examples of the use of energy from respiration in the human body.

4. EXTENDED Write the balanced symbol equation for aerobic respiration.

End of topic checklist

Key terms

aerobic respiration, alveoli, bronchiole, bronchitis, bronchus, carcinogenic, cellular respiration, cilia, COPD, emphysema, gas exchange, larynx, lungs, mucus, synthesis, trachea, ventilation

During your study of this topic you should have learned:

○ EXTENDED That humans exchange gases with the environment by diffusion, so the lungs need a large surface area, a short distance for diffusion to the blood, and continual ventilation of the inside of the lungs.

○ To describe ventilation as the breathing in and out of air to the lungs, through the larynx, where sound may be produced, down the trachea to the two bronchi, and through the bronchioles to the alveoli, where the gases are exchanged with the many capillaries that lie next to the alveoli.

○ That expired air contains more carbon dioxide, less oxygen and more water vapour than inspired air.

○ EXTENDED The differences in gas concentration in inspired air and expired air are the result of respiration in cells using oxygen and producing carbon dioxide, and the evaporation of water vapour from surfaces inside the lungs.

○ That an increased level of activity increases the rate and depth of breathing.

○ EXTENDED The increased rate and depth of breathing during activity removes the increased amount of carbon dioxide produced by respiration and so stops the pH of body tissues and blood falling.

○ That goblet cells in the linings of the lungs secrete mucus that traps pathogens and particles. The mucus is swept out of the lungs by cilia, which protect the lungs from damage and infection.

○ EXTENDED Tobacco smoking can cause chronic obstructive pulmonary disease, lung cancer and coronary heart disease.

○ EXTENDED The effects of tobacco smoke on the gas exchange system.

○ To define respiration as the process in which energy is released from nutrient molecules in the cells of living organisms, and is controlled by enzymes.

○ How to define aerobic respiration.

○ That in aerobic respiration, glucose is broken down using oxygen from the air:
glucose + oxygen → carbon dioxide + water (+ energy)
EXTENDED $C_6H_{12}O_6 + 6O_2 \rightarrow 6CO_2 + 6H_2O$ (+ energy)

○ That in humans, energy from respiration is used in muscle contraction, protein synthesis, cell division, active transport, growth, the passage of nerve impulses and the maintenance of a constant body temperature.

End of topic questions

Note: The marks given for these questions indicate the level of detail required in the answers. In the examination, the number of marks given to questions like these may be different.

1. a) Define the terms *diffusion* and *gas exchange.* **(2 marks)**

   b) Describe the role of diffusion in gas exchange in humans. **(2 marks)**

   c) EXTENDED Explain how the tissues and organs of the lungs are adapted to maximise the rate of gas exchange. **(4 marks)**

2. EXTENDED Fig. 8.10 shows a few cells of the epithelium lining the trachea.

   a) State where the trachea is found and explain its role in the body. **(2 marks)**

   b) Name the type of cell shown by cell A and describe its function. **(2 marks)**

   c) Name the type of cell shown by cell B and describe its function. **(2 marks)**

   d) Explain the role of these cells in protecting the body. **(3 marks)**

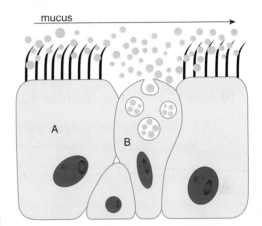

△ Fig. 8.10. Cells of the trachea epithelium

3. a) List the body systems in a human that are involved in supplying the reactants of cellular respiration. **(3 marks)**

   b) List the body systems in a human that are involved in removing the products of cellular respiration. **(3 marks)**

4. Draw up a table to summarise the similarities and differences between aerobic and anaerobic respiration. **(8 marks)**

5. Students were studying the results of respiration in some woodlice. They set up two identical sets of apparatus: a boiling tube fitted with a bung and linked to a tube of limewater through a delivery tube. They placed some woodlice in one boiling tube, and no woodlice in the other tube. The boiling tubes were fitted with their bungs so that no additional air could enter the apparatus and then were left overnight.

   a) What was the role of the second set of apparatus? Explain your answer. **(2 marks)**

   b) Suggest what happened to the limewater in the two sets of apparatus. **(2 marks)**

   c) Explain as fully as you can your answer to part **b).** **(3 marks)**

6. A whale takes a deep breath of air and then dives for half an hour. Suggest how energy would be generated in the whale's muscles over the period of the dive.

(4 marks)

7. In bread-making, the yeast is first mixed with a solution containing sugar and kept in a warm place until a froth forms on the surface. The yeast mixture is then added to flour and any additional constituents, and mixed thoroughly. The dough is placed in a warm place for a while, until it has doubled in size. It is then baked as a loaf.

a) Explain why a sugar solution is added to the yeast at the start. **(2 marks)**

b) 'A warm place' means around 25–30 °C.

i) Explain what would happen if the temperature was lower than this.

(3 marks)

ii) Explain what would happen if the temperature was higher than this.

(3 marks)

Organisms need to be able to respond to a changing environment. Hormones are one of the ways of sending a message to change the activity of cells or organs to help organisms respond. Hormones affect everything from how excited or scared we feel, to how tall we grow, to whether we are male or female.

In plants these responses are called tropisms, and it is how plants can respond to gravity and light.

SECTION CONTENTS

9

Coordination and response

△ A micrograph showing adrenal tissue from the adrenal glands. This is where the hormone adrenaline is made.

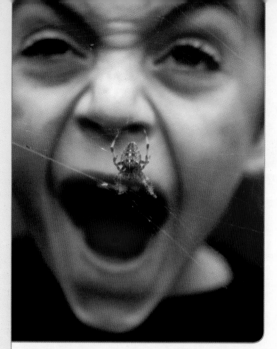

Coordination and response

INTRODUCTION
Has a spider every made you jump?

Adrenaline is part of the flight or flight response. When something makes us scared, such as seeing a spider, adrenaline is the hormone that gets us ready to run away or fight. Our heart rate increases, we breathe faster and our pupils widen.

△ Fig. 9.1 Adrenaline is released as part of a fight or flight response.

KNOWLEDGE CHECK

✓ Plants and animals detect the environment with specialised sense organs.
✓ Animals respond to changes in the environment using nervous and hormonal systems.
✓ Plants respond to changes in the environment, for example, by growth.

LEARNING OBJECTIVES

✓ Describe a hormone as a chemical substance produced by a gland that changes the activity of one or more target organs.
✓ Describe some of the effects of adrenaline, which is produced at times of increased action or stress.
✓ Describe investigations of gravitropism and phototropism.
✓ **EXTENDED** Discuss the role of the hormone adrenaline in the chemical control of metabolic activity including increasing the blood glucose concentration and pulse rate.
✓ **EXTENDED** Explain phototropism and gravitropism as a result of the plant hormone auxin.

HORMONES IN HUMANS

Hormones are chemical messengers used in the **hormonal system**. They are produced in **endocrine glands**. Endocrine glands do not have ducts (tubes) to carry away the hormones they make: the hormones are secreted directly into the blood to be carried around the body dissolved in the blood plasma. Hormones change the activity of other specific parts of the body, called the **target organs**. Most hormones affect several target organs; others may only affect one target organ.

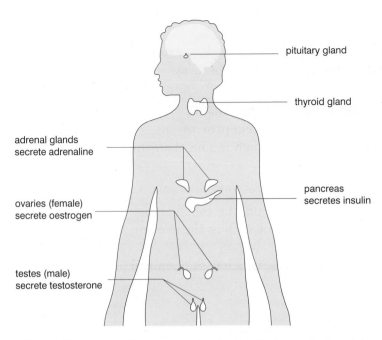

labels:
- pituitary gland
- thyroid gland
- adrenal glands secrete adrenaline
- pancreas secretes insulin
- ovaries (female) secrete oestrogen
- testes (male) secrete testosterone

△ Fig. 9.2 The position of some endocrine glands in the human body and the hormones they secrete.

Adrenaline

Adrenaline is a hormone that is produced in the adrenal glands just above the kidneys. This hormone is released in the crucial moments when an animal must instantly decide whether to attack or run for its life.

Some of the effects of adrenaline are:

- increased pulse (heart) rate to circulate blood more rapidly around the body and deliver glucose and oxygen to muscle cells to allow more rapid contraction
- increased depth of breathing and breathing rate to take more oxygen into the body and remove carbon dioxide more rapidly from the body
- dilated pupils for better vision.

All these changes prepare the body for action.

EXTENDED

Adrenaline also causes liver and muscle cells to release glucose, which increases blood glucose concentration. This means that there is more glucose available for increased muscle cell respiration, which will release more energy for muscle contraction.

END OF EXTENDED

Sensitivity

Sensitivity is the ability to recognise and respond to changes in external and internal conditions, and is one of the characteristics of living organisms.

A change in conditions is called a **stimulus**. For a coordinated response to occur to that stimulus there must be a **receptor organ**, which recognises the stimulus, and an **effector**, which is a mechanism to carry out the response.

There are two systems involved in coordination and response in humans.

- One is the *nervous system*, which includes the brain, the spinal cord, the peripheral **nerves** and specialist sense organs such as the eye and the ear.
- The other is the *hormonal* (or endocrine) *system*, which uses chemical communication by means of **hormones**.

QUESTIONS

1. Explain the meaning of the following terms:

   a) hormone

   b) endocrine gland

   c) target organ.

2. In what conditions might adrenaline be released in the body?

3. Explain the advantages of adrenaline in preparing the body for action.

TROPIC RESPONSES

Plants generally respond to changes in the environment by a change in the way that they are growing. For example, a shoot will grow towards light and in the opposite direction to the force of gravity, whereas a root will grow away from light but towards moisture and in the direction of the force of gravity. These growth responses to a stimulus in plants are called **tropisms**. They help the plant to produce leaves where there is the most light, and roots that can supply the water that the plant needs.

△ Fig. 9.3 Growing towards light helps a plant get more light for photosynthesis.

- Growth in response to the direction of light is called **phototropism**. If the growth is towards light, it is called *positive* phototropism, as occurs in shoots. If the growth is away from light, it is called *negative* phototropism, as occurs in roots.
- Growth in response to gravity is called **gravitropism** (sometimes also called *geotropism*). Plant shoots show negative gravitropism and plant roots show positive gravitropism.

Developing investigative skills

Figure 9.4 shows apparatus that can be used to investigate the effect of light on the growth of seedlings.

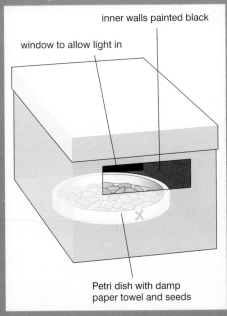

inner walls painted black

window to allow light in

Petri dish with damp paper towel and seeds

△ Fig. 9.4 Apparatus for investigating the effect of light on the growth of seedlings.

Devise and plan investigations

❶ Explain how the apparatus could be used for this investigation.

❷ Explain how you would set up a control for this investigation.

Make observations and measurements

❸ If this investigation were set up correctly, what result would you expect to see in the seedlings from the windowed box, compared with your control? Explain your answer.

Evaluate data and methods

❹ Suggest how this investigation could be extended to investigate whether roots also show a phototropic response.

Control of tropic responses

Tropisms are controlled by plant hormones called **auxins**. Auxin is made in the tips of shoots and roots. It dissolves in water in the cells and diffuses away from the tip. Further back along a shoot, auxin *stimulates* cells to elongate so that the shoot or root grows longer.

 SCIENCE IN CONTEXT **GARDENER'S TIP**

One effect of auxin is to inhibit the growth of side shoots. This is why a gardener who wants a plant to stop growing taller and encourage it to become more bushy will take off the shoot tip, so removing a source of auxin.

The growth of shoots toward light can be explained by the response of auxin to light.

- When all sides of a shoot receive the same amount of light, equal amounts of auxin diffuse down all sides of the shoot. So cells all around the shoot are stimulated equally to grow longer. This means the shoot will grow straight up.
- When the light on the shoot comes mainly from the side, auxin on that side of the shoot appears to move across the shoot to the shaded side. The cells on the shaded side of the shoot will receive more auxin, and so grow longer, than those on the bright side. This causes the shoot to curve as it grows, so that it grows toward the light.

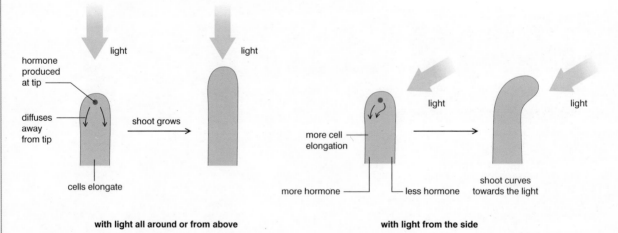

Δ Fig. 9.5 The effect of light on the growth of shoots.

In roots, auxin has the opposite effect on cells, so that it *reduces* how much the cells elongate.

- When roots are pointing straight downward, all sides of the root receive the same amount of auxin, so all cells elongate by the same amount.

- When the root is growing at an angle to the force of gravity, gravity causes the auxin to collect on the lower side. This reduces the amount of elongation of cells on the lower side of the root, so that the root starts to curve as it grows until it is in line with the force of gravity.

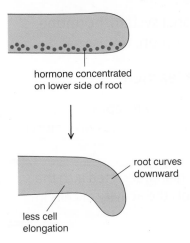

hormone concentrated on lower side of root

root curves downward

less cell elongation

△ Fig. 9.6 The effect of gravity on the growth of roots.

REMEMBER

- A full understanding of the phototropic responses in stems is needed.
- Auxin causes curvature in shoots by the elongation of existing cells, not by the production of more cells.

END OF EXTENDED

QUESTIONS

1. Define the term *tropism* in your own words.

2. Give one example of:
   a) positive phototropism
   b) positive gravitropism.

3. EXTENDED Describe the action of auxin in a shoot growing in one-sided light.

End of topic checklist

Key terms

auxin, effector, endocrine gland, gravitropism, hormonal system, hormone, nerve, phototropism, receptor organ, stimulus, target organ, tropism

During your study of this topic you should have learned:

○ A coordinated response requires a stimulus that is sensed by receptor cells, which results in a change in the organism brought about by an effector (usually muscles or glands in an animal).

○ A hormone is a chemical that is made in an endocrine gland, secreted into the blood so that it can move around the body, and controls the activity of cells in one or more target organs.

○ Adrenaline is the hormone that prepares the body for action or flight at times of stress, by increasing pulse rate and breathing and dilating the pupils.

○ EXTENDED Adrenaline also causes muscle and liver cells to release glucose, and so increase blood glucose concentration.

○ Plants respond to stimuli often by growth responses called tropisms.

○ Plant shoots show positive phototropism when they grow toward light, and negative gravitropism as they grow away from gravity.

○ Plant roots show positive gravitropism when they grow toward the force of gravity, and negative phototropism when they grow away from light.

○ EXTENDED Auxins are plant growth hormones that control the responses of shoots and roots to light and gravity.

End of topic questions

Note: The marks given for these questions indicate the level of detail required in the answers. In the examination, the number of marks given to questions like these may be different.

1. Explain the survival advantage to plants of having:

   a) shoots that are positively phototropic   (2 marks)

   b) roots that are positively gravitropic.   (2 marks)

2. Adrenaline is a hormone.

   a) Define the word hormone, including where they are produced, and how they are carried.   (3 marks)

   b) Describe two examples in which adrenaline secretion would increase.   (2 marks)

   c) EXTENDED Discuss the role of adrenaline in the chemical control of pulse rate.   (2 marks)

Scientists believe that there has been life on Earth for over 3500 million years. Nobody knows yet what triggered non-living molecules to become organised into living things that can reproduce themselves, but scientists have found traces of bacteria-like structures that may represent the earliest forms of life, in very ancient rocks.

Reproduction leads to different combinations of characteristics in offspring, and mutation produces new characteristics. The environment in which early organisms lived determined which of these combinations of characteristics would be the most successful, and so which individuals were likely to survive and pass on those characteristics to their offspring through reproduction. This process led to the evolution of new species and eventually to the millions of species that are alive on Earth today.

SECTION CONTENTS

10 Reproduction

△ Each poppy plant can produce up to 60 000 seeds, as a result of sexual reproduction.

Reproduction

INTRODUCTION

Most multicellular organisms reproduce sexually, requiring the transfer of gametes from the male to the female for fertilisation. Some flowering plants and a very few animals can reproduce asexually, where there is no transfer of gametes and females produce new individuals (more females) without fertilisation. Until recently, it was thought that asexual reproduction in animals was something that only happened in addition to sexual reproduction. However, DNA evidence suggests that some species of stick insect have not reproduced sexually for over 1 million years. Males of these species don't exist.

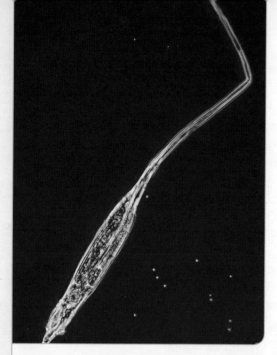

△ Fig. 10.1 Scientists think that this species of rotifer has not reproduced sexually for over 40 million years.

KNOWLEDGE CHECK

✓ The flower is the reproductive structure in flowering plants.
✓ The human reproductive system consists of organs, tissues and cells that are specially adapted for their role in reproduction.
✓ Sexual reproduction is the production of new individuals as a result of fertilisation; asexual reproduction is the production of new individuals without fertilisation.

LEARNING OBJECTIVES

✓ Describe the differences between sexual and asexual reproduction.
✓ Identify asexual reproduction in a range of organisms.
✓ Describe fertilisation as the fusion of gamete (sex cell) nuclei in sexual reproduction to produce a zygote.
✓ EXTENDED Identify the structures in wind-pollinated and insect-pollinated flowers and describe how they are adapted to their functions in reproduction.
✓ Define pollination of a flower as the transfer of pollen from an anther to a stigma.
✓ State that fertilisation occurs when a pollen nucleus fuses with a nucleus in an ovule.
✓ Investigate the conditions required for germination.
✓ Identify the organs, tissues and cells in the male and female human reproductive systems and describe their functions in reproduction.
✓ Describe the menstrual cycle in terms of changes in the ovaries and the lining of the uterus.
✓ EXTENDED Relate the size, structure, motility and number of sperm and egg cells to their role in reproduction.
✓ Describe how the embryo implants into the uterus wall and how structures in the uterus and the placenta support the developing embryo.

ASEXUAL AND SEXUAL REPRODUCTION

Asexual reproduction

Some organisms increase in number by **asexual reproduction**. For this type of reproduction it is not necessary to have two parents. During asexual reproduction, cells from an adult organism divide to produce the offspring. This means that offspring produced by asexual reproduction are genetically identical to their parent and to each other.

Asexual reproduction is used by many different organisms. Bacteria reproduce asexually using binary fission. When they are large enough their genetic material copies itself exactly and then the cell splits in half. The process then begins all over again. This can occur very rapidly to produce large numbers of identical bacteria.

△ Fig. 10.2 The toadstools we see growing are specialised spore-producing bodies of fungi.

Almost all fungi can reproduce asexually. Different types of fungi use different means of asexual reproduction but by far the most important type is that of spore formation. This can be seen in *Mucor,* the common pin mould, which often grows on bread. When this fungus has a plentiful supply of nutrients a **hypha** grows up vertically and the tip swells with cytoplasm containing many nuclei. This tip releases many spores into the atmosphere. If they find the right conditions for growth, each spore can develop into a new **mycelium**.

Another form of asexual reproduction is seen in plants, such as potatoes, that produce tubers. Tubers form from the end of stems that grow underneath the soil surface. The stems swell into storage organs filled with starch.

△ Fig. 10.3 Each of the potato tubers formed by this plant could produce a new plant in the next growing season.

When the leaves and stems of the plant die back at the end of the growing season the tubers stay dormant until the next season. Each tuber then produces several potato plants from the buds on the side of the tuber. Each potato plant gives rise to several tubers and each tuber produces a number of plants, so several new plants are formed from one parent.

QUESTIONS

1. Define the term *asexual reproduction* in your own words.

2. Explain why binary fission of bacteria is an example of asexual reproduction.

Sexual reproduction

Sexual reproduction is the most common method of reproduction for the majority of larger organisms, including almost all animals and plants. It occurs when there is **fertilisation**, which is when the nucleus of a male **gamete** (sex cell) fuses with the nucleus of a female gamete to form a **zygote**. The zygote will contain some of the genetic information of each of its parents. So it will be genetically different from each of the parents. It will also be genetically different from all other offspring produced by those parents (unless it has an identical twin).

QUESTION

1. Define the following terms in your own words:

a) *fertilisation*

b) *sexual reproduction.*

SEXUAL REPRODUCTION IN PLANTS

The most successful group of plants is the flowering plants. These are the only plants to have true flowers and produce **seeds** with a tough protective coat. During sexual reproduction, flowering plants:

- produce male and female gametes – some species may produce male and female gametes in the same flowers; other species may have male-only flowers and female-only flowers on the same plant; and in other species male flowers and female flowers are produced on different plants
- male pollen is transferred to the female part of the flower so that **pollination** can take place
- the male gamete and female gamete fuse during fertilisation to form a zygote
- the zygote develops to form an embryo within a seed, which protects the embryo and provides food during germination of the seed

- seeds are dispersed, so that they germinate and grow away from the parent.

Structure of flowers

All flowers have a similar basic arrangement. They have structures stacked one on top of each other along a short stem, arranged either in a spiral or in separate rings.

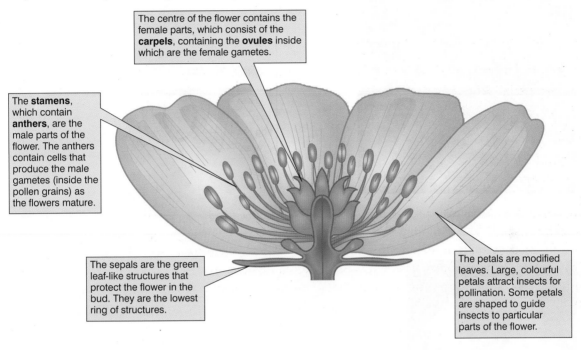

The centre of the flower contains the female parts, which consist of the **carpels**, containing the **ovules** inside which are the female gametes.

The **stamens**, which contain **anthers**, are the male parts of the flower. The anthers contain cells that produce the male gametes (inside the pollen grains) as the flowers mature.

The sepals are the green leaf-like structures that protect the flower in the bud. They are the lowest ring of structures.

The petals are modified leaves. Large, colourful petals attract insects for pollination. Some petals are shaped to guide insects to particular parts of the flower.

△ Fig. 10.4 Structure of an insect-pollinated flower.

The male part of a flower is the ring of **stamens**. There may be up to 100 stamens, or fewer than a dozen. Each stamen consists of two parts – the **anther** at the top and a stalk called the *filament*. Pollen grains develop inside the anthers. Inside each pollen grain is a male gamete. As a grain matures, it develops a thick outer wall to protect the delicate male gamete inside. When all the pollen grains in the anther are mature, the anther splits open to release them.

The female part of the flower is the **carpel**. A flower can contain more than one carpel, each with its own **style** and **stigma**. The stigma is the part of the carpel where the pollen lands during pollination. The **ovary** at the base of the carpel protects the female gamete from the dry air outside. The ovary contains one or more **ovules**, and each ovule contains an egg sac that surrounds the egg cell (female gamete).

△ Fig. 10.5 Alder catkins contain flowers that shed pollen into the air to be transported to other flowers.

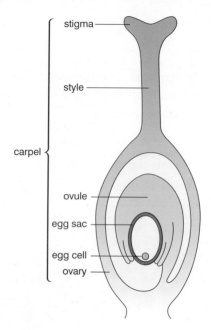

△ Fig. 10.6 The carpel.

◁ Fig. 10.7 Even complex flowers like daisies, which contain thousands of male and female parts, have carpels surrounded by stamens.

QUESTIONS

1. Name the female parts of a flower, and describe the function of each part.

2. Name the male parts of a flower, and describe the function of each part.

Pollination

Before fertilisation can take place, the male gametes have to reach the female gametes. This involves transferring the pollen to the stigma, in a process known as *pollination*. In many plants this means transferring the pollen from one flower to another. Some plants use the wind to transfer their pollen between flowers; others use animals, especially insects, to carry the pollen. Flowers have different features depending on whether they are pollinated by wind or by insects.

Wind-pollinated plants	Insect-pollinated plants
small petals, which do not obstruct pollen dispersal	large petals for insects to land on
green or inconspicuous petals	brightly coloured petals to attract insects
no scent	often scented to attract insects
no nectaries	nectaries present at the base of the flower produce a sugary liquid to attract insects, for example, bees and butterflies
many anthers, which are often large and hang outside the flower so that pollen is easily dispersed	a few small anthers, usually held inside the flower
pollen grains have smooth outer walls	pollen grains have sticky or spiky outer walls
stigmas are large and feathery, often hanging outside the flower to trap pollen	stigmas are small and held inside the flower
produce large amounts of pollen	produce smaller amounts of pollen
pollen is lightweight	pollen is heavier

△ Table 10.1 Comparison of wind-pollinated and insect-pollinated flowers.

△ Fig. 10.8 In insect-pollinated plants, nectaries secrete a sugary liquid to attract insects.

PHYSICS – PROPERTIES OF WAVES

- Insects that transfer pollen often have eyes that are sensitive to different wavelengths of the electromagnetic spectrum in comparison to humans. The colours of flowers are adapted to match this – flowers can appear very differently when viewed as the insects would view them.

PHYSICS – FORCES

- Several of the adaptations of pollen from wind-pollinated plants relate to forces and motion – shapes that reduce drag, masses that are easy to accelerate on the wind – and there are links to the motion of parachutes, for example.

EXTENDED

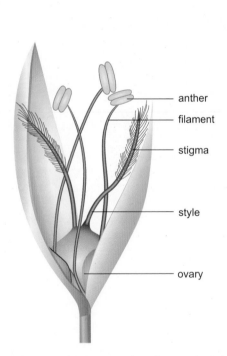

△ Fig. 10.9 These grass plants have anthers that hang outside the flowers and release large amounts of pollen to the wind. The stigmas also hang outside the flower to collect pollen from other grass plants.

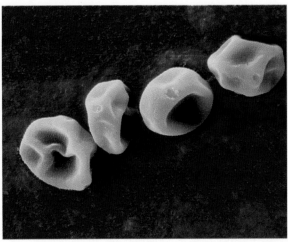

△ Fig. 10.10 Pollen grains. Left: from a wind-pollinated plant (birch) with a simple smooth outer wall. Right: from an insect-pollinated flower (daisy), with a spiky coat that helps the grains stick to the hairs on an insect's body.

END OF EXTENDED

SCIENCE IN CONTEXT

FLOWERS AND POLLINATORS

Different features of animal-pollinated flowers attract different pollinators. Tube-shaped flowers attract insects with a long tongue, such as butterflies, or birds with a long bill, such as hummingbirds. Blue and violet flowers are more attractive to bees, whereas butterflies often prefer red. Plants pollinated by moths or bats tend to open at night and may not be brightly coloured but instead produce a strong sweet scent. Plants that rely on flies to pollinate them often smell like rotting flesh.

One of the most bizarre partnerships between flowers and insects occurs between a particular species of orchid and a wasp. Male wasps are attracted to the flowers to mate with what they think are female wasps. During the 'mating' the flowers deposit pollen on the insect, which then carries it to the next flower that it is attracted to.

△ Fig. 10.11 A male wasp receiving pollen while 'mating' with an orchid flower.

REMEMBER

Be very careful not to confuse *pollination* with *fertilisation*.

SCIENCE IN CONTEXT

THE PROBLEM WITH BEES

About a third of all plants that we use for food or other uses depend on bees for pollination. This includes plants such as oilseed rape, cotton, coffee, apples and pears. If there are few bees, the crop harvest can be reduced by up to 75%. During the flowering season of crop plants, farmers and growers may place bee hives close to the crop to encourage successful pollination of most flowers. This helps to ensure a good harvest.

△ Fig. 10.12 By encouraging bees to build their hives in portable boxes, the farmer can move the hives to where the flowers of a crop are ready for pollination.

Recently, people have become concerned about a large decrease in bee populations. There are many possible reasons for this. In some places, it has been suggested that the lack of a range of food plants, including weeds, has been the cause. An increase in the use of pesticides that also kill bees may be another cause of the fall in their numbers.

Without bees, food production will be greatly affected. So, there are many studies being done to identify why bee numbers are decreasing and to work out how to improve the environment for bees.

QUESTIONS

1. Distinguish between *pollination* and *fertilisation* in a plant.

2. Describe *three* differences in structure between wind-pollinated and insect-pollinated flowers.

3. **EXTENDED** **a)** Explain the advantage to a flower of having adaptations for attracting insects rather than relying on wind for pollination.

   b) Describe one disadvantage for an insect-pollinated plant that relies on one or just a small number of insect species for pollination.

Germination

Germination is when the seed coat breaks open and the embryo starts to grow and develop into a new plant.

There are three environmental conditions that need to be right for seeds to germinate:

- temperature
- moisture
- oxygen.

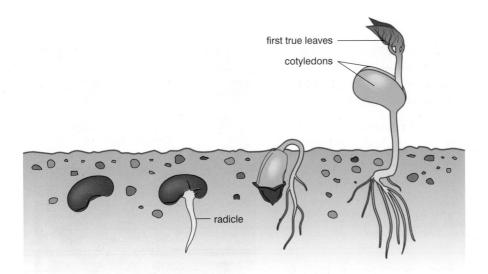

first true leaves

cotyledons

radicle

△ Fig. 10.13 Germination of a bean seed.

The presence of light is not usually needed for germination. This is because most seeds germinate below ground, so they cannot get their food from photosynthesis.

Temperature

A seed will not start to germinate until the conditions around it reach a suitable temperature. Many seeds lie dormant for long a time during cold periods, such as winter, and start to grow as the earth warms. However, if the temperature becomes too hot, the seed may be killed. This is why it is very important to store seeds in the correct conditions and to control the temperature in glasshouses carefully, for example, through the use of ventilation and shading.

 SCIENCE IN CONTEXT

CONDITIONS FOR GERMINATION

Different plants are suited to different climates. Those that are adapted to colder climates will germinate at lower temperatures. They may also need a very cold period followed by an increase in temperature before they will germinate.

Other seeds will not germinate until they have been exposed to very high temperatures, such as the heat from a forest fire. The extreme heat weakens the seed coat so that water can enter the seed and germination can begin.

△ Fig. 10.14 Fire clears the ground of competing plants, and stimulates these seeds to germinate in ideal conditions.

Germinating after a fire means that there is likely to be less competition with other species that usually cover the ground. Also, the ash left from the burning acts as a natural fertiliser for the new plants.

Water

Water is required to swell the seed and burst the seed coat. All seeds contain some moisture, but during germination metabolic reactions are being carried out rapidly. More water is needed for:

- activation of hormones and enzymes
- hydrolysis of storage compounds, for example, conversion of starch to glucose
- transport of materials to be used for respiration and growth
- metabolic reactions and enzyme actions that occur in solution.

Oxygen

Active living cells respire and the most useful form of respiration, aerobic respiration, requires oxygen. Seeds can use anaerobic respiration for a short while, but the rate at which energy is released is very slow (not useful in an actively growing organism) and the by-products are toxic. That is why most seeds will only germinate successfully if there is plenty of oxygen in the soil.

△ Fig. 10.15 Waterlogged soil excludes oxygen, making it difficult for these seeds to germinate and grow.

Developing investigative skills

We can investigate the particular conditions for germination.

Devise and plan investigations

❶ Using the apparatus shown, write a plan to investigate the effect of (a) light, (b) water and (c) temperature on the germination of seeds. Think carefully about what controls to use in each case.

Make observations and measurements

An investigation was carried out using two petri dishes containing 20 seeds of the same species. Both dishes received the same amount of light and moisture, but they were kept at different temperatures. The table shows the number of seeds that germinated over a period of 8 days.

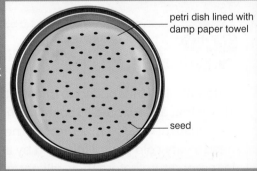

△ Fig. 10.16 A simple set-up for investigating seed germination.

petri dish lined with damp paper towel

seed

| Day | Total number germinated seedlings | |
	Cool/10°C	Warm/20°C
1	0	0
2	0	0
3	0	5
4	1	11
5	6	15
6	16	17
7	18	17
8	18	17

❷ Display the results of this investigation in a suitable way.

Analyse and interpret data

❸ Describe the patterns shown by these results.

❹ Draw a conclusion from these results.

❺ Explain the results using your scientific knowledge.

QUESTIONS

1. What is *germination*?

2. What effect do the following conditions have on the germination of a seed? Explain why they have these effects.

 a) oxygen

 b) moisture

 c) warmth

SEXUAL REPRODUCTION IN HUMANS

Male reproductive system

A human male has two **testes** (singular: testis) in which **sperm** are produced. The testes are supported outside the body in the scrotum to keep them cooler, because at higher temperatures fewer sperm are produced.

Sperm ducts carry the sperm from the testes to the penis, through the prostate gland and seminal vesicles. The prostate gland and seminal vesicles together produce the liquid in which the sperm are able to swim. Semen is the mixture of sperm cells and fluids.

Semen passes along the sperm duct to the urethra to outside the body. The urethra also carries urine from the

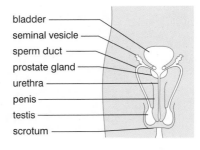

bladder
seminal vesicle
sperm duct
prostate gland
urethra
penis
testis
scrotum

△ Fig. 10.17 The male reproductive system. (Note that the bladder is not part of the reproductive system.)

bladder to outside the body. When the man is sexually excited, large spaces in the penis fill with blood. This causes the penis to become larger and stiffer causing an erection. At the same time a muscle ring (sphincter) at the top of the urethra contracts, preventing urine entering the urethra from the bladder.

The erection makes it possible for the man to insert his penis into the vagina of the woman for sexual intercourse. Rapid contractions of muscles in the penis during ejaculation send the sperm shooting out into the vagina.

Female reproductive system

The two **ovaries** are the organs in human females that produce the eggs. They are positioned within the abdominal cavity, either side of the **uterus** and joined to it by the **oviducts**.

Every month from puberty until menopause, when a woman is around 50 years old, one ovary usually releases one egg, which travels down the oviduct to the uterus (womb). If it is not fertilised, the egg will be flushed from the uterus during the monthly period (bleed). At the lower end of the uterus is the cervix. This canal produces mucus which changes during the menstrual cycle, allowing sperm to pass through at some times and not others. It also keeps the developing baby secure in the uterus until birth. The cervix leads into the vagina. The **vagina** is an elastic muscular tube where sperm are received from the penis during sexual intercourse.

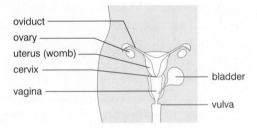

△ Fig. 10.18 The female reproductive system. (Note that the bladder is not part of the reproductive system.)

Human gametes

The human gametes are specialised for their roles in reproduction. The adaptive features of the sperm include the flagellum and the presence of enzymes. The adaptive features of the egg cells include the jelly coating, which changes after fertilisation, and the energy stores. Figure 10.19 shows the main adaptive features of the human sperm cell and egg cell.

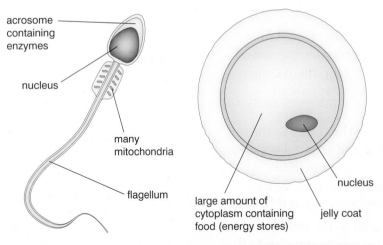

△ Fig. 10.19 Left: human sperm. Right: human egg. (Not to scale: the volume of an egg cell is 20 hundreds of times bigger than the volume a single sperm cell.)

REPRODUCTION

Sperm are among the smallest cells in the human body, at about 45 micrometres long. Over 100 million sperm cells are produced each day. There are many mitochondria in the part of the sperm between the head and the flagellum (tail). The mitochondria provide energy from respiration that allows the flagellum to beat back and forth to move the sperm cell.

Most of the cell is made up of the flagellum, which propels the sperm through the female uterus to the egg for fertilisation. At the front tip of the sperm is a small sac of enzymes called the **acrosome**. When a sperm reaches an egg cell, the acrosome bursts open to release the enzymes. The enzymes digest through the jelly coat and cell membrane of the egg cell, allowing the male nucleus to enter the egg cell.

The egg cell is one of the largest human cells, at about 0.2 mm in diameter. It cannot move on its own, but is wafted along the oviduct by cilia on the inside of the tube. An ovary may contain thousands of egg cells, but only one is usually released from one ovary at **ovulation** each month. Within the egg cell is the nucleus and a large amount of cytoplasm. The cytoplasm provides nutrients for the dividing zygote after fertilisation. Surrounding the cell membrane is a jelly coat that protects the cell. Immediately after fertilisation by one sperm, the jelly coat changes to an impenetrable barrier. This prevents other sperm nuclei entering the egg cell.

END OF EXTENDED

QUESTIONS

1. a) Sketch a diagram of the human male reproductive system.

   b) Add labels to your sketch to name the main parts of the system.

   c) Describe the role of each of the main parts of the system in human reproduction.

2. a) Sketch a diagram of the human female reproductive system.

   b) Add labels to your sketch to name the main parts of the system.

   c) Describe the role of each of the main parts of the system in human reproduction.

3. EXTENDED Draw a table to compare the size, numbers and mobility of human egg and sperm cells.

The menstrual cycle

The **menstrual cycle** is a sequence of changes that occur in a woman's body every month. The average cycle is 28 days long, but it is normal for it to vary in different women.

The cycle begins with the monthly period, or bleeding, which is produced from the breakdown of the thickened lining of the uterus. After this, the uterus lining starts to thicken again. Ovulation occurs about halfway through the cycle, when an egg is released from one of the ovaries. The egg travels along the oviduct to the uterus.

If the egg is fertilised during this time, the egg will implant in the uterus lining and the lining will continue to develop for pregnancy. If the egg is not fertilised, the cell and the uterus lining are shed during the monthly period at the start of the next cycle.

Fertilisation and development of the fetus

During sexual intercourse, sperm deposited near the cervix swim up into the uterus, and then along the oviduct to the egg. Many sperm fail to make the journey, but some will reach the oviducts at the top end of the uterus.

The egg travels along the oviduct while the sperm swim up from the uterus. Fertilisation takes place in the oviduct. The nucleus of one sperm cell fuses with the nucleus of the egg cell, forming a fertilised egg, or zygote.

After fertilisation in the oviduct, the fertilised egg (zygote) travels on towards the uterus. The journey takes about three days, during which time the zygote will divide several times to form a ball of 64 cells, which is now called an **embryo**.

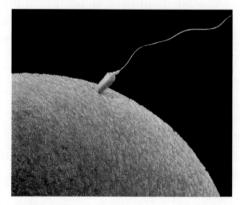

△ Fig. 10.20 The moment just before fertilisation: sperm approaching an egg.

In the uterus, the embryo embeds in the thickened lining (**implantation**) and cell division and growth continue. For the first three months, the embryo gets nutrients from the mother by diffusion through the uterus lining.

By the end of three months, the placenta has developed, and the embryo has become a **fetus** in which all the main organs of the body can be identified.

Over the next 28 weeks the fetus will increase its mass roughly 8 million times. At no other point in an individual's lifetime will growth occur at such a high rate. This period of development in the uterus is known as *gestation*, and it lasts about 40 weeks in humans, measured from the time of the woman's last period. The rapid growth during gestation depends on a good supply of food and oxygen, provided by the mother.

The fetus develops inside a bag of fluid called **amniotic fluid**. This fluid is produced from the amniotic membrane that forms the outer layer of the bag (**amniotic sac**). The fluid protects the fetus from mechanical damage, for example, if the mother moves suddenly. It also reduces the effect of large temperature variations that would affect the rate of development of the fetus. One of the signs that birth will happen soon is when this bag bursts shortly before labour. Once the placenta has formed, until birth, it is the only way that the developing fetus exchanges materials with the outside world. Birth usually occurs when all the organs of the fetus are fully developed and ready to carry out the life processes on their own.

The **placenta** is an organ that is produced by the growing fetus. The placenta allows a constant exchange of materials between the mother and fetus. The fetus is joined to the placenta by the umbilical cord, which carries the blood vessels of the fetus.

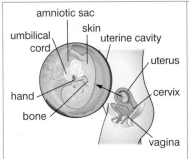
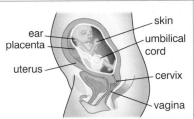
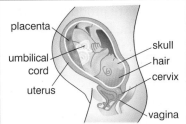

11 weeks:
- fetus about 4 cm long
- most of the main body structures have formed

23 weeks:
- fetus about 29 cm long and weighs about 500 g
- fetus hears sound from outside, and moves

40 weeks:
- fetus is about 51 cm long and weighs about 3.4 kg
- all organs fully developed ready for birth

△ Fig. 10.21 The fetus in the uterus at three time points during gestation.

SCIENCE IN CONTEXT — ULTRASOUND SCANS

Ultrasound is very high frequency sound, far above the frequency that can be heard. It is used in medical imaging for showing soft tissues inside the body. It is particularly useful for looking at the developing fetus in the uterus, because it does not harm either the fetus or the mother.

An ultrasound scan is commonly done about halfway through gestation, to make sure that the fetus is developing normally. At about this stage, if the fetus is lying at the right angle, it may even be possible to tell if it is a male fetus because the testes can be distinguished at this age.

Ultrasound scans may be done at other times during gestation if there is any concern about the development of the fetus.

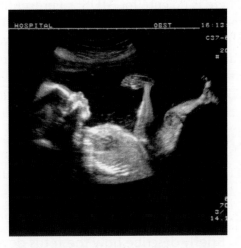

△ Fig. 10.22 This ultrasound scan was taken in the 20th week of gestation and shows that the fetus is developing normally.

The placenta and the uterus wall have a large number of blood vessels that run very close to each other, but do not touch. So maternal and fetal bloods do not mix. If they did, the higher blood pressure in the mother could damage the fetus. The structure of the placenta also helps to prevent many pathogens and some chemicals getting into the blood of the fetus.

Dissolved food molecules, oxygen and other nutrients that the fetus needs for growth diffuse from the mother's blood into the blood of the fetus. Waste products from metabolism, such as carbon dioxide and urea, in the fetus's blood diffuse across into the mother's blood.

Sexually transmitted infections (STIs)

Sexual intercourse is a method by which infection can spread, because of the exchange of body fluids, which may contain pathogens. There are many sexually transmitted infections (STIs), including HIV, which usually leads to acquired immunodeficiency disease (AIDS).

AIDS

AIDS is a disease of the immune system caused by a virus called HIV. The virus in an infected person is present in sexual fluids such as semen and vaginal fluids, and so can be transmitted during sexual intercourse. It may also be passed to another person in blood, either through a scratch, or through the sharing of needles for intravenous injection of drugs such as heroin. Infection can also pass from a mother to her fetus, through the placenta, or to her baby through breast-feeding after birth.

There is no cure for AIDS, so prevention of infection is essential. The methods are the same for all STIs. This is most easily done by abstinence from sex, or by limiting sexual partners to those who do not carry the virus. As a person may have no obvious symptoms early in infection, barrier methods such as the condom or femidom are most effective in reducing the risk of infection during intercourse.

QUESTIONS

1. Define the following terms in your own words:
   a) *zygote*, b) *embryo*, c) *fetus*.

2. Where in the human body does fertilisation of the egg cell occur?

3. Briefly describe how the fetus develops up to the point of birth.

4. EXTENDED Describe the role of the placenta during the development of a fetus.

End of topic checklist

Key terms

acrosome, amniotic fluid, amniotic sac, anther, asexual reproduction, carpel, embryo, fertilisation, fetus, gamete, germination, hypha, implantation, menstrual cycle, mycelium, ovary, oviduct, ovulation, ovule, placenta, pollination, seed, sexual reproduction, sperm, stamen, stigma, style, testis, uterus, vagina, zygote

During your study of this topic you should have learned:

○ How to define asexual reproduction as the production of new individuals without fertilisation. It is the division of the body cells of one parent. It produces offspring that are genetically identical to the parent and to each other.

○ How to define sexual reproduction as the production of new individuals from the fusion of a male gamete and a female gamete during fertilisation. It requires two parents, and produces offspring that are genetically different to their parents and to each other.

○ That fertilisation of a male gamete and female gamete produces a zygote that develops into an embryo by cell division.

○ How to identify the male and female parts of a flower.

○ An insect-pollinated flower has features such as coloured petals, scent and nectaries to attract insects to feed at the flower. The insects pick up pollen, which they transfer to other flowers that they move on to.

○ A wind-pollinated flower is usually small, without colour, scent or nectaries. It produces a large amount of lightweight pollen, which is scattered over a large distance in the wind.

○ Seeds need moisture, oxygen and warmth for successful germination.

○ How to identify and name the human male reproductive system, including the testes where sperm are made; the sperm ducts, which carry the sperm to the urethra; the prostate gland and seminal vesicles, which produce liquid in which the sperm swim; the penis, which when erect delivers sperm into the vagina of the woman; and the urethra, which carries the sperm from the sperm ducts to the outside of the body.

○ How to identify and name the human female reproductive system, including the ovaries, where the egg cells are made; the oviducts, which carry the eggs to the uterus and where fertilisation takes place with sperm cells; the uterus, where the embryo embeds and develops into a fetus; the cervix, where sperm are deposited at the base of the uterus; and the vagina, where the penis is inserted during sexual intercourse.

End of topic checklist continued

○ The menstrual cycle and egg is released from the ovary, and the uterus lining thickens. If fertilisation does not take place, the uterus lining and egg are shed at the start of the next cycle.

○ Once the embryo has embedded in the uterus wall, it develops the placenta. This is where nutrients and waste materials are exchanged between the blood of the mother and of the fetus.

○ During development, the embryo (and later the fetus) is protected from mechanical damage and temperature fluctuations by amniotic fluid that surrounds it in the amniotic sac.

End of topic questions

Note: The marks given for these questions indicate the level of detail required in the answers. In the examination, the number of marks given to questions like these may be different.

1. The photograph in Fig. 10.23 shows a catkin on a goat willow tree. A catkin is formed from a group of flowers.

a) What is the purpose of the flowers on a goat willow tree? **(1 mark)**

b) Name the yellow parts of the flowers shown in this photograph. **(1 mark)**

c) Describe their purpose in a flower. **(1 mark)**

d) Are goat willow flowers pollinated by the wind or by insects? Explain your answer using clues from the photograph. **(3 marks)**

△ Fig. 10.23 A goat willow catkin.

2. a) Explain the advantage to a flower of having adaptations for attracting insects rather than relying on wind for pollination. **(1 mark)**

b) Describe one disadvantage for an insect-pollinated plant of relying on one or just a small number of insect species for pollination. **(1 marks)**

3. **EXTENDED** In many flowers that have stamens and stigmas, the stamens mature and shed their pollen before the stigma matures and accepts pollen.

a) Describe and explain one advantage for the plant of doing this. **(2 marks)**

b) Describe and explain one disadvantage for the plant of doing this. **(2 marks)**

4. A gardener has some packets of seeds for planting. The packets explain how to plant the seeds to get the best germination.

a) What is meant by *germination*? (1 mark)

b) All the packets say that the seeds need to be planted in moist compost and kept warm. Explain why the seeds need these conditions. (2 marks)

c) Explain why the seeds will not germinate successfully in waterlogged soil. (2 marks)

d) The larger seeds need to be planted deeper in the compost, and the tiniest seeds need to be scattered on the surface of the compost. Explain why different seeds need to be planted at different depths. (Hint: think about food reserves.) (3 marks)

e) Some seeds that come from plants in high-latitude regions (for example, Canada or Russia) need to be placed in the freezer for a few weeks before they will germinate. This makes them respond as if they had been through a cold winter. Explain the survival advantage of this adaptation. (2 marks)

5. a) Where are sperm cells made in the human body? (1 mark)

b) Where are egg cells made in the human body? (1 mark)

c) Where is an egg cell fertilised by a sperm cell? (1 mark)

d) Starting from the point of their formation, explain how a sperm cell reaches the egg cell at fertilisation. (4 marks)

6. a) What is the *placenta*? (1 mark)

b) What role does the placenta play in supporting the fetus? (1 mark)

c) EXTENDED How are substances exchanged across the placenta? (2 marks)

d) EXTENDED What is the advantage of keeping the mother's blood separated from the blood of the fetus? (1 mark)

Human activity can have a huge impact on food chains and food webs, and even lead to the extinction of whole species. The introduction of just 24 wild rabbits into Australia in 1859 resulted in a rapidly spreading population that has decimated huge areas of diverse environments, causing havoc to native animal and plant species. Various attempts to keep the rabbit population in Australia in check have been attempted, for example, by introducing the rabbit disease myxomatosis. Even so, scientists estimate that rabbits still cost Australian farmers more than $200 million annually in lost production, in addition to the ongoing damage that they wreak on the natural environment.

SECTION CONTENTS
Organisms and their environment

11
Organisms and their environment

△ Plants make their food by using light energy from the Sun.

△ Fig. 11.1 Alligators can survive for months without food, although they are always on the lookout for a good meal.

Organisms and their environment

INTRODUCTION

All animals need to eat, to provide the fuel for respiration. Some animals such as the common shrew need to consume two or three times their body weight of insects, slugs and worms every day in order to survive. They live life quickly, being on the hunt for food for most of the time, especially at night. By contrast, alligators only need to feed about once a week, and can live for months without food. They live life much more slowly than shrews, waiting in ambush for prey to get close before attacking. Most animals eat on average somewhere between these extremes, although adult mayflies have no mouthparts and never eat. They live their brief lives of a few days using energy stored from earlier stages in their life cycle, as their only purpose is to reproduce, after which they die.

KNOWLEDGE CHECK

✓ Respiration is the release of energy from food molecules.
✓ During photosynthesis, light energy is transferred by plants to sugars as chemical energy.
✓ Energy released by respiration is used for a range of purposes, including making new body tissue.
✓ Organisms that feed on one another can be displayed in a food chain that shows who eats what.
✓ Food chains within a habitat can be combined to produce a food web.

LEARNING OBJECTIVES

✓ Identify the principal source of energy input to biological systems.
✓ Define the terms *food chain*, *food web*, *producer*, *consumer*, *decomposer*, *herbivore* and *carnivore*.
✓ Interpret food chains and food webs, and use them to describe the effect of humans on habitats.
✓ Draw, describe and interpret *pyramids of numbers*.
✓ **EXTENDED** Describe how energy is transferred between trophic levels.
✓ **EXTENDED** Explain why food chains usually have fewer than five trophic levels.
✓ **EXTENDED** Define the terms *trophic level* and *ecosystem*.

ORGANISMS AND THEIR ENVIRONMENT

Energy flow

Plants make their food (sugars) from carbon dioxide and water using light energy from the Sun. In systems terminology, sunlight is the energy input for plants.

Most food chains on the surface of the Earth begin with photosynthesising plants. This means that the Sun is the main input of energy into biological systems, such as food chains and food webs.

ENERGY INPUT FROM THE SUN

As a result of the curvature of the Earth, the amount of light energy from the Sun that falls on every square metre is greatest at the equator, and decreases as you move towards the poles. The tilt of Earth's axis in relation to the Sun causes variation in the amount of sunlight energy received by high latitude regions at different times of the year, causing seasons.

These differences in energy received have major effects on the ecosystems in each region. Parts of the world near the equator that receive sufficient rainfall, such as tropical rainforests, have a greater productivity of plants in a year than other regions. This greater productivity supplies more food for animals, leading to a greater productivity of animals – some of these areas are the most biodiverse on the planet.

The seasonal effects in high latitude regions result in rapid plant growth in summer months and virtually no growth during the winter, although some of this effect is the result of lack of heat energy from the Sun as much as lack of light energy.

Food chains

You should be familiar with food chains from your earlier work. A **food chain** shows 'who eats what' in a habitat. For example, in Fig.11.2, owls eat shrews, shrews eat grasshoppers, grasshoppers eat grass. (Remember, the arrows in a food chain show the direction of energy flow.)

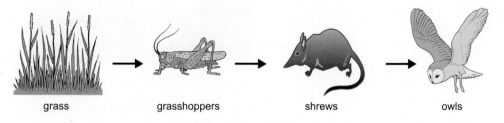

| grass | grasshoppers | shrews | owls |

Δ Fig. 11.2 An example of a food chain.

Each level in a food chain shows a separate level at which that species is feeding.

- Grass – this is the **producer** level, because grass is a plant and produces its own food (organic nutrients) using light energy during photosynthesis. All food chains start with a producer level.
- Grasshoppers – these are the primary consumers, '**consumer**' because they eat the grass and 'primary' because they are the first eaters of other organisms in the food chain. This level may also be called **herbivores**, because they eat plant material.
- Shrews – these are consumers too, but they are specifically secondary consumers because they eat the primary consumers. They are also called **carnivores**, because they eat meat.
- Owls – these are also consumers, but they are specifically tertiary consumers because they eat the secondary consumers. They are also carnivores.

If anything ate owls, they would be quaternary consumers, but food chains often don't reach that level. Animals at the highest trophic level in a food chain may also be called the top consumers, or top predators. All animals are consumers, because they eat other organisms to get their food, in contrast to plants, which are producers.

EXTENDED

Each feeding level of a food chain is called a **trophic level**. So producers are one trophic level, primary consumers (or herbivores) are the trophic level that feeds on producers, and so on.

What isn't shown in a food chain is what happens to all the dead plant and animal material that isn't scavenged. This material decays as a result of the action of **decomposers**, such as fungi and bacteria. Fungi digest their food by secreting enzymes outside their hyphae; they then absorb the dissolved food materials. Many bacteria also do this. However, only some of the digested food materials are absorbed – the rest are released into the environment. Decomposers play an essential role in ecosystems, as you will see later in the nitrogen cycle.

END OF EXTENDED

△ Fig. 11.3 The hyphae of this fungus are growing through the dead tree and secreting enzymes that cause the wood to break down into simpler chemicals.

OTHER PRODUCERS

Not all producers are plants, and not all producers use light energy. There are species, mainly bacteria, which produce their own food without the presence of light energy from the Sun. Instead they get the energy they need for the formation of sugars from chemical reactions.

These bacteria are the source of food for food chains and webs that exist where there is no sunlight, such as deep in oceans and in underground caves. Be careful to avoid the statement that 'all life on Earth depends on the Sun', as this is an oversimplification and not totally accurate.

Energy in food chains

We can look at food chains in terms of energy. Plants use energy from light to build new substances. These substances act as stores of energy. A herbivore gains this store of energy, by ingestion, when it eats the plant. Some of that energy becomes stored in the substances in the animal's body, and so can be ingested by any animal that eats it. So, we can define a food chain also as the transfer of energy between organisms by ingestion.

Food webs

If we look more closely at food chains, it is rare to find an organism that is eaten by just one other species, or a predator that feeds on just one type of prey. It may also be the case that a predator may feed on different kinds of organisms – an **omnivore**, for example, is a primary consumer when feeding on plants, but a secondary or tertiary consumer when eating other animals. So food chains within a habitat are linked together to form a **food web**. A food web is a better description of the feeding relationships in a habitat and shows how living organisms are interconnected.

Food webs still usually group the organisms according to their feeding level. For example, in the simplified food web shown in Fig.11.4, the rabbit, squirrel, mouse, seed-eating bird and herbivorous insect are all primary consumers and are placed just above the producer level.

There are usually many more species in a habitat than shown in Fig.11.4, and linking them all in one food web can get confusing. So food web diagrams may focus on the relationships between key organisms rather than all of them. For example, they may only include the most numerous species, or focus on the most vulnerable species. This can be helpful if you want to use the food web to predict what would happen to the ecosystem if the food web were changed in some way, such as by human activity.

You could use the food web shown to predict what would happen if the plants were sprayed with an insecticide. This would kill the herbivorous

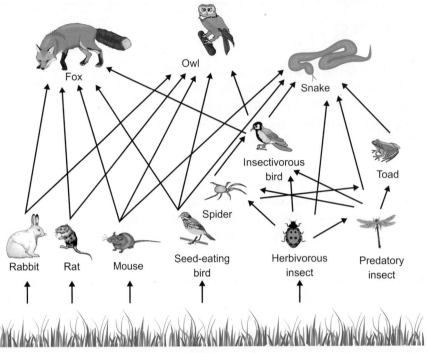

△ Fig. 11.4 A simplified food web.

insects and so reduce the amount of food available to all the animals that feed on them.

Trophic levels describe the feeding level of organisms in a food chain or food web, starting with the producers which are trophic level one.

Energy is lost between each trophic level, very little energy is passed from one trophic level to the next and so food chains and food webs usually have fewer than five trophic levels.

Interpreting human impact on food chains and food webs

We can use food chains and food webs to help us understand the wider impact on habitats that we have when we affect particular organisms. For example, along the Atlantic coast of the USA there has been overfishing of large shark species. These species are predators of smaller fish, such as skate and rays. These smaller fish feed on shellfish, including scallops.

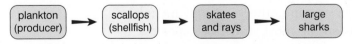

△ Fig. 11.5 A marine food chain.

As the numbers of large sharks have decreased, there has been less predation of the smaller fish. So, the numbers of skates and rays have greatly increased. This has had a major impact on the shellfish, with scallops becoming nearly extinct in several areas.

Many of the fish that we eat are predators higher up food chains, and so this effect of changing the population sizes of organisms lower in the food chain is being seen in many parts of the ocean.

We also affect food chains and webs when we introduce species from one area to another. This may happen intentionally, such as:

- to provide more food (for example, goats provide meat and milk)
- to control a pest species, for example, introducing cane toads to Australia to control beetles that are pests of sugarcane plantations
- because the species is a pet (for example, cats and dogs).

Sometimes the introduction is accidental, for example, the introduction of rats to some places because they were onboard the ships that transported humans to those places.

Goats have become a pest in many places, particularly on islands, because they eat much of the vegetation. This prevents new trees growing, and many of the local plants that are not adapted to being browsed like this die out. Changing the plants that grow in the area will also change the animals that can live there.

Cats and rats have also become pests on islands because they eat many birds and their eggs. In New Zealand, many species of ground-nesting birds have become extinct because of these introductions.

Cane toads were introduced to Australia to control beetles that were attacking the sugarcane plantations in the northern regions. Unfortunately the toads didn't stay in the plantations, because they needed more shelter. So they moved out, and started eating small animals in other areas. This left less food for the predators of the local food web, including many species of small lizards. Population sizes of these lizards have decreased and some are at risk of extinction. The toad is now considered a pest in these areas.

◁ Fig. 11.6 Cane toads have glands in their skin that produce chemicals that are toxic to many animals. So, there are few predators in Australia that can eat them.

EXTINCTION IN HAWAII

A deep hole on one of the Hawaiian islands provides a 10 000 year record of the effects of humans on the plants and animals that lived there. Before humans arrived, the only organisms must have arrived by chance on the wind or water. Only a limited number of species could travel the thousands of miles from the mainland to the islands. Since the time that they arrived, they evolved into a range of new species that were found nowhere else. The birds, in particular, evolved into a wide range of forms, including some that were too large to fly and behaved more like pigs and goats, grazing the plants.

About 900 years ago the first bones of a rat appear in the deep hole. It arrived with people on a boat. More rat bones were found in the hole since that time. Since 900 years ago, many island species have disappeared from the bone collection. Many species of birds, including the large species, became extinct. Only birds that nest on the islands and then leave for the rest of the year are still found in Hawaii. All the species of land snails, which were an important food for predators on the islands, became extinct, and this affected other species in the island food web.

△ Fig. 11.7 The Laysan albatross is now protected when it nests on Midway Island in Hawaii, although cats are still a threat to its eggs.

The rats, and other species that humans brought to the island, including cats and goats, changed the island food web forever.

QUESTIONS

1. Use your own words to define the following terms: *producer*, *consumer*, *herbivore*, *carnivore*.

2. **EXTENDED** Use your own words to define the following terms: *decomposer*, *trophic level*.

3. Name the principal source of energy to an ecosystem. Explain your answer.

4. Distinguish between a food chain and food web.

5. Describe how food webs can be **a)** useful and **b)** difficult to draw.

EXTENDED

An ecosystem includes the interactions between the living and the non-living factors in an area. Many of the interactions involve energy transfers.

Energy transfers in biological systems

Plants transfer the energy from light into stored energy inside them. This energy is transferred to animals when they digest and assimilate plant food to make new substances in their body tissue.

The energy that a plant receives from light, or that an animal gets in its food, is always greater than the amount of energy it stores in the substances in its tissues. This is because some of the energy that it takes in is transferred to the environment in various forms. The energy losses from plants and animals differ in some ways.

Energy losses from plants

The amount of energy from sunlight that falls on the Earth's surface varies at different times of the day and year, and varies in different parts of the world (with places near the equator receiving more light energy than places nearer the poles). On average, tropical areas receive between 3 and 5 kWh/m² per day (which is about the same energy as a one-bar electric heater left on for 3–5 hours).

Plants use only a tiny proportion of this for many reasons, as shown in Fig. 11.8. It has been estimated that most plants only transfer about 1% to 2% of the energy in the light that falls on them into chemical energy in their tissues (**biomass**). This is the energy available to a herbivore that eats the plant.

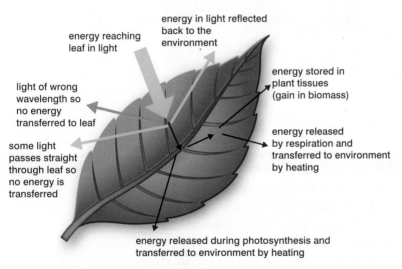

energy reaching leaf in light

energy in light reflected back to the environment

energy stored in plant tissues (gain in biomass)

light of wrong wavelength so no energy transferred to leaf

energy released by respiration and transferred to environment by heating

some light passes straight through leaf so no energy is transferred

energy released during photosynthesis and transferred to environment by heating

△ Fig. 11.8 Energy gains and losses of a plant.

Energy losses in animals

When an animal eats, the food is digested in the alimentary canal and the soluble food molecules are absorbed into the body. The undigested and unabsorbed food in the alimentary canal is egested as faeces.

Absorbed food molecules may be used for different purposes in the body:

• to produce new animal tissue or gametes for reproduction

- as a source of energy for respiration
- converted to waste products in chemical reactions.

The energy stored in the food molecules may stay in the body, stored in body tissue, it may be transferred to the environment, or stored in waste chemicals, such as urine. When food molecules are broken down during respiration and other reactions, some of the energy released from the molecules is transferred to the environment by heating, through the processes of conduction, convection and radiation. So, only a small proportion of the energy stored in the animal's food is converted into energy stored in its body tissues as an increase in the animal's biomass.

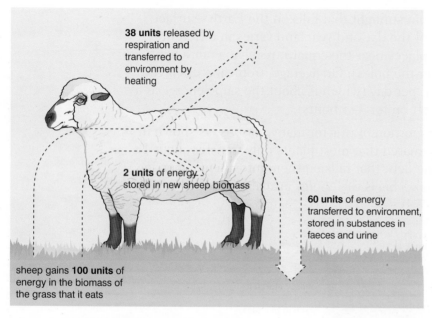

38 units released by respiration and transferred to environment by heating

2 units of energy stored in new sheep biomass

60 units of energy transferred to environment, stored in substances in faeces and urine

sheep gains **100 units** of energy in the biomass of the grass that it eats

Δ Fig. 11.9 The energy flow through a sheep.

REMEMBER

Energy transfer efficiency is the amount of energy stored in the body tissue at a particular trophic level compared with the amount in the previous level. Calculating the energy transfer efficiency between trophic levels involves the estimation of many values. This means that the transfer efficiencies you may find in textbooks and on the internet are only best estimates and must not be taken as exact.

In addition, many sources quote a value of 10% as the efficiency of energy transfer between any trophic level and the one above. Calculations of efficiency vary from about 0.2% to around 20% for different organisms in different ecosystems. This gives an *average* of 10%, but over such a large range this is not very reliable. It is better to prepare to explain how energy is gained and lost between trophic levels, in order to explain the shape of pyramids of energy and lengths of food chains, than to quote specific values for energy transfer efficiency.

QUESTIONS

1. Draw a flowchart to show the energy gains by and losses from a plant leaf.

2. Draw a flowchart to show the energy gains by and losses from a herbivore.

3. Explain why the amount of energy stored in an organism's tissues is always less than the amount of energy that it gained.

END OF EXTENDED

SCIENCE LINK

PHYSICS – ENERGY TRANSFER

- Food chains and webs describe which organisms consume others, but a key feature is that they show how energy is transferred from each trophic level to the next and this links to the law of conservation of energy.

- Energy ideas can be used to explain why the number of organisms in a predator level are much lower than the number of organisms in the level below – energy transfers are never 100% efficient and not all energy is stored and available for the next level up.

- Energy ideas also indicate why food chains have a limited length – the useful energy is reduced to a point where an additional level could not be supported.

End of topic checklist

Key terms

biomass, carnivore, consumer, decomposer, food chain, food web, herbivore, omnivore, producer, trophic level

During your study of this topic you should have learned:

◯ Producers are organisms that make their own organic nutrients (food), such as plants that use sunlight to produce sugars through photosynthesis.

◯ Consumers are organisms that gain energy by feeding on other organisms. There are different levels of consumer, depending on their position in a food chain.

◯ A herbivore is an animal that eats plants, and a carnivore is an animal that eats other animals.

◯ A food chain shows the transfer of energy between organisms as a result of feeding, starting with a producer.

◯ A food web is an interconnection of food chains that share some organisms.

◯ How to define decomposer as an organism that gets its energy from dead or waste organic material.

◯ EXTENDED A trophic level is a feeding level within a food chain, food web or pyramid of numbers or biomass.

◯ EXTENDED Energy is gained at each trophic level as it is transferred from light in plants or from food in animals to make new body tissue.

◯ EXTENDED Energy is transferred to the environment at each trophic level as heat energy from respiration, and also as chemical energy from animals in the form of faeces and urine.

◯ EXTENDED That a food chain is rarely more than five trophic levels in length because the top trophic level within the chain contains too little energy to support another trophic level.

End of topic questions

Note: The marks given for these questions indicate the level of detail required in the answers. In the examination, the number of marks given to questions like these may be different.

1. The photograph in Fig.11.10 shows lions eating a dead wildebeest. Before the lions killed the wildebeest, it had been feeding on grass.

△ Fig. 11.10 Lions eating their kill.

   a) Is the lion a carnivore or herbivore? Explain your answer. **(2 marks)**

   b) At which level of a food chain does the wildebeest feed? **(1 mark)**

   c) Draw a food chain for the organisms shown in the photograph. **(2 marks)**

   d) Lions also feed on the herbivores, gazelle and zebra. Use all these organisms to draw a food web for the African grassland. **(3 marks)**

2. **EXTENDED** In a tropical forest, the layer of dead leaves (called the leaf litter) on the forest floor is usually very thin at all times of the year. In temperate woodlands (where there are seasons of summer and winter), many trees drop their leaves in the autumn and grow new ones in the spring.

   a) Tropical trees drop a few leaves at a time, at any time of year. What happens to the leaves on the ground? Explain your answer as fully as possible. **(2 marks)**

   b) The leaf litter in a temperate woodland is deep all through winter, when it may be cold enough for snow, until it gets warm again in spring. Then the leaf litter disappears. Explain these observations as fully as you can. **(3 marks)**

△ Fig. 11.11 A temperate woodland in winter.

End of topic questions continued

3. Use the food web in Fig.11.4 to predict what would happen to the following species if all the herbivorous insects were killed by insecticide. Explain your answers.

    a) predatory insects   **(2 marks)**

    b) insectivorous birds   **(2 marks)**

    c) mice   **(2 marks)**

    d) snakes   **(2 marks)**

4. EXTENDED Explain as fully as you can why a food chain is unlikely to include more than five trophic levels. **(6 marks)**

5. EXTENDED Food chains in northern regions on Earth may be much shorter than food chains in tropical rainforests. Thinking only in terms of energy, try to explain this difference. **(4 marks)**

There is probably no place on Earth that is not affected by human activity. An estimated 40% of the land surface is cultivated to produce food, either from crops or animals. Changing land use, so that it produces food or provides places for us to live and work, destroys habitats for other organisms. Then, the waste we produce causes pollution of land, water and air unless it is properly controlled. As the human population grows, we must find ways of reducing our impact on ecosystems, so that we conserve resources and the many organisms with which we share the Earth.

SECTION CONTENTS

Human influences on ecosystems

12
Human influences on ecosystems

△ Deforestation destroys the habitats of many species of plants and animals that depend on the trees.

Human influences on ecosystems

INTRODUCTION

The Earth has many natural cycles; one of these is the carbon cycle. Processes such as photosynthesis and respiration move carbon around the Earth.

Humans are changing the balance of these processes and this is having major impacts on ecosystems.

△ Fig. 12.1 Humans have destroyed parts of the Amazon rainforest to make room for cattle farming.

Deforestation, to make room for agriculture, is one of the ways humans are changing ecosystems. Removing trees leads to habitat destruction and possible extinction of the animals and plants that live there. The removal of tree roots leads to loss of soil, which makes it difficult for new plants to grow. With fewer trees there is less photosynthesis and so more carbon dioxide is left in the atmosphere.

KNOWLEDGE CHECK

✓ Water is essential for many life processes.
✓ Carbohydrates, proteins and lipids all contain carbon.
✓ Proteins also contain nitrogen.
✓ Plants lose water to the environment through their leaves, in transpiration.
✓ Decomposers digest dead organic material releasing some of the products of digestion into the environment.
✓ Earth's atmosphere is affected by human activity such as deforestation and combustion of fuels.

LEARNING OBJECTIVES

✓ Describe stages in the carbon cycle, including respiration, photosynthesis, decomposition and combustion.
✓ EXTENDED Discuss the effects of the combustion of fossil fuels and the cutting down of forests on the oxygen and carbon dioxide concentrations in the atmosphere.
✓ List the undesirable effects of deforestation as an example of habitat destruction.
✓ EXTENDED Explain the process of eutrophication of water.

HUMAN INFLUENCES ON ECOSYSTEMS

Nutrient cycles

Unlike energy, which is transferred through organisms and eventually to the environment in a way that is not useful to the organisms, nutrients continually transfer between the environment and organisms and back again, in what are described as **nutrient cycles**. One example is the carbon cycle.

The carbon cycle

Carbon is continually cycled through the living and non-living parts of ecosystems, in different forms at different stages of the **carbon cycle**. Carbon dioxide from the atmosphere is converted to complex carbon compounds in plants during photosynthesis. This is often called the 'fixing' of carbon by plants. Respiration in plants returns some of this fixed carbon back to the atmosphere as carbon dioxide. Carbon in the form of complex carbon compounds passes along the food chain. At each stage, some of this carbon is released as carbon dioxide to the atmosphere as a result of respiration.

When organisms die, their bodies decay as they are digested by decomposers. Some of the complex carbon compounds are taken into the bodies of the decomposers, where some may be converted to carbon dioxide during respiration. Carbon dioxide may also be released directly into the atmosphere during decay.

Fig. 12.2 Water excludes air from the ground, which prevents decaying organisms from respiring. So, dead plant material in waterlogged ground builds up over time, forming peat. Peat can be burnt as a fuel, although this is being discouraged so that peat bog habitats can be protected.

Combustion

If dead organic material is buried by sediment or water too quickly for decomposers to cause decay, and if it remains buried, then it may be converted to other complex carbon compounds. Peat is formed when mosses and other plants are buried in swampy ground for hundreds of years. Over many millions of years, where there were once huge forests growing in swampy regions, heat and pressure have turned the organic material into coal. Heat and pressure over many millions of years also produces oil from the decaying bodies of tiny marine organisms that were buried in sediment at the bottom of oceans. Peat, coal and oil are **fossil fuels**. We can release the carbon from the complex carbon compounds in fossil fuels into the air as carbon dioxide during **combustion**, when we burn them.

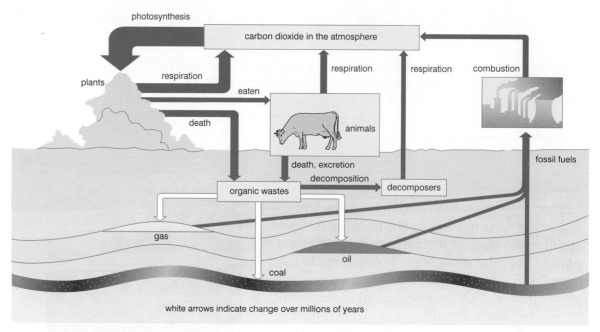

photosynthesis

carbon dioxide in the atmosphere

plants

respiration

respiration

respiration

combustion

eaten

death

animals

fossil fuels

death, excretion
decomposition

organic wastes

decomposers

gas

oil

coal

white arrows indicate change over millions of years

△ Fig. 12.3 A summary of the carbon cycle.

REMEMBER

Make sure you are certain what form carbon is in (carbon dioxide or complex carbon compounds such as carbohydrates) at each stage of the carbon cycle.

The effects of large-scale deforestation and combustion

Deforestation is the permanent destruction of large areas of forests and woodlands. It usually happens in areas that provide quality wood for furniture, such as the tropical hardwood forests of Malaysia, or to create farming or grazing land (all over the world).

Deforestation can result in many kinds of damage to the environment and the organisms that live there, including:

- extinction of organisms when there is nowhere left that is suitable for them to live
- loss of soil
- flooding
- carbon dioxide build-up in the atmosphere because there are not enough trees to store the carbon after photosynthesis.

EXTENDED

Forests act as a major carbon store because carbon dioxide is taken up from the atmosphere during photosynthesis and used to produce the chemical compounds that make up trees. When forests are cleared, and the trees are either burnt or

△ Fig. 12.4 Radar image of the island of Sumatra showing the extent of deforestation. Native forest appears in green and pink areas represent deforestation.

left to rot, this carbon is released quickly into the air as carbon dioxide. This rapidly increases the proportion of carbon dioxide compared with oxygen in the air surrounding the forest. In addition, the amount of oxygen removed from the local atmosphere by plants for photosynthesis may also drop, changing the balance between carbon dioxide and oxygen in the atmosphere locally.

On the scale of deforestation in the Amazon Basin, the amount of carbon dioxide released is so great that it cannot be brought back into balance as a result of photosynthesis. This additional carbon dioxide remains in the atmosphere.

Over the past 10 000 years or so, as a result of photosynthesis and respiration and other physical processes, the exchange of carbon between organisms and the atmosphere resulted in little change in the amount of carbon dioxide in the atmosphere. On average, over one year about 120 billion tonnes of carbon dioxide are removed from the atmosphere by photosynthesis, and a similar amount is returned by respiration.

During the past 250 years, however, deforestation and the combustion of fossil fuels as a result of human activity has added increasing amounts of CO_2 to the atmosphere. Today about 5.5 billion tonnes of carbon dioxide are added to the atmosphere every year through human activity, particularly through combustion of fossil fuels.

Compared with 120 billion tonnes through natural processes, this may not seem a lot, but there is no process that balances this addition. So, the concentration of carbon dioxide in the atmosphere is increasing. It is this additional carbon dioxide in the atmosphere that most people believe is causing global warming and climate change.

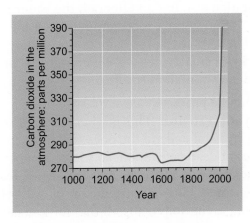

△ Fig. 12.5 Atmospheric carbon dioxide concentration from 1000 CE to recent times.

END OF EXTENDED

QUESTIONS

1. Describe the role of the following in the carbon cycle:

   a) respiration, b) photosynthesis, c) decomposition.

2. In what form is carbon when it is in the following stages of the carbon cycle?

   a) Earth's atmosphere b) plant tissue c) fossil fuels

3. EXTENDED Describe the effect of large-scale deforestation on the oxygen and carbon dioxide concentrations of the atmosphere.

CHEMISTRY – CHEMICAL REACTIONS

- The carbon cycle connects together a number of chemical reactions, each of these can be described in terms of particles and using word equations and symbol equations.

- Combustion (reacting with oxygen) leads to a number of processes, for example, acid rain or the production of carbon dioxide, that impact on the environment and the organisms that can survive in a particular habitat.

- Since many aspects of an ecosystem are interconnected, human activity in one situation (such as producing fertilisers or burning fuels to provide energy) can have an impact in a much wider range of situations and contexts.

EXTENDED

Deforestation also has an effect on the water cycle. Trees draw ground water up through their roots and release it into the atmosphere by transpiration. As forest trees are removed, the amount of water that can be held in an area decreases, which in turn can cause either increasing or decreasing rainfall in the area.

Removing the protective cover of vegetation from the soil can also result in **soil erosion**. This is where the soil is washed away by rain. The top layers of soil are the ones that contain the most nutrients, from the decay of dead vegetation, so soil erosion removes essential nutrients from the land. Soil nutrients are also lost by **leaching**, which is the soaking away of soluble nutrients in soil water because there are few plant roots in the soil to absorb the nutrients and lock them away in plant tissue. This loss of nutrients from the soil is permanent, and makes it very difficult for forest trees to regrow in the area, even if the land is not cultivated.

◁ Fig. 12.6 This satellite image of a river estuary in Madagascar shows large amounts of soil in the water (orange). This is a result of deforestation near the river.

Loss of plant species due to deforestation will result in a loss of animal species in the same **community**, because of their feeding relationships in the food web. Many tropical rainforests are areas of high **biodiversity**, where many organisms live. They also contain many species found nowhere else because they have evolved together in an energy-rich and relatively unchanging environment. Destruction of tropical rainforests, such as in the Amazon Basin, is causing a high rate of extinction of species.

END OF EXTENDED

Fertilisers and eutrophication

Fertilisers are chemicals that farmers use on fields to add nutrients, for example nitrates, that help the crops to grow better and so produce greater yields. However, if a farmer adds more fertiliser to a field than the crop plants can absorb, the remaining nutrients will soak away in ground water into nearby streams and rivers. Also, if there is heavy rainfall soon after the fertiliser has been spread on a field, the nutrients will dissolve in the rainwater and run off the surface of the field into streams and rivers.

This adding of nutrients to water is called **eutrophication**. The nutrients in the water will have the same effect on plants and algae in the water as they have on plants that grow on land, and will encourage them to grow faster. As they grow faster, they respire more rapidly, taking oxygen from the water. This leaves less oxygen in the water for other organisms, such as fish, and those organisms may die.

EXTENDED

The effects of eutrophication on an aquatic ecosystem can be explained in the following way.

- Eutrophication increases the rate of growth of photosynthesising organisms (producers) in the water, in particular algae and plants that grow at the surface of the water.

- If the plants and algae at the surface grow so much that they block light to plants that grow deeper in the water, the deep-water plants will die because they cannot photosynthesise.
- This will provide more food for decomposers, such as bacteria, which will increase in numbers rapidly.
- The bacteria respire more rapidly in order to make new materials for growth and reproduction. Respiration takes dissolved oxygen from the water, reducing the oxygen concentration of the water.
- Other aquatic (water-living) organisms find it increasingly difficult to get the oxygen they need from the water for respiration.
- If the amount of oxygen dissolved in the water falls too low, many organisms, particularly active animals such as fish, will die.
- The decay of dead organisms in the water provides more nutrients, so more bacteria grow, respire and take more oxygen from the water.
- Eventually, most of the large aquatic plants and animals in the water may die.

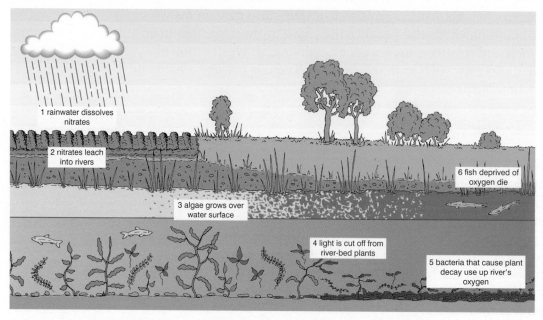

1 rainwater dissolves nitrates

2 nitrates leach into rivers

3 algae grows over water surface

4 light is cut off from river-bed plants

5 bacteria that cause plant decay use up river's oxygen

6 fish deprived of oxygen die

△ Fig. 12.7 Eutrophication can lead to the death of water organisms.

REMEMBER

Eutrophication is often wrongly defined as the pollution of water and death of aquatic organisms. This is incorrect – eutrophication is simply the adding of nutrients. It comes from the Greek word *eutrophia*, meaning 'healthy or adequate nutrition'. Adding nutrients that the ecosystem can use normally may be an advantage, but adding them in excess may lead to the death of aquatic organisms as a result of the depletion of dissolved oxygen in the water. So, excess nutrients can cause pollution.

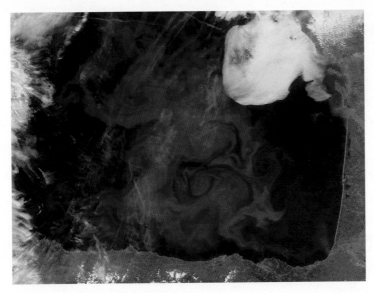

△ Fig. 12.8 Large-scale algal growth can be seen in satellite photos. This algal bloom occurred in the Baltic Sea in 2010 as a result of fertilisers being washed off the surrounding land.

END OF EXTENDED

QUESTIONS

1. **EXTENDED** Explain how deforestation may affect:

   a) the water cycle

   b) soil fertility

   c) atmospheric carbon dioxide.

2. Explain what we mean by *eutrophication*.

3. Give two reasons why the use of artificial fertiliser on a field could cause eutrophication of a nearby stream.

4. Describe how eutrophication can lead to the death of fish in a stream.

5. **EXTENDED** Draw a flow diagram to explain how sewage can cause eutrophication and water pollution.

End of topic checklist

Key terms

biodiversity, carbon cycle, combustion, community, deforestation, eutrophication, fossil fuel, leaching, nutrient cycle, soil erosion

During your study of this topic you should have learned:

◯ The carbon cycle can be represented as a diagram that shows how photosynthesis, respiration, decomposition and combustion contribute to the transfer of carbon between organisms and the environment.

◯ EXTENDED Combustion and deforestation can rapidly increase the carbon dioxide concentration in the atmosphere.

◯ Deforestation can damage the environment by causing species to become extinct, increasing loss of soil and the risk of flooding, and by resulting in more carbon dioxide in the atmosphere.

◯ Why deforestation has undesirable effects on the environment.

◯ EXTENDED Eutrophication caused by the addition of nutrients to water sources, is the increase in nutrients in water from fertilisers or untreated sewage, which may lead to the death of organisms, such as fish, due to lack of oxygen in the water.

◯ EXTENDED Eutrophication caused by the addition of nutrients to water sources, increases the rate of growth of surface plants and algae that block light to plants lower in the water. Lower plants die and are decayed by bacteria that remove oxygen from the water for respiration. This leaves little oxygen in the water for other organisms such as fish.

End of topic questions

Note: The marks given for these questions indicate the level of detail required in the answers. In the examination, the number of marks given to questions like these may be different.

1. The graph in Fig.12.9 shows the change in carbon dioxide concentration above a forest over 2 days, and the light intensity just above the top of the trees.

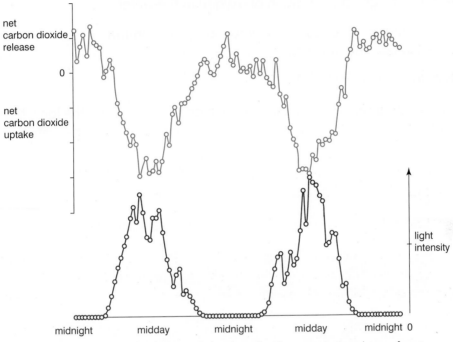

△ Fig. 12.9 Changes in light intensity and carbon dioxide concentration above a forest.

a) Explain the changes in light intensity shown in the graph. **(2 marks)**

b) Explain the changes in carbon dioxide concentration shown in the graph. (Remember there are more organisms than just the trees in the forest.) **(4 marks)**

2. **EXTENDED** If rainforest is cleared on a large scale and then left to recover, it is rare that the same species of plants and animals return to the area, even after many years. Explain why this happens, referring to soil fertility, climate and biodiversity in your answer. **(4 marks)**

3. Rivers and lakes that are used for water supplies may be monitored to make sure that the water in them is safe for use. One way of monitoring is to measure the amount of oxygen that is used by the water (the oxygen demand) over a period of 5 days.

a) Why might concentration of oxygen decrease in the water? Explain your answer. **(2 marks)**

b) In this test would polluted water use more oxygen than unpolluted water? Explain your answer. **(2 marks)**

Another way of monitoring the water is to sample the small organisms that live in it. Some species, such as worms, are better adapted for living in water that has a low oxygen concentration. Other species, such as mayfly larvae, need a high concentration of oxygen in the water.

c) Which of the two species above would be more common in polluted water? Explain your answer. **(2 marks)**

d) What does *adapted* mean? **(1 mark)**

e) Why might sampling the organisms be a better measure of the long-term health of the water than measuring the oxygen demand of the water? **(2 marks)**

This section provides the basic ideas that the rest of your course is built on. You may have covered some aspects in your previous work, but it is important to understand the key principles thoroughly before seeing how these can be applied across all the other sections. The section covers some of the experimental techniques you will meet in your course.

First you will look at the existing evidence for the particulate nature of matter. Next, you will consider the structure of an atom and why the atoms of different elements have different properties. You will look at the different ways that atoms of elements join together when they form compounds, and how the method of combination will determine the properties of the compound formed. You will develop your skills in writing word and symbol equations; and also use an equation to work out the products of a reaction.

STARTING POINTS

1. What is an atom?

2. Name some of the particles that are found in an atom.

3. What name is given to a particle formed when two atoms combine together?

4. You will be learning about the states of matter. Do you know what these states are?

5. One type of chemical bonding you will study is called ionic bonding. Find out what an ion is.

SECTION CONTENTS

a) The particulate nature of matter

b) Experimental techniques

c) Atoms, elements and compounds

d) Ions and ionic bonds

e) Molecules and covalent bonds

f) Stoichiometry

1
Principles of chemistry

△ Diamond and graphite are both forms of carbon but have quite different properties.

△ Fig. 1.1 Water in all its states of matter.

The particulate nature of matter

INTRODUCTION

Nearly all substances may be classified as solid, liquid or gas – the states of matter. In science these states are often shown in shorthand as (s), (l) and (g) after the formula or symbol (these are called **state symbols**). The kinetic particle theory is based on the idea that all substances are made up of extremely tiny particles. The particles in these three states are arranged differently and have different types of movement and different energies. In many cases, matter changes into different states quite easily. The names of many of these processes are in everyday use, such as **melting** and condensing. Using simple models of the particles in solids, liquids and gases can help to explain what happens when a substance changes state.

KNOWLEDGE CHECK

✓ Be able to classify substances as solid, liquid or gas.
✓ Be familiar with some of the simple properties of solids, liquids and gases.
✓ Know that all substances are made up of particles.

LEARNING OBJECTIVES

✓ Be able to state the distinguishing properties of solids, liquids and gases.
✓ Be able to describe the structure of solids, liquids and gases in terms of particle separation, arrangement and types of motion.
✓ Be able to describe the changes of state in terms of melting, boiling, evaporation, freezing and condensation.
✓ Be able to describe qualitatively the pressure and temperature of a gas in terms of the motion of its particles.
✓ Be able to show an understanding of the terms *atom*, *molecule* and *ion*.
✓ EXTENDED Be able to explain changes of state in terms of particle theory.

△ Fig. 1.2 Water covers nearly four-fifths of the Earth's surface. In this photo you can see that all three states of matter can exist together: solid water (the ice) is floating in liquid water (the ocean), and the surrounding air contains water vapour (clouds).

HOW DO SOLIDS, LIQUIDS AND GASES DIFFER?

The three states of matter each have different properties, depending on how strongly the particles are held together.

• **Solids** have a fixed volume and shape.
• **Liquids** have a fixed volume but no definite shape. They take up the shape of the container in which they are held.

- **Gases** have no fixed volume or shape. They spread out to fill whatever container or space they are in.

Substances don't always exist in the same state; depending on the physical conditions, they change from one state to another (interconvert).

Some substances can exist in all three states in the natural world. A good example of this is water.

QUESTIONS

1. What is the state symbol for a liquid?

2. Which is the only state of matter that has a fixed shape?

3. In what ways does fine sand behave like a liquid?

Why do solids, liquids and gases behave differently?

The behaviour of solids, liquids and gases can be explained if we think of all matter as being made up of very small particles that are in constant motion. This idea has been summarised in the **particle theory** of matter.

In solids, the particles are held tightly together in a fixed position, so solids have a definite shape. However, the particles are vibrating about their fixed positions because they have energy.

In liquids, the particles are held tightly together but have enough energy to move around. Liquids have no definite shape and will take on the shape of the container they are in.

In gases, the particles are further apart with enough energy to move apart from each other and are constantly moving. Gas particles can spread apart to fill the container they are in.

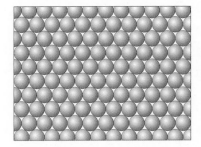

Δ Fig. 1.3 Particles in a solid.

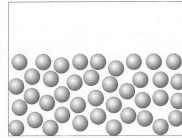

Δ Fig. 1.4 Particles in a liquid.

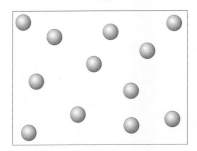

Δ Fig. 1.5 Particles in a gas.

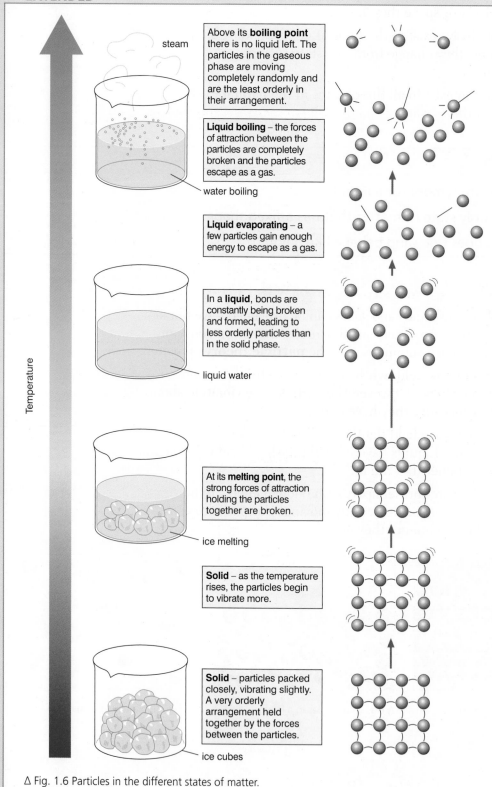

Above its **boiling point** there is no liquid left. The particles in the gaseous phase are moving completely randomly and are the least orderly in their arrangement.

Liquid boiling – the forces of attraction between the particles are completely broken and the particles escape as a gas.

water boiling

steam

Liquid evaporating – a few particles gain enough energy to escape as a gas.

In a **liquid**, bonds are constantly being broken and formed, leading to less orderly particles than in the solid phase.

liquid water

At its **melting point**, the strong forces of attraction holding the particles together are broken.

ice melting

Solid – as the temperature rises, the particles begin to vibrate more.

Solid – particles packed closely, vibrating slightly. A very orderly arrangement held together by the forces between the particles.

ice cubes

Temperature

Δ Fig. 1.6 Particles in the different states of matter.

END OF EXTENDED

HOW DO SUBSTANCES CHANGE FROM ONE STATE TO ANOTHER?

To change solids into liquids and then into gases, heat energy must be put in. The heat provides the particles with enough energy to overcome the forces holding them together.

To change gases into liquids and then into solids involves cooling, so removing heat energy. This makes the particles come closer together as the substance changes from gas to liquid and the particles bond together as the liquid becomes a solid.

The temperatures at which one state changes to another have specific names:

Name of temperature	Change of state
Melting point	Solid to liquid
Boiling point	Liquid to gas
Freezing point	Liquid to solid
Condensation point	Gas to liquid

△ Table 1.1 Changes of state.

The particles in a liquid can move around. They have different energies, so some are moving faster than others. The faster particles have enough energy to escape from the surface of the liquid and it changes into the gas state (also called **vapour** particles). This process is **evaporation**. The rate of evaporation increases with increasing temperature because heat gives more particles the energy to be able to escape from the surface.

Fig. 1.7 summarises the changes in states of matter. Note that melting and freezing happen at the same temperature – as do boiling and condensing.

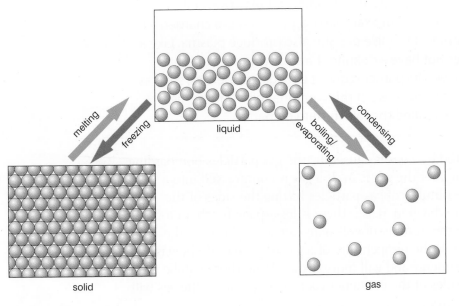

△ Fig. 1.7 Changes of state. Note that melting and freezing happen at the same temperature – as do boiling and condensing.

THE STATES OF MATTER

There are three states of matter – or are there? To complicate this simple idea, some substances show the properties of two different states of matter. Some examples are given below.

Liquid crystals

Liquid crystals are commonly used in displays in computers and televisions. Within particular temperature ranges the particles of the liquid crystal can flow like a liquid, but remain arranged in a pattern in which the particles cannot rotate.

△ Fig. 1.8 An LCD (liquid crystal display) television.

Superfluids

When some liquids are cooled to very low temperatures they form a second liquid state, described as a *superfluid* state. Liquid helium at just above absolute zero has infinite fluidity and will 'climb out' of its container when left undisturbed – at this temperature the liquid has zero viscosity. (You may like to look up 'fluidity' and 'viscosity'.)

Plasma

Plasmas, or ionised gases, can exist at temperatures of several thousand degrees Celsius. An example of a plasma is the charged air produced by lightning. Stars like our Sun also produce plasma. Like a gas, a plasma does not have a definite shape or volume but the strong forces between its particles give it unusual properties, such as conducting electricity. Because of this combination of properties, plasma is sometimes called the fourth state of matter.

The **pressure** exerted by a gas is caused by the gas particles bombarding the sides of the container the gas is in. If a gas is compressed into a smaller volume, the number of gas particles hitting the sides of the container every second will increase (there is less space for them to move in). The increase in the number of collisions per second causes the increase in pressure. If the temperature of a gas is increased, the gas particles have more energy and will move faster. Again there will be more collisions with the sides of the container each second and so the gas will exert a greater pressure.

1. What type of movement do the particles in a solid have?

2. In which state are the particles held together more strongly: in solid water, liquid water or water vapour?

3. What is the name of the process that occurs when the faster-moving particles in a liquid escape from its surface?

4. What name is given to the temperature at which a solid changes into a liquid?

SCIENCE LINK

PARTICLES

- Particles make up the structure of all living things in Biology and everything in the Universe that we study in Physics.

- Biological processes in cells happen through the movement of particles.

- Larger-scale processes such as digestion, respiration and photosynthesis are driven by the interactions of particles.

- In Physics, ideas about particles help explain the structure of buildings, how heat energy is transferred, how electrical circuits work, what happens to cause the weather, and so on.

- Particle ideas – the different sizes, how particles join together, how particles are arranged and how they move – are ideas that return again and again.

ELEMENTS, ATOMS AND MOLECULES

All matter is made from **elements**. A Periodic Table of elements is shown on page 327 of this book and you will recognise the names of some of the more common elements that you know about, for example, carbon (C), oxygen (O), aluminium (Al) and iron (Fe). Elements are substances that cannot be broken down into anything simpler, as they are made up of one kind of the same small particle. These small particles are called **atoms**.

Almost always, the atoms in an element combine with other atoms to form molecules. For example, two atoms of hydrogen combine with one atom of oxygen to form a molecule of water. So the formula of water is H_2O. One atom of carbon combines with two atoms of oxygen to form carbon dioxide – the chemical formula is CO_2. Water and carbon dioxide are molecules. There are also particles called ions. Unlike atoms and molecules these particles are not neutral, they are charged. (see page 234).

△ Fig. 1.9 Model of a water molecule.

QUESTIONS

1. What is the name of the particle that is found in all elements?

2. The molecule methane has the chemical formula CH_4. Which atoms does it contain and how many of each?

3. Glucose is a molecule with the chemical formula of $C_6H_{12}O_6$. Explain how this molecule is made up.

4. What is the key difference between an ion and an atom?

End of topic checklist

Key terms

atom, boiling, boiling point, condensation, element, evaporation, freezing, freezing point, gas, liquid, melting, melting point, particle theory, pressure, solid, state symbols, vapour

During your study of this topic you should have learned:

○ About the different properties of solids, liquids and gases.

○ How to describe the structure of solids, liquids and gases in terms of particle separation, arrangement and types of motion.

○ How to describe changes of state in terms of melting, boiling, evaporation, freezing and condensation.

○ How to describe the pressure and temperature of a gas in terms of the motion of its particles.

○ About atoms, molecules and ions.

○ EXTENDED How to explain changes of state in terms of the particle theory.

End of topic questions

Note: The marks given for these questions indicate the level of detail required in the answers. In the examination, the number of marks given to questions like these may be different.

1. In which of the three states of matter are the particles moving fastest? **(1 mark)**

2. Describe the arrangement and movement of the particles in a liquid. **(2 marks)**

3. In which state of matter do the particles just vibrate about a fixed point? **(1 mark)**

4. Sodium (melting point 98 °C) and aluminium (melting point 660 °C) are both solids at room temperature. From their melting points, what can you conclude about the forces of attraction between the particles in the two metals? **(1 mark)**

5. What is the name of the process involved in each of the following changes of state:

   a) $Fe(s) \rightarrow Fe(l)$? **(1 mark)**

   b) $H_2O(l) \rightarrow H_2O(g)$? **(1 mark)**

   c) $H_2O(g) \rightarrow H_2O(l)$? **(1 mark)**

   d) $H_2O(l) \rightarrow H_2O(s)$? **(1 mark)**

6. Ethanol liquid turns into ethanol vapour at 78 °C. What is the name of this temperature? **(1 mark)**

7. A student wrote in her exercise book, 'The particle arrangement in a liquid is more like the arrangement in a solid than in a gas'. Do you agree with this statement? Explain your reasoning. **(2 marks)**

Experimental techniques

INTRODUCTION

Practical work is a very important part of studying chemistry. In your practical work you will need to develop your skills so that you can safely, correctly and methodically use and organise techniques, apparatus and materials. This involves being able to use appropriate apparatus for measurement to give readings to the required degree of accuracy. It is important to be able to use techniques that will determine the purity of a substance and, if necessary, techniques that can be used to purify mixtures of substances.

△ Fig. 1.10 Using the neutralisation method for a titration.

KNOWLEDGE CHECK

✓ Be familiar with some simple equipment for measuring time, temperature, mass and volume.
✓ Know that some substances are mixtures of a number of different components.

LEARNING OBJECTIVES

✓ Be able to name appropriate apparatus for accurate measurement of time, temperature, mass and volume.
✓ Be able to interpret simple chromatograms.
✓ **EXTENDED** Be able to use R_f values in interpreting simple chromatograms.
✓ Be able to describe methods of separation and purification by the use of: a suitable solvent; filtration; crystallisation; distillation; fractional distillation; paper chromatography.
✓ Be able to suggest suitable purification techniques, given information about the substances involved.

MEASUREMENT

In your study of chemistry you will carry out practical work. It is essential to use the right apparatus for the task.

Time is measured with clocks, such as a wall clock. The clock should be accurate to about 1 second. You may be able to use your own wristwatch or a stopclock.

Temperature is measured using a thermometer. The range of the thermometer is commonly $-10\,°C$ to $+110\,°C$ with intervals of $1\,°C$.

Mass is measured with a balance or a set of scales.

Volume of liquids can be measured with burettes, pipettes and measuring cylinders.

△ Fig. 1.11 Measuring equipment.

CRITERIA OF PURITY

Paper chromatography

Paper **chromatography** is a way of separating solutions or liquids that are mixed together.

Black ink is a mixture of different coloured inks. The diagrams in Fig. 1.12 show how paper chromatography is used to find the colours that make up a black ink.

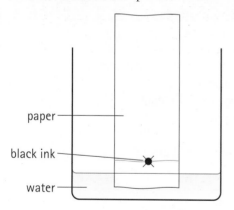
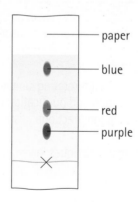

△ Fig. 1.12 Paper chromatography being used to separate a solution to find the colours in black ink. The left part of the diagram shows the paper before the inks have been separated, and the right part shows the paper after the inks have been separated.

A spot of ink is placed on the × mark and the paper is suspended in water. As the water rises up the paper, the different dyes travel different distances and so are separated on the **chromatogram**.

Paper chromatography can be used to identify what an unknown liquid is made of. This involves interpreting a chromatogram.

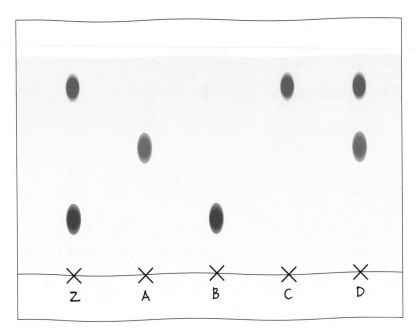

△ Fig. 1.13 A chromatogram.

The unknown liquid Z is compared with known liquids – in this case A to D.

Z must be made of B and C because the pattern of their dots matches the pattern shown by Z.

△ Fig. 1.14 A piece of filter paper is marked with black ink and dipped into water in a beaker.

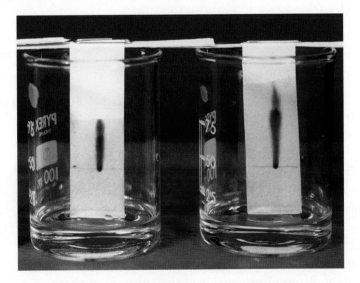

△ Fig. 1.15 After a few minutes the chromatogram has been created by the action of the water on the ink.

Retention factors

Substances can also be identified using chromatography by measuring their **retention factor** on the filter paper. The retention factor (R_f) for a particular substance compares the distance the substance has travelled up the filter paper with the distance travelled by the **solvent**. The retention factor can be calculated using the following formula:

$$R_f = \frac{\text{Distance moved by a substance from the baseline}}{\text{Distance moved by the solvent from the baseline}}$$

As the solvent will always travel further than the substance, R_f values will always be less than 1.

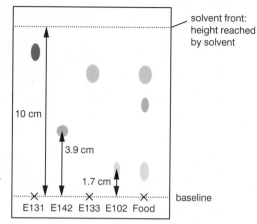

△ Fig. 1.16 The R_f value for the food additive E102 is 0.17.

END OF EXTENDED

QUESTIONS

1. The start line, or baseline, in chromatography should be drawn in pencil. Explain why.

2. In a chromatography experiment, why must the solvent level in the beaker be below the baseline?

3. In a chromatography experiment to compare the dyes in two different inks, one of the inks does not move at all from the baseline. Suggest a reason for this.

4. EXTENDED Look at the diagram in Fig. 1.16. Explain why the retention factor for the food additive E102 is 0.17.

METHODS OF SEPARATION AND PURIFICATION

Many substances exist in mixtures with other substances. To obtain the pure substance it is first necessary to separate the mixture into its components and then purify the components. Techniques for separating and purifying solids and liquids rely on finding different properties of the substances that make up the impure mixture.

There are some important terms you will need to be familiar with:

A **solution** is formed when a **solute** (a solid) dissolves in a **solvent** (a liquid). The more solute that dissolves, the more concentrated the solution will become.

△ Fig. 1.17 Filtration of copper(II) hydroxide.

Purifying impure solids

The method is:

1. Add a solvent that the required solid is **soluble** in, and dissolve it.

2. Filter the mixture to remove the insoluble impurity.

3. Heat the solution to remove some solvent and leave it to crystallise.

4. Filter off the crystals, wash with a small amount of cold solvent and dry them – this is the pure solid.

An example of using this technique would be separating salt from 'rock salt' (the impure form of sodium chloride). Water is added to dissolve the salt but leave the other solids undissolved. Filter off the insoluble impurities, warm the salt solution and leave it to crystallise to form salt crystals.

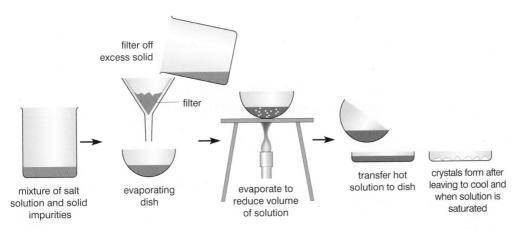

mixture of salt solution and solid impurities

filter off excess solid

filter

evaporating dish

evaporate to reduce volume of solution

transfer hot solution to dish

crystals form after leaving to cool and when solution is saturated

Δ Fig. 1.18 Separating impurities in rock salt.

Purifying impure liquids

There are two methods:

1. Liquids contaminated with soluble solids dissolved in them.

The method is **distillation**.

The solution is heated, the solvent boils and turns into a vapour. It is condensed back to the pure liquid and collected.

This is the technique used in **desalination** plants, which produce pure drinking water from sea water. The solids are left behind after boiling off the water.

Δ Fig.1.19 Distillation apparatus.

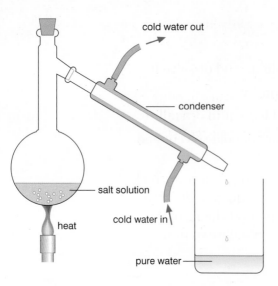

△ Fig. 1.20 Distillation of salt water.

2. Liquids contaminated with other liquids.

In this case the technique is **fractional distillation**, which uses the difference in boiling points of the different liquids mixed together.

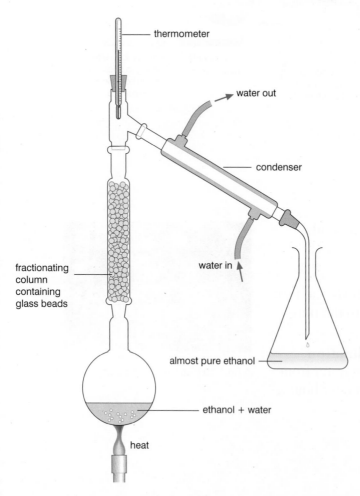

△ Fig. 1.21 Apparatus for fractional distillation of an alcohol–water mixture.

The mixture is boiled, and the liquid with the lowest boiling point turns to a vapour first, rises up the fractionating column and is condensed back to liquid in the condenser. The next lowest boiling point liquid comes off, and so on until all the liquids have been separated. You can identify the fraction you want to collect by the temperature reading on the thermometer. The fractionating column increases the purity of the distilled product by reducing the amount of other substances in the vapour when it condenses.

Fractional distillation is the method used in the separation of crude oil and collecting ethanol from the fermentation mixture.

QUESTIONS

1. What is a *solvent*?

2. What does the term *soluble* mean?

3. What method would you use to separate a pure liquid from a solution of a solid and the liquid?

4. Complete this sentence selecting the correct term below.
   To separate two liquids by fractional distillation they must have different ...

   a) melting points

   b) boiling points

   c) colours

   d) viscosities

End of topic checklist

Key terms

chromatogram, desalination, distillation, fractional distillation, chromatography, retention factor, soluble, solute, solution, solvent

During your study of this topic you should have learned:

◯ About the appropriate apparatus for the measurement of time, temperature, mass and volume, including burettes, pipettes and measuring cylinders.

◯ How to define the terms solvent, solute, solution and concentration.

◯ About the technique of paper chromatography.

◯ How to interpret simple chromatograms.

◯ **EXTENDED** How to interpret simple chromatograms, including the use of R_f values.

◯ How to describe methods of purification by the use of:

- a suitable solvent – to separate a soluble solid from an insoluble solid
- filtration – to separate a solid from a liquid
- crystallisation – to separate a solid from its solution
- distillation – to separate a solid and a liquid from a solution
- fractional distillation – to separate liquids with different boiling points.

◯ How to suggest suitable purification techniques given information about the substances involved.

End of topic questions

Note: The marks given for these questions indicate the level of detail required in the answers. In the examination, the number of marks given to questions like these may be different.

1. You are provided with four samples of black water-soluble inks. Two of the ink samples are identical. Describe how you would use paper chromatography to identify which two ink samples are the same. **(3 marks)**

2. You are trying to separate the dyes in a sample of ink using paper chromatography. You set up the apparatus as shown in Fig. 1.14. After 20 minutes the black spot is unchanged and the water has risen nearly to the top of the filter paper.

   a) Suggest a reason why the black spot has remained unchanged. **(1 mark)**

   b) What could you change that might lead to a successful separation of the dyes? **(1 mark)**

3. In the fractional distillation of ethanol and water, why does the ethanol vapour condense in the condenser? **(1 mark)**

4. Describe how you would produce crystals of sodium chloride from a sodium chloride solution. **(2 marks)**

5. What process could be used to separate the following mixtures:

   a) sand from a sand–water mixture? **(1 mark)**

   b) petrol from a petrol–diesel mixture? **(1 mark)**

   c) pure water from salt solution? **(1 mark)**

6. EXTENDED Look at the chromatogram produced when testing four food colouring compounds A, B, C and D.

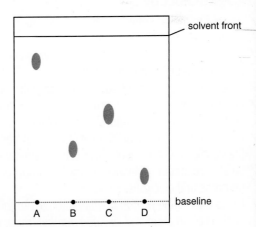

   a) Which compound has the largest retention factor (R_f)? **(1 mark)**

   b) Which compound has the smallest R_f? **(1 mark)**

   c) Estimate the R_f for compound C. Explain how you made the estimate. **(2 marks)**

   d) Why are all R_f values less than 1.0? **(1 mark)**

△ Fig. 1.22

Atoms, elements and compounds

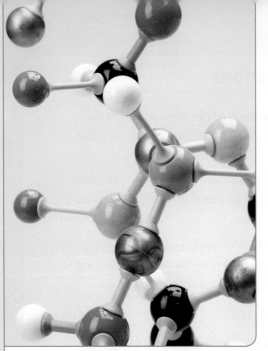

△ Fig. 1.23 A model showing molecular structure.

INTRODUCTION

This topic is about the structure, or the makeup, of all substances. Some substances exist in nature as elements, others as compounds that are formed when elements combine chemically. The topic starts by considering the structure of the atoms that make up elements. It shows how the arrangement of elements in the Periodic Table is determined by the structure of their atoms. The properties of metals and non-metals are explained and an introduction to the combination of atoms forming compounds is provided. The following topics look in more detail at how atoms combine together to form ions and molecules, and the structure of metals.

KNOWLEDGE CHECK

✓ Know the three states of matter and how to use the particle theory to explain the conversion of one state into another.
✓ Understand how diffusion experiments provide evidence for the existence of particles.
✓ Know that compounds are formed when elements combine together chemically.

LEARNING OBJECTIVES

✓ Be able to identify physical change and chemical change and understand the differences between them.
✓ Be able to describe the differences between elements, mixtures and compounds, and between metals and non-metals.
✓ Be able to use the terms solvent, solute, solution and concentration.
✓ Be able to describe the structure of an atom in terms of a central nucleus, containing protons and neutrons, and 'shells' of electrons.
✓ Be able to describe the build-up of electrons in 'shells' around the nucleus and understand the significance of the noble gas electronic structures and of the outer shell electrons.
✓ Be able to state the relative charges and approximate relative masses of protons, neutrons and electrons.
✓ Be able to define *proton number* (atomic number) as the number of protons in the nucleus of an atom and *nucleon number* (mass number) as the total number of protons and neutrons in the nucleus of an atom.
✓ EXTENDED Be able to use proton number and the simple structure of atoms to explain the basis of the Periodic Table, with special reference to the elements with proton numbers 1 to 20.

PHYSICAL AND CHEMICAL CHANGES

A chemical change, or chemical reaction, is quite different from physical changes that occur, for example, when sugar dissolves in water.

In a chemical change, one or more new substances are produced. In many cases an observable change is apparent, for example, the colour changes or a gas is produced.

An apparent change in mass can occur. This change is often quite small and difficult to detect unless accurate balances are used. Mass is conserved in *all* chemical reactions – the apparent change in mass usually occurs because one of the reactants or products is a gas (whose mass may not have been measured).

An energy change is almost always involved. In most cases energy is released and the surroundings become warmer. In some cases energy is absorbed from the surroundings, and so the surroundings become colder. Note: Some physical changes, such as evaporation, also have energy changes.

THE STRUCTURE OF MATTER

All matter can be classified into the three categories of elements, mixtures and compounds. As you have seen, the elements can be ordered in the Periodic Table. You will be learning how elements combine together to form compounds in the next two topics.

A **mixture** contains more than one substance (elements or compounds). In a mixture, the individual substances can be separated by simple means. This is because the substances in a mixture have not combined chemically.

Most elements can be classified as either **metals** or **non-metals**. In the Periodic Table, the metals are arranged on the left and in the middle, and the non-metals are on the right.

Metals and non-metals have quite different physical and chemical properties.

Δ Fig. 1.24 Non-metals: from left: silicon, chlorine, sulfur.

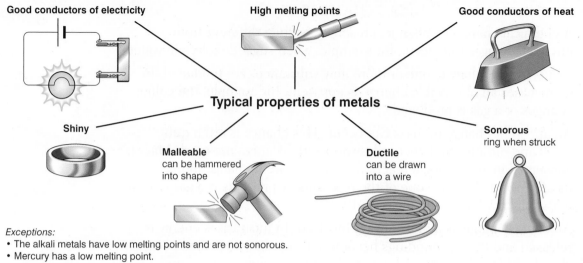

Δ Fig. 1.25 Typical properties of metals.

Exceptions:
- The alkali metals have low melting points and are not sonorous.
- Mercury has a low melting point.

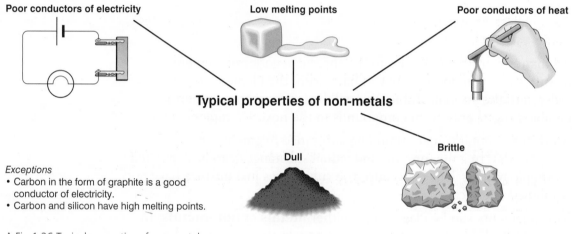

Exceptions
- Carbon in the form of graphite is a good conductor of electricity.
- Carbon and silicon have high melting points.

Δ Fig.1.26 Typical properties of non-metals.

QUESTION

1. What is the difference between a *mixture* and a *compound*?

SCIENCE LINK

BIOLOGY – BIOLOGICAL MOLECULES

- The behaviour of the different types of particles – atoms, molecules and ions – is important in describing how the different life processes happen.

- The combination of atoms of different elements into molecules leads to the chemicals required for life to exist.

- Although some of the molecules required for life processes are very complicated, they arise through the same rules of combination that apply to the simplest compounds.

- Particular elements must be present for living things to survive successfully, for example, plants need particular 'nutrients', a number of which are simply chemical elements.

ATOMIC THEORY

In 1808, the British chemist John Dalton published a book outlining his theory of atoms. These were the main points of his theory:

- All matter is made of small, indivisible spheres called atoms.
- All the atoms of a given element are identical and have the same mass.
- The atoms of different elements have different masses.
- Chemical compounds are formed when different atoms join together.

All the molecules of a chemical **compound** have the same type and number of atoms.

An element is the smallest part of a substance that can exist on its own. When two or more elements combine together a compound is formed.

Since 1808, atomic theory has developed considerably and yet many of Dalton's ideas are still correct. Modern theory is built on an understanding of the particles that make up atoms – the so-called sub-atomic particles.

Sub-atomic particles

The smallest amount of an element that still behaves like that element is an atom. Each element has its own unique type of atom. Atoms are made up of smaller, sub-atomic particles. The three main sub-atomic particles are **protons**, **neutrons** and **electrons**.

These particles are very small and have very little mass. However, it is possible to compare their masses using a relative scale. Their charges may also be compared in a similar way. The proton and neutron have the same mass, and the proton and electron have equal but opposite charges.

Sub-atomic particle	Relative mass	Relative charge
Proton	1	+1
Neutron	1	0
Electron	about $\frac{1}{2000}$	−1

△ Table 1.2 Relative masses and charges of sub-atomic particles.

Protons and neutrons are found in the centre of the atom in a cluster called the **nucleus**. The electrons form a series of 'shells' around the nucleus.

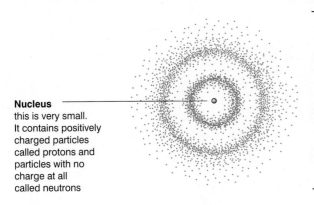

Nucleus
this is very small.
It contains positively charged particles called protons and particles with no charge at all called neutrons

Electrons
are negatively charged particles that form a series of 'shells' around the nucleus

△ Fig. 1.27 Structure of an atom.

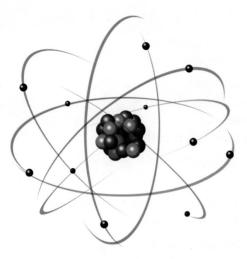

△ Fig. 1.28 Diagrams are another way of representing the structure of an atom.

ARRANGEMENTS OF ELECTRONS IN THE ATOM

An atom's electrons are arranged in **shells** around the nucleus. These do not all contain the same number of electrons – the shell nearest to the nucleus can take only two electrons, whereas the next one out from the nucleus can take eight.

Electron shell	Maximum number of electrons
1	2
2	8
3	8 (initially, with up to 18 after element 20)

△ Table 1.3 Maximum number of electrons in a shell.

Oxygen has a proton number of 8, so it has 8 electrons. Of these, two are in the first shell and six are in the second shell. This arrangement is written 2,6. A phosphorus atom with a proton number of 15 has 15 electrons, arranged 2,8,5. The electrons in the outer electron shell that are involved in chemical bonding are known as the **valency electrons**.

Atom diagrams

The atomic structure of an atom can be shown simply in a diagram.

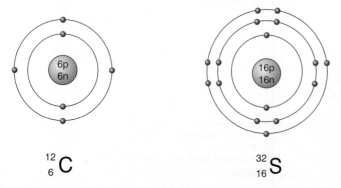

$^{12}_{6}\text{C}$ $\qquad$ $^{32}_{16}\text{S}$

△ Fig. 1.29 Atom diagrams for carbon and sulfur showing the numbers of protons and neutrons and the electron arrangements.

The arrangement of electrons in an atom is called its **electronic configuration**.

Periodicity and electronic configuration

In the Periodic Table lithium, sodium and potassium are placed on the left, and neon and argon are placed on the right. The proton number increases from lithium to neon, moving through a section, or **period,** of the Periodic Table. The number of electrons in the outer shell increases. This is called **periodicity**.

ELECTRONIC CONFIGURATION AND CHEMICAL PROPERTIES

Elements that have similar electronic configurations have similar chemical properties.

Lithium (2,1), sodium (2,8,1) and potassium (2,8,8,1) all have one electron in their outer shell. These are all highly reactive metals. They are called Group I elements.

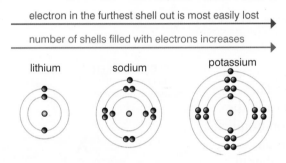

△ Fig. 1.30 Electronic configurations of lithium, sodium and potassium.

Fluorine (2,7), chlorine (2,8,7), bromine (2,8,18,7) and iodine (2,8,18,18,7) all have seven electrons in their outer shell. These elements are all highly reactive non-metals. They are called Group VII elements, or halogens.

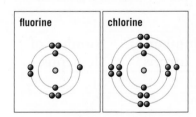

△ Fig. 1.31 Electronic configuration of fluorine and chlorine.

Similarly, all the elements in Group III of the Periodic Table have three electrons in their outer electron shell.

The elements helium (2), neon (2,8), argon (2,8,8), krypton (2,8,18,8) and xenon (2,8,18,18,8) either have a full outer shell or have eight electrons in their outer shell and therefore the atoms do not lose or gain electrons easily. This means that these gases are unreactive. They are called **noble gases** and are in Group VIII or 0 in the Periodic Table.

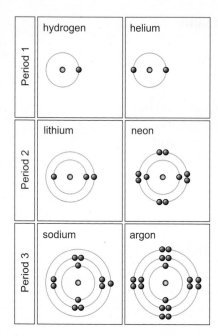

△ Fig. 1.32 Electronic configurations of helium, neon and argon.

△ Fig. 1.33 Neon lighting in Hong Kong.

Proton number and nucleon number

In order to describe the numbers of protons, neutrons and electrons in an atom, scientists use two numbers. These are called the **proton number** (or **atomic number**) and the **nucleon number** (or **mass number**). The proton number, as you might expect, describes the number of protons in the atom. The nucleon number describes the number of particles in the nucleus of the atom – that is, the total number of protons and neutrons.

Proton numbers are used to arrange the elements in the **Periodic Table**. The atomic structures of the first ten elements in the Periodic Table are shown in Table 1.4.

Hydrogen is the only atom that has no neutrons.

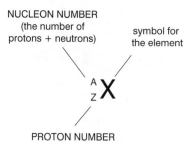

NUCLEON NUMBER
(the number of protons + neutrons)

symbol for the element

$$^A_Z X$$

PROTON NUMBER
(the number of protons, which equals the number of electrons)

△ Fig. 1.34 Chemical symbol showing nucleon number and proton number.

Element	Proton number	Nucleon number	Number of protons	Number of neutrons	Number of electrons
Hydrogen	1	1	1	0	1
Helium	2	4	2	2	2
Lithium	3	7	3	4	3
Beryllium	4	9	4	5	4
Boron	5	11	5	6	5
Carbon	6	12	6	6	6
Nitrogen	7	14	7	7	7
Oxygen	8	16	8	8	8
Fluorine	9	19	9	10	9
Neon	10	20	10	10	10

△ Table 1.4 Atomic structures of the first ten elements.

QUESTIONS

1. Which sub-atomic particle has the smallest relative mass?

2. Why do atoms have the same number of protons as electrons?

3. An aluminium atom can be represented as $^{27}_{14}$Al.

    a) What is aluminium's nucleon number?

    b) How many neutrons does this atom of aluminium have?

SCIENCE IN CONTEXT

SUB-ATOMIC PARTICLES

△ Fig. 1.35 The Large Hadron Collider at CERN in Switzerland.

Protons, neutrons and electrons are the particles from which atoms are made. However, in the past 20 years or so scientists have discovered a number of other sub-atomic particles: quarks, leptons, muons, neutrinos, bosons and gluons. The properties of some of these particles have become well known, but there is still much to learn about the others. Finding out about these, and possibly other sub-atomic particles, is one of the challenges of the twenty-first century.

To study the smallest known particles, a particle accelerator has been built underground at CERN near Geneva, Switzerland. This giant instrument, called the Large Hadron Collider (LHC), has a circumference of 27 km. It attempts to recreate the conditions that existed just after the 'Big Bang' by colliding beams of particles at very high speed – only about 5 m/s slower than the speed of light. It promises to revolutionise scientific understanding of the nature of atoms. Who knows – school science in 10 or 20 years' time may be very different from your lessons today!

Electronic configuration: The first 20 elements of the Periodic Table

There are over 100 different elements. They are arranged in the Periodic Table according to their chemical and physical properties.

The chemical properties of elements depend on the arrangement of electrons in their atoms. The electronic structure of the first 20 elements is shown in Table 1.5.

Element	Symbol	Proton number	Electron number	Electronic configuration
Hydrogen	H	1	1	1
Helium	He	2	2	2
Lithium	Li	3	3	2,1
Beryllium	Be	4	4	2,2
Boron	B	5	5	2,3
Carbon	C	6	6	2,4
Nitrogen	N	7	7	2,5
Oxygen	O	8	8	2,6
Fluorine	F	9	9	2,7
Neon	Ne	10	10	2,8
Sodium	Na	11	11	2,8,1
Magnesium	Mg	12	12	2,8,2
Aluminium	Al	13	13	2,8,3
Silicon	Si	14	14	2,8,4
Phosphorus	P	15	15	2,8,5
Sulfur	S	16	16	2,8,6
Chlorine	Cl	17	17	2,8,7
Argon	Ar	18	18	2,8,8
Potassium	K	19	19	2,8,8,1
Calcium	Ca	20	20	2,8,8,2

△ Table 1.5 Electronic structure of first 20 elements.

QUESTIONS

1. a) How many electrons does magnesium have in its outer electron shell?

b) In which group of the Periodic Table is magnesium?

2. Draw atom diagrams for:

a) aluminium

b) calcium.

3. Why are noble gases (Group VIII or 0) unreactive?

End of topic checklist

Key terms

compound, electron, electronic configuration, metal, mixture, neutron, noble gases, non-metal, nucleon number (mass number), nucleus, period, Periodic Table, periodicity, proton, proton number (atomic number), shell, valency electrons

During your study of this topic you should have learned:

○ How to identify physical and chemical changes and understand the differences between them.

○ About the differences between elements, mixtures and compounds, and between metals and non-metals.

○ How to define the terms solvent, solute, solution and concentration.

○ To describe the structure of an atom in terms of a central nucleus, containing protons and neutrons, and 'shells' of electrons.

○ To describe the build-up of electrons in 'shells' and understand the significance of the noble gas electronic structures and of valency or outer electrons.

○ About the relative charges and approximate relative masses of protons, neutrons and electrons.

○ How to define *proton number* (atomic number) and *nucleon number* (mass number).

○ EXTENDED How to use proton numbers and the simple structure of atoms to explain the basis of the Periodic Table, with special reference to the elements with proton numbers 1 to 20.

End of topic questions

Note: The marks given for these questions indicate the level of detail required in the answers. In the examination, the number of marks given to questions like these may be different.

1. What is the relative mass of a proton? (1 mark)

2. Explain the meaning of:

    a) *proton number* (atomic number) (1 mark)

    b) *nucleon number* (mass number). (1 mark)

3. Copy and complete the table. (4 marks)

Atom	Number of protons	Number of neutrons	Number of electrons	Electron arrangement
$^{28}_{14}Si$				
$^{24}_{12}Mg$				
$^{32}_{16}S$				
$^{40}_{18}Ar$				

4. The table shows information about the structure of six particles (A–F).

Particle	Protons	Neutrons	Electrons
A	8	8	10
B	12	12	10
C	6	6	6
D	8	10	10
E	6	8	6
F	11	12	11

    a) In each of questions **i)** to **iv)**, choose one of the six particles A–F. Each letter may be used once, more than once or not at all.

    Choose a particle that:

      i) has a nucleon number of 12 (1 mark)

     ii) has the highest nucleon number (1 mark)

    iii) has no overall charge (1 mark)

    iv) has an overall positive charge. (1 mark)

    b) Draw an atom diagram for particle E. (2 marks)

5. Draw an atom diagram for:

   a) oxygen   **(2 marks)**

   b) potassium.   **(2 marks)**

6. EXTENDED For each of parts **a)** to **d)** say whether the statement is TRUE or FALSE.

There is a relationship between the group number of the first 20 elements in the Periodic Table and:

   a) the number of protons in an atom of the element   **(1 mark)**

   b) the number of neutrons in an atom of the element   **(1 mark)**

   c) the number of electrons in an atom of the element   **(1 mark)**

   d) the number of electrons in the outer electron shell of the element.   **(1 mark)**

Ions and ionic bonds

INTRODUCTION

When the atoms of elements react and join together, they form compounds. When one of the reacting atoms is a metal, the compound formed is called an ionic compound. They do not contain molecules; instead they are made of particles called ions. Ionic compounds have similar physical properties, many of which are quite different from the properties of substances made up of atoms or molecules.

△ Fig. 1.36 Sodium chloride is an example of an ionic compound.

KNOWLEDGE CHECK

✓ Understand that compounds are formed when the atoms of two or more elements combine together.
✓ Know that protons have a positive charge and are found in the nucleus of the atom.
✓ Know that electrons have a negative charge and are found in shells around the nucleus.
✓ Know that the number of outer electrons in an atom depends on its group in the Periodic Table.

LEARNING OBJECTIVES

✓ Be able to describe the formation of ions by electron loss or gain.
✓ Be able to use dot-and-cross diagrams to describe the formation of ionic bonds between elements from Groups I and VII.
✓ EXTENDED Be able to describe the formation of ionic bonds between metallic and non-metallic elements.
✓ EXTENDED Be able to describe the lattice structure of ionic compounds as a regular arrangement of alternating positive and negative ions.

THE FORMATION OF IONS

Atoms bond with other atoms in a **chemical reaction** to make a compound. For example, sodium reacts with chlorine to make sodium chloride. **Ionic compounds** contain a metal combined with one or more non-metals. They are not made up of molecules – they are made up of **ions**.

Ions are formed from atoms by the gain or loss of electrons. Both metals and non-metals try to achieve complete (filled) outer electron shells or the electron configuration of the nearest noble gas.

Metals lose electrons from their outer shells and form positive ions. Non-metals gain electrons in their outer shells and form negative ions.

The bonding process can be represented in dot-and-cross diagrams.
Look at the reaction between sodium and chlorine as an example.

Sodium is a metal. It has a proton number of 11 and so has 11 electrons, arranged 2,8,1. Its atom diagram looks like this:	Chlorine is a non-metal. It has a proton number of 17 and so has 17 electrons, arranged 2,8,7. Its atom diagram looks like this:

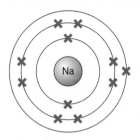 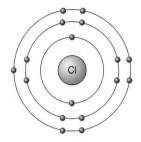

△ Fig. 1.37 Dot-and-cross diagrams for sodium and chlorine.

Sodium has one electron in its outer shell. It can achieve a full outer shell by losing this electron. The sodium atom transfers its outermost electron to the chlorine atom.	Chlorine has seven electrons in its outer shell. It can achieve a full outer shell by gaining an extra electron. The chlorine atom accepts an electron from the sodium.

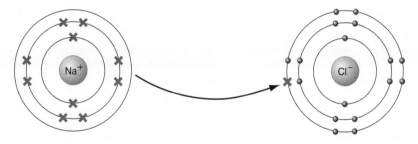

△ Fig. 1.38 Dot-and-cross diagram for sodium chloride, NaCl.

The sodium is no longer an atom; it is now an ion. It does not have equal numbers of protons and electrons, so it is no longer neutral. It has one more proton than it has electrons, so it is a positive ion with a charge of 1+. The ion is written as Na^+.	The chlorine is no longer an atom; it is now an ion. It does not have equal numbers of protons and electrons, so it is no longer neutral. It has one more electron than it has protons, so it is a negative ion with a charge of 1−. The ion is written as Cl^-.

EXTENDED

METALS CAN TRANSFER MORE THAN ONE ELECTRON TO A NON-METAL

Magnesium combines with oxygen to form magnesium oxide. The magnesium (electron arrangement 2,8,2) transfers two electrons to the oxygen (electron arrangement 2,6). Magnesium therefore forms an Mg^{2+} ion and oxygen forms an O^{2-} ion.

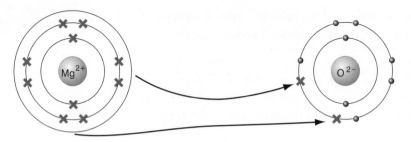

△ Fig. 1.39 Dot-and-cross diagram for magnesium oxide, MgO.

Aluminium has an electron arrangement 2,8,3. When it combines with fluorine with an electron arrangement 2,7, three fluorine atoms are needed for each aluminium atom. The formula of aluminium fluoride is therefore AlF_3.

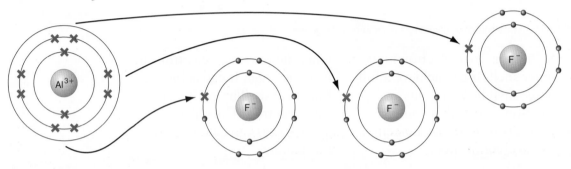

△ Fig. 1.40 Dot-and-cross diagram for aluminium fluoride, AlF_3.

REMEMBER

It is important to remember the difference between oxidation and reduction. In ionic bonding the atom that loses electrons is said to be *oxidised*. The atom that gains the electrons is said to be *reduced*. So, aluminium is oxidised and fluorine is reduced when aluminium fluoride is made.

END OF EXTENDED

QUESTIONS

1. Draw a dot-and-cross diagram to show how lithium and fluorine atoms combine to form lithium fluoride. You must show the starting atoms and the finishing ions. (Proton numbers: Li 3; F 9)

2. EXTENDED Draw a dot-and-cross diagram to show how calcium and sulfur atoms combine to form calcium sulfide. You must show the starting atoms and finishing ions. (Proton numbers: Ca 20; S 16)

3. EXTENDED How do you know that phosphorus oxide is not an ionic compound?

ELECTRONIC CONFIGURATION AND IONIC CHARGE

When atoms form ions, they are trying to achieve the electronic configuration of their nearest noble gas (Group VIII or 0). Some common ions and their electronic configurations are shown in Table 1.6.

Ion	Electronic configuration
Li^+	2
Na^+	2,8
Mg^{2+}	2,8
F^-	2,8
Cl^-	2,8,8
O^{2-}	2,8

△ Table 1.6 Electronic configurations of some ions.

PROPERTIES OF IONIC COMPOUNDS

Ionic compounds have high melting points and high boiling points because of strong electrostatic forces between the ions.

The strong electrostatic attraction between oppositely charged ions is called an **ionic bond**.

Ionic compounds form giant lattice structures. For example, when sodium chloride is formed by ionic bonding, the ions do not pair up. Each sodium ion is surrounded by six chloride ions, and each chloride ion is surrounded by six sodium ions.

The electrostatic attractions between the ions are very strong. The properties of sodium chloride can be explained using this model of its structure.

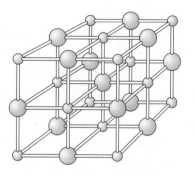

○ chloride ion   ○ sodium ion

△ Fig. 1.41 In solid sodium chloride, the ions are held firmly in place. All ionic compounds have giant ionic lattice structures like this.

△ Fig. 1.42 Crystals of sodium chloride.

Properties of sodium chloride	Explanation in terms of structure
Hard crystals	Strong forces of attraction between the oppositely charged ions
High melting point (801 °C)	Strong forces of attraction between the oppositely charged ions
Dissolves in water	The water is also able to form strong electrostatic attractions with the ions – the ions are 'plucked' off the lattice structure
Does not conduct electricity when solid	Strong forces between the ions prevent them from moving
Conducts electricity when molten or dissolved in water	The strong forces between the ions have been broken down and so the ions are able to move

△ Table 1.7 Properties of sodium chloride.

Magnesium oxide is another ionic compound. Its ionic formula is $Mg^{2+}O^{2-}$.

MgO has a much higher melting point and boiling point than NaCl because of the increased charges on the ions. The forces holding the ions together are stronger in MgO than in NaCl.

END OF EXTENDED

SCIENCE IN CONTEXT

MAGNESIUM OXIDE

Magnesium oxide is a very versatile compound. It is used extensively in the construction industry, both in making cement and in making fire-proof construction materials. The fact that it has a melting point of over 2800 °C makes it ideal for this use.

▷ Fig. 1.43 The heat resistance of magnesium oxide means that it is used to line furnaces.

IONIC CRYSTALS

△ Fig. 1.44 Gemstones are examples of ionic crystals.

All ionic compounds form giant structures, and all have relatively high melting and boiling points. The charges on the ions determine the strength of the electrostatic attraction between the ions, and hence the melting and boiling points of the compound compared to others.

Another factor that affects the strength of the electrostatic attraction is the relative sizes of the positive and negative ions and how well they are able to pack together. The overall arrangement of the ions is determined by attractive forces between oppositely charged ions and repulsive forces between similarly charged ions. In sodium chloride, for example, six chloride ions fit around one sodium ion without the chloride ions getting too close together and repelling one another. Similarly, six sodium ions can fit around one chloride ion. This structure is sometimes called a 6:6 lattice (see Fig. 1.41 for a diagram of this structure).

Caesium is a metal in the same group of the Periodic Table as sodium, but caesium ions are much bigger than sodium ions. In the structure of caesium chloride, eight chloride ions can fit around each caesium ion. So, although sodium and caesium are in the same group, their chlorides have different structures.

Some of the most valuable gemstones are ionic compounds. Rubies and sapphires, for example, are both aluminium oxide. The different colours of the gemstones are due to traces of other metals such as iron, titanium and chromium.

QUESTIONS

1. **EXTENDED** Why does an ionic compound such as magnesium oxide not conduct electricity when it is solid?

2. **EXTENDED** Suggest a reason why magnesium oxide has a higher melting point than sodium chloride.

End of topic checklist

Key terms

chemical reaction, ion, ionic bond, ionic compound

During your study of this topic you should have learned:

○ How to describe the formation of ions by electron loss or gain.

○ How to describe the formation of ionic bonds between elements from Groups I and VII.

○ **EXTENDED** How to describe the formation of ionic bonds between metallic and non-metallic elements.

○ **EXTENDED** How to describe the lattice structure of ionic compounds as a regular arrangement of alternating positive and negative ions.

End of topic questions

Note: The marks given for these questions indicate the level of detail required in the answers. In the examination, the number of marks given to questions like these may be different.

1. For each of the following reactions, say whether the compound formed is ionic or not:

    a) hydrogen and chlorine (1 mark)

    b) carbon and hydrogen (1 mark)

    c) sodium and oxygen (1 mark)

    d) chlorine and oxygen (1 mark)

    e) calcium and bromine. (1 mark)

2. Write down the formulae of the ions formed by the following elements:

    a) potassium (1 mark)

    b) aluminium (1 mark)

    c) sulfur (1 mark)

    d) fluorine. (1 mark)

3. The table below shows the electronic arrangement of three atoms, X, Y and Z. Copy and complete the table to show the electronic arrangements and charges of the ions these atoms will form. **(3 marks)**

Atom	Electronic arrangement of the atom	Electronic arrangement of the ion	Charge on the ion
X	2,6		
Y	2,8,8,2		
Z	2,1		

4. EXTENDED Draw dot-and-cross diagrams to show how the following atoms combine to form ionic compounds. (You must show the electronic arrangements of the starting atoms and the finishing ions.)

    a) potassium and oxygen (proton numbers: K 19; O 8) (2 marks)

    b) magnesium and chlorine (proton numbers: Mg 12; Cl 17) (2 marks)

5. EXTENDED Explain why an ionic substance such as potassium chloride:

    a) has a high melting point (2 marks)

    b) can conduct electricity. (2 marks)

6. EXTENDED Explain why magnesium oxide has a higher melting point and boiling point than sodium chloride. (2 marks)

Molecules and covalent bonds

△ Fig. 1.45 All plastics are covalent substances.

INTRODUCTION

Unlike ionic compounds, covalent substances are formed when atoms of non-metals combine. Although covalent substances all contain the same type of bond, their properties can be quite different – some are gases, others are very hard solids with high melting points. Plastics are a common type of covalent substance. Because chemists now understand how the molecules form and link together, they can produce plastics with almost the perfect properties for a particular use, from soft and flexible (as in contact lenses) to hard and rigid (as in electrical sockets).

HOW COVALENT BONDS ARE FORMED

Covalent bonding involves electron sharing and occurs between atoms of non-metals. It results in the formation of a **molecule**.

The non-metal atoms try to achieve complete outer electron shells or the electron arrangement of the nearest noble gas by sharing electrons.

A single **covalent bond** is formed when two atoms each contribute one electron to a shared pair of electrons. For example, hydrogen gas exists as H_2 molecules. Each hydrogen atom needs to fill its electron shell. They can do this by sharing electrons.

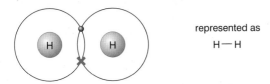

represented as

H—H

△ Fig. 1.46 The dot-and-cross diagram and displayed formula of H_2.

A covalent bond is the result of attraction between the bonding pair of electrons (negative charges) and the nuclei (positive charges) of the atoms involved in the bond. A single covalent bond can be represented by a single line. The formula of a hydrogen molecule can be written as a displayed formula, H—H. The hydrogen atoms and oxygen atoms in water are also held together by single covalent bonds.

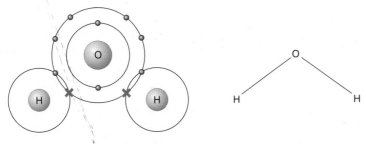

△ Fig. 1.47 Water contains single covalent bonds.

The hydrogen and carbon atoms in methane are held together by single covalent bonds.

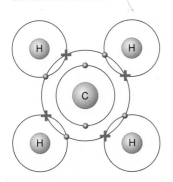

△ Fig. 1.48 Methane contains four single covalent bonds.

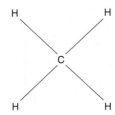

△ Fig. 1.49 The displayed formula for methane.

The hydrogen chloride molecule, HCl, is also held together by a single covalent bond.

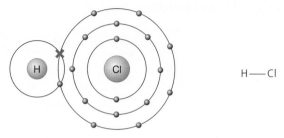

H——Cl

△ Fig. 1.50 Hydrogen chloride has a single covalent bond.

Ethane has a slightly more complex electron arrangement.

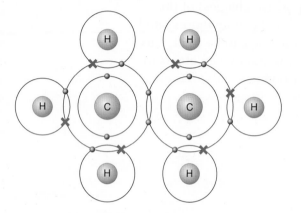

△ Fig. 1.51 Covalent bonds in ethane.

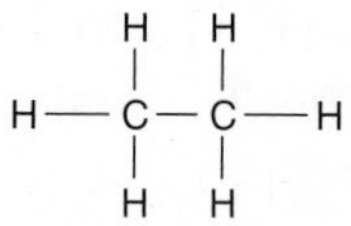

H   H
| |
H—C—C—H
| |
H   H

△ Fig. 1.52 Displayed formula for ethane.

EXTENDED

The alcohol methanol is covalently bonded as shown in Fig. 1.53.

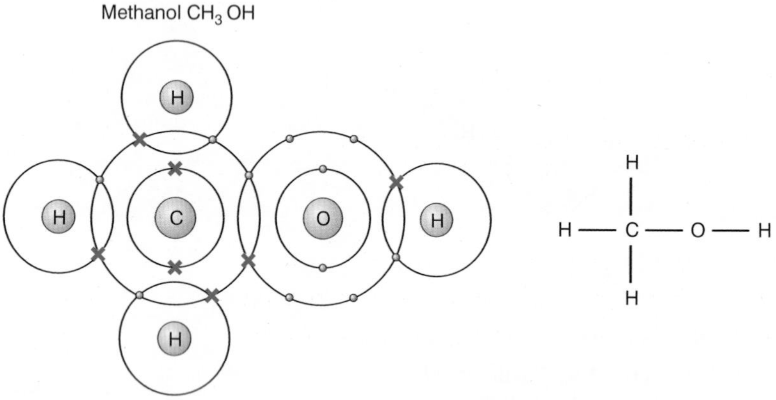

Methanol CH_3OH

       H
       |
H —— C —— O —— H
       |
       H

△ Fig. 1.53 Covalent bonds in methanol.

Some molecules contain double covalent bonds. In carbon dioxide, the carbon atom has an electron arrangement of 2,4 and needs an additional four electrons to complete its outer electron shell. It needs to share its four electrons with four electrons from oxygen atoms (electron arrangement 2,6). So two oxygen atoms are needed, each sharing two electrons with the carbon atom.

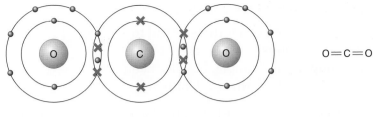

$O=C=O$

△ Fig. 1.54 Carbon dioxide contains double bonds.

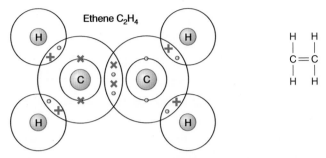

Ethene C₂H₄

△ Fig. 1.55 Ethene contains a double bond.

Some molecules contain triple covalent bonds. In the nitrogen molecule, each nitrogen atom has an electron arrangement of 2,5 and needs an additional three electrons to complete its outer electron shell. It needs to share three of its outer electrons with another nitrogen atom. This forms a triple bond, which is shown as N≡N.

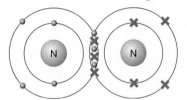

◁ Fig. 1.56 A nitrogen molecule contains a triple bond.

END OF EXTENDED

QUESTIONS

1. Draw a dot-and-cross diagram and displayed formula to show how the covalent bonds are formed in chlorine gas (Cl_2). The proton number of chlorine is 17.

2. **EXTENDED** Draw a dot-and-cross diagram and displayed formula to show how the covalent bonds are formed in the gas ammonia (NH_3). The proton number of hydrogen is 1; the proton number of nitrogen is 7.

3. **EXTENDED** Draw a dot-and-cross diagram and displayed formula to show the double bond in an oxygen molecule (O_2). The proton number of oxygen is 8.

4. **EXTENDED** Draw a dot-and-cross diagram and displayed formula to show the covalent bonds in ethene (C_2H_4). The proton number of hydrogen is 1; the proton number of carbon is 6.

5. **EXTENDED** Draw a dot-and-cross diagram and displayed formula to show the covalent bonds in hydrazine (N_2H_4). The proton number of hydrogen is 1; the proton number of nitrogen is 7.

HOW MANY COVALENT BONDS CAN AN ELEMENT FORM?

The number of covalent bonds a non-metal atom can form is linked to its position in the Periodic Table. Metals (Groups I, II, III) do not form covalent bonds. The noble gases in Group 0, for example, helium, neon and argon, are unreactive and also do not usually form covalent bonds.

Group in the Periodic Table	I	II	III	IV	V	VI	VII	VIII or 0
Covalent bonds formed	–	–	–	4	3	2	1	–

△ Table 1.8 Group number and number of covalent bonds formed.

MOLECULAR CRYSTALS

Covalent compounds can form simple molecular crystals. Many covalent crystals exist only in the solid form at low temperatures. Some simple molecular crystals are ice, solid carbon dioxide and iodine.

EXTENDED

PROPERTIES OF COVALENT COMPOUNDS

Substances with molecular structures are usually gases, liquids or solids with low melting points and boiling points.

Covalent bonds are strong bonds. They are **intramolecular bonds** – formed *within* each molecule. Much weaker **intermolecular forces** attract the individual molecules to each other.

The properties of covalent compounds can be explained using a simple model involving these two types of bond or forces.

Properties of hydrogen	Explanation in terms of structure
Hydrogen is a gas with a very low melting point (−259 °C).	The intermolecular forces of attraction between the molecules are weak.
Hydrogen does not conduct electricity.	There are no ions or free electrons present. The covalent bond (intramolecular bond) is a strong bond and the electrons cannot be removed from it easily.

△ Table 1.9 Properties of hydrogen.

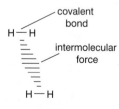

△ Fig. 1.57 Force in and between hydrogen molecules.

END OF EXTENDED

COMPARING THE PROPERTIES OF COVALENT AND IONIC COMPOUNDS

Simple covalent compounds typically have very different properties to ionic compounds. A comparison can be seen in Table 1.10. The volatility of a compound is a measure of how easily it forms a vapour. Compounds with low melting and boiling points are often described as being **volatile.**

Property	Ionic compounds	Simple covalent compounds
Volatility	Non-volatile (high melting and boiling points)	Volatile (low melting and boiling points)
Solubility	Often soluble in water	Mostly insoluble in water
Electrical conductivity	Conduct electricity only when dissolved in water or molten (the ions separate and are free to move, carrying their electric charge)	Low electrical conductivity – are non-electrolytes (do not contain ions and so cannot carry an electrical current; however, some covalent compounds do form ions when dissolved in water)

△ Table 1.10 Comparison of simple covalent compounds with ionic compounds.

EXTENDED

Type of compound	Intermolecular force	Property
Ionic	Strong	High melting and boiling points
Simple covalent	Weak	Low melting and boiling points

END OF EXTENDED

QUESTIONS

1. **EXTENDED** Why does a covalently bonded compound such as carbon dioxide have a relatively low melting point?

2. Would you expect a covalently bonded compound such as ethanol to conduct electricity? Explain your answer.

End of topic checklist

Key terms

covalent bond, molecule, intermolecular force, intramolecular bond, volatile

During your study of this topic you should have learned:

○ How to describe the formation of single covalent bonds in NH_3, H_2, Cl_2, H_2O, CH_4 and HCl as the sharing of pairs of electrons leading to the noble gas configuration.

○ How to describe the differences in volatility, solubility and electrical conductivity between ionic and covalent compounds.

○ EXTENDED How to describe the electron arrangements in more complex covalent molecules such as N_2, C_2H_4, CH_3OH and CO_2.

○ EXTENDED How to explain the difference in melting point and boiling point of ionic and covalent compounds in terms of intermolecular forces.

End of topic questions

Note: The marks given for these questions indicate the level of detail required in the answers. In the examination, the number of marks given to questions like these may be different.

1. Draw dot-and-cross diagrams to show the bonding in the following compounds:

   a) hydrogen fluoride, HF   **(2 marks)**

   b) carbon disulfide, CS_2 **(2 marks)**

   c) ethanol, C_2H_5OH.   **(2 marks)**

2. Candle wax is a covalently bonded compound. Explain why candle wax has a relatively low melting point.   **(2 marks)**

3. Ozone (O_3) is a gas found in the Earth's atmosphere. How do you know that ozone is covalently bonded and not ionically bonded?   **(2 marks)**

4. EXTENDED Explain why methane (CH_4), which has strong covalent bonds between the carbon atom and the hydrogen atoms, is a gas at room temperature and pressure, and has a very low melting point.   **(2 marks)**

5. EXTENDED Substance X has a simple molecular structure.

   a) In which state(s) of matter might you expect it to exist in at room temperature and pressure? Explain your answer.   **(2 marks)**

   b) How would you expect the boiling point of X to compare with the boiling point of an ionic compound such as sodium chloride? Explain your answer.   **(2 marks)**

Stoichiometry

△ Fig. 1.58 When this reaction is described as $S(s) + O_2(g) \rightarrow SO_2(g)$, it is understood by chemists all over the world.

INTRODUCTION

Stoichiometry is the branch of chemistry concerned with the relative quantities of reactants and products in a chemical reaction. A study of stoichiometry depends on balanced chemical equations which, in turn, depend on knowledge of the chemical symbols for the elements and the formulae of chemical compounds. This topic starts by considering how simple chemical formulae are written and then looks in detail at chemical equations. The topic then focuses on how chemical equations can be used to work out how much reactant is needed to make a certain amount of product.

KNOWLEDGE CHECK

✓ Know that elements are made up of atoms.
✓ Know that compounds are formed when atoms combine together.
✓ Know that molecules are formed in covalent bonding and that ions are formed in ionic bonding.

LEARNING OBJECTIVES

✓ Be able to use the symbols of the elements and write the formulae of simple compounds.
✓ Be able to deduce the formula of a simple compound from the relative number of atoms present.
✓ Be able to deduce the formula of a simple compound from a model or a diagrammatic representation.
✓ Be able to construct word equations.
✓ Be able to interpret and balance simple symbol equations.
✓ EXTENDED Be able to determine the formula of an ionic compound from the charges on the ions present.
✓ EXTENDED Be able to construct and use symbol equations with state symbols, including ionic equations.

HOW ARE CHEMICAL FORMULAE WRITTEN?

When elements chemically combine, they form compounds.
A compound can be represented by a **chemical formula**.

All substances are made up from simple building blocks called elements. Each element has a unique **chemical symbol**, containing one or two letters. Elements discovered a long time ago often have symbols that don't seem to match their name. For example, silver has the chemical symbol Ag. This is derived from *argentum*, the Latin name for silver.

'COMBINING POWERS' OF ELEMENTS

There are a number of ways of working out chemical formulae. In this topic you will start with the idea of a 'combining power' for each element and then look at how the charges on ions can be used for ionic compounds. Later in the course you will be introduced to oxidation states and how these can be used to work out chemical formulae.

There is a simple relationship between an element's *group number* in the Periodic Table and its combining power. Groups are the vertical columns in the Periodic Table. The combining power is linked to the *number of electrons* in the outer shell of atoms of the element.

Group number	I	II	III	IV	V	VI	VII	VIII or 0
Combining power	1	2	3	4	3	2	1	0

Δ Table 1.11 Combining powers of elements.

Groups I–IV: combining power = group number

Groups V–VII: combining power = 8 – (group number)

If an element is not in one of the main groups, its combining power is included in the name of the compound containing it. For example, copper is a transition metal and is in the middle block of the Periodic Table. In copper(II) oxide, copper has a combining power of 2.

Sometimes an element does not have the combining power you would predict from its position in the Periodic Table. The combining power of these elements is also included in the name of the compound containing it. For example, phosphorus is in Group V, so you would expect it to have a combining power of 3, but in phosphorus(V) oxide its combining power is 5.

The only exception is hydrogen. Hydrogen is not included in a group, nor is its combining power given in the name of compounds containing hydrogen. It has a combining power of 1.

SIMPLE COMPOUNDS

Many compounds contain just two elements. For example, when magnesium burns in oxygen, a white ash of magnesium oxide is formed. To work out the chemical formula of magnesium oxide:

1. Write down the name of the compound.

2. Write down the chemical symbols for the elements in the compound.

3. Use the Periodic Table to find the 'combining power' of each element. Write the combining power of each element under its symbol.

4. If the numbers can be cancelled down, do so.

5. Swap the combining powers. Write them after the symbol, slightly below the line (as a 'subscript').

6. If any of the numbers are 1, you do not need to write them.

Magnesium oxide has the chemical formula you would have probably guessed: MgO.

The chemical formula of a compound is not always immediately obvious, but if you follow these rules you will have no problems.

Compounds containing more than two elements

Some elements exist bonded together in what is called a **radical**. For example, in copper(II) sulfate, the sulfate part of the compound is a radical.

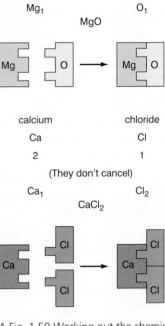

△ Fig. 1.59 Working out the chemical formulae for magnesium oxide and calcium chloride.

There are a number of common radicals, each having its own combining power. You cannot work out these combining powers easily from the Periodic Table – you have to learn them. Notice that all the radicals exist as ions.

Combining power = 1	Combining power = 2	Combining power = 3
Hydroxide OH^-	Carbonate CO_3^{2-}	Phosphate PO_4^{3-}
Hydrogencarbonate HCO_3^-	Sulfate SO_4^{2-}	
Nitrate NO_3^-		
Ammonium NH_4^+		

△ Table 1.12 Combining compounds for common radicals.

The same rules for working out formulae apply to radicals as to elements. For example:

Copper(II) sulfate		Potassium nitrate	
Cu	SO_4	K	NO_3
2	2	1	1
$CuSO_4$		KNO_3	

△ Table 1.13 Combining elements and radicals.

If the formula contains more than one radical unit, the radical must be put in brackets. For example:

Calcium hydroxide	
Ca	OH
2	1
$Ca(OH)_2$	

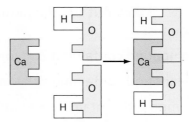

△ Fig. 1.60 Working out the chemical formula for calcium hydroxide.

The brackets are used just as they are used in mathematics: the number outside a bracket multiplies everything inside it. Be careful how you use the brackets, for example, do not be tempted to write calcium hydroxide as $CaOH_2$ rather than $Ca(OH)_2$. This is wrong.

$CaOH_2$ contains one Ca, one O, two H ✗

$Ca(OH)_2$ contains one Ca, two O, two H ✓

EXTENDED

The formula of an ionic compound can be worked out from the ions present. For example, sodium chloride is an ionic compound.

Sodium is in Group I and forms an ion with a charge of 1+, Na^+.

Chlorine is in Group VII and forms an ion with a charge of 1−, Cl^-.

When these ions combine, the charges must cancel each other out:

NaCl (the 1+ and 1− charges cancel)

What is the formula of lead(II) bromide, which contains Pb^{2+} and Br^- ions?

To cancel the 2+ charge, two 1− charges are needed, so the formula is $PbBr_2$.

END OF EXTENDED

QUESTIONS

1. Work out the chemical formulae of the following compounds:

   a) potassium bromide

   b) calcium oxide

   c) aluminium chloride

   d) carbon hydride (methane).

2. Work out the chemical formulae of the following compounds:

   a) copper(II) nitrate

   b) aluminium hydroxide

   c) ammonium sulfate

   d) iron(III) carbonate.

3. **EXTENDED** Work out the chemical formulae of the following compounds:

   a) a compound containing Zn^{2+} ions and Cl^- ions

   b) a compound containing Cr^{3+} ions and O^{2-} ions

   c) a compound containing Fe^{2+} and OH^- ions.

WRITING CHEMICAL EQUATIONS

In a chemical equation the starting chemicals are called the **reactants** and the finishing chemicals are called the **products**.

Follow these rules to write a chemical equation.

1. Write down the word equation.

2. **EXTENDED** Write down the symbols (for elements) and formulae (for compounds).

3. Balance the equation, to make sure there are the same number of each type of atom on each side of the equation.

4. **EXTENDED** Include the **state symbols** solid (s); liquid (l); gas (g); solution in water (aq).

State	State symbol
Solid	(s)
Liquid	(l)
Gas	(g)
Solution	(aq)

△ Table 1.14 States and their symbols.

Remember that some elements are **diatomic**. They exist as molecules containing two atoms.

Element	Formula
Hydrogen	H_2
Oxygen	O_2
Nitrogen	N_2
Chlorine	Cl_2
Bromine	Br_2
Iodine	I_2

△ Table 1.15 Some diatomic elements.

WORKED EXAMPLES

1. When a lighted splint is put into a test tube of hydrogen, the hydrogen burns with a 'pop'. In fact the hydrogen reacts with oxygen in the air (the reactants) to form water (the product). Write the chemical equation for this reaction.

Word equation: hydrogen + oxygen → water

EXTENDED

Symbols and formulae: H_2 + O_2 → H_2O

END OF EXTENDED

Balance the equation: $2H_2$ + O_2 $\rightarrow$ $2H_2O$

For every two molecules of hydrogen that react, one molecule of oxygen is needed and two molecules of water are formed.

EXTENDED

Add the state symbols: $2H_2(g)$ + $O_2(g)$ $\rightarrow$ $2H_2O(l)$

END OF EXTENDED

2. What is the equation when sulfur burns in air?

Word equation: sulfur + oxygen $\rightarrow$ sulfur dioxide

EXTENDED

Symbols and formulae: S + O_2 $\rightarrow$ SO_2

END OF EXTENDED

Balance the equation: S + O_2 $\rightarrow$ SO_2

EXTENDED

Add the state symbols: $S(s)$ + $O_2(g)$ $\rightarrow$ $SO_2(g)$

END OF EXTENDED

◁ Fig. 1.61 Methane is burning in the oxygen in the air to form carbon dioxide and water.

BALANCING EQUATIONS

Balancing equations can be quite tricky. It is essentially done by trial and error. However, the golden rule is that *balancing numbers can only be put in front of the formulae*.

For example, to balance the equation for the reaction between methane and oxygen:

	Reactants	**Products**
Start with the unbalanced equation	$CH_4 + O_2$	$CO_2 + H_2O$
Count the number of atoms on each side of the equation	1C ✓, 4H, 2O	1C ✓, 2H, 3O
There is a need to increase the number of H atoms on the products side of the equation. Put a '2' in front of the H_2O	$CH_4 + O_2$	$CO_2 + 2H_2O$
Count the number of atoms on each side of the equation again	1C ✓, 4H ✓, 2O	1C ✓, 4H ✓, 4O
There is a need to increase the number of O atoms on the reactant side of the equation. Put a '2' in front of the O_2	$CH_4 + 2O_2$	$CO_2 + 2H_2O$
Count the atoms on each side of the equation again	1C ✓, 4H ✓, 4O ✓	1C ✓, 4H ✓, 4O ✓

△ Table 1.16 Steps in balancing the equation for the reaction between methane and oxygen.

No atoms have been created or destroyed in the reaction. The equation is balanced.

$$CH_4(g) \ + \ 2O_2(g) \ \rightarrow \ CO_2(g) \ + \ 2H_2O(l)$$

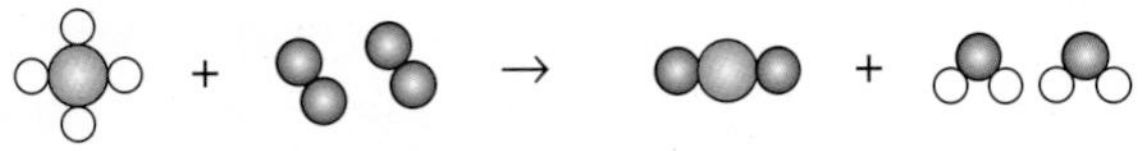

△ Fig. 1.62 The number of each type of atom is the same on the left and right sides of the equation.

In balancing equations involving radicals such as sulfate, hydroxide and nitrate, you can use the same procedure. For example, when lead(II) nitrate solution is mixed with potassium iodide solution, lead(II) iodide and potassium nitrate are produced (Fig. 1.63).

1. Words:

lead(II) nitrate	+	potassium iodide	→	lead(II) iodide	+	potassium nitrate

2. Symbols:

$$Pb(NO_3)_2 \quad + \quad KI \quad \rightarrow \quad PbI_2 \quad + \quad KNO_3$$

3. Balance the nitrates:

$$Pb(NO_3)_2 \quad + \quad KI \quad \rightarrow \quad PbI_2 \quad + \quad 2KNO_3$$

4. Balance the iodides:

$$Pb(NO_3)_2(aq) \quad + \quad 2KI(aq) \quad \rightarrow PbI_2(s) \quad + \quad 2KNO_3(aq)$$

△ Fig. 1.63 This reaction occurs simply on mixing the solutions of lead(II) nitrate and potassium iodide. Lead iodide is an insoluble yellow solid.

QUESTIONS

1. Balance the following chemical equations:

a) $Ca(s) + O_2(g) \rightarrow CaO(s)$

b) $H_2S(g) + O_2(g) \rightarrow SO_2(g) + H_2O(l)$

c) $Pb(NO_3)_2(s) \rightarrow PbO(s) + NO_2(g) + O_2(g)$

2. **EXTENDED** Write balanced equations for the following word equations:

a) sulfur + oxygen → sulfur dioxide

b) magnesium + oxygen → magnesium oxide

c) copper(II) oxide + hydrogen → copper + water

As mentioned earlier, the general method for balancing equations is by trial and error, but it helps if you are systematic – always start on the left-hand side with the reactants. Sometimes you can balance an equation using fractions. In more advanced study, such balanced equations are perfectly acceptable. Getting rid of the fractions is not difficult though. Look at this example:

WORKED EXAMPLE

Ethane (C_2H_6) is a hydrocarbon fuel and burns in air to form carbon dioxide and water.

Unbalanced equation: $C_2H_6(g) + O_2(g) \rightarrow CO_2(g) + H_2O(l)$

Balancing the carbon and hydrogen atoms gives:

$$C_2H_6(g) \quad + \quad O_2(g) \quad \rightarrow \quad 2CO_2(g) \quad + \quad 3H_2O(l)$$

The equation can then be balanced by putting 3½ in front of the O_2. By doubling every balancing number, the equation is then balanced using whole numbers.

$$2C_2H_6(g) \quad + \quad 7O_2(g) \quad \rightarrow \quad 4CO_2(g) \quad + \quad 6H_2O(l)$$

SCIENCE LINK — BIOLOGY – BIOLOGICAL MOLECULES, CHARACTERISTICS OF LIVING THINGS

- Being able to describe chemical changes using balanced equations allows us to check that all the starting chemicals (the reactants) have been accounted for after the change (forming the products) – this applies to all the chemical reactions in living things, such as respiration and photosynthesis.

- The energy transfer through a series of chemical changes can also be tracked.

PHYSICS – CONSERVATION LAWS

- The idea of balancing equations – that all the 'starting' particles must be accounted for – is an example of a conservation law.

During your course you will become familiar with balancing equations and become much quicker at doing it. Try balancing the equations below. The third is the chemical reaction often used for making chlorine gas in the laboratory.

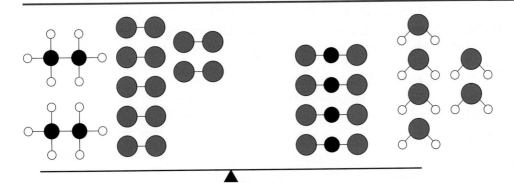

△ Fig. 1.64 Balancing the equation for burning ethane in air.

QUESTION

1. Balance the following equations:

   a) $C_5H_{10}(g) + O_2(g) \rightarrow CO_2(g) + H_2O(l)$

   b) $Fe_2O_3(s) + CO(g) \rightarrow Fe(s) + CO_2(g)$

   c) $KMnO_4(s) + HCl(aq) \rightarrow KCl(aq) + MnCl_2(aq) + H_2O(l) + Cl_2(g)$

Ionic equations

Ionic equations show reactions involving ions (atoms or radicals that have lost or gained electrons). The size of the charge on an ion is the same as its combining power – whether it is positive or negative depends on which part of the Periodic Table the element is placed in.

In many ionic reactions some of the ions play no part in the reaction. These ions are called **spectator ions**. A simplified ionic equation can then be written, using only the important, reacting ions. In these equations, state symbols are often used and appear in brackets.

The equation must balance in terms of chemical symbols and charges.

WORKED EXAMPLES

1. In the reaction to produce lead(II) iodide, the potassium ions and nitrate ions are spectators – the important ions are the lead(II) ions and the iodide ions.

   The simplified ionic equation is:

   $Pb^{2+}(aq) + 2I^-(aq) \rightarrow PbI_2(s)$

	Reactants	Products
	$Pb^{2+}(aq) + 2I^-(aq)$	$PbI_2(s)$
Symbols	1Pb ✓, 2I ✓	1Pb ✓, 2I ✓
Charges	2+ and 2− = 0 ✓	0 ✓

The equation shows that *any* solution containing lead(II) ions will react with *any* solution containing iodide ions to form lead(II) iodide.

2. Any solution containing copper(II) ions and any solution containing hydroxide ions can be used to make copper(II) hydroxide, which appears as a solid:

$$Cu^{2+}(aq) + 2OH^-(aq) \rightarrow Cu(OH)_2(s)$$

	Reactants	Products
	$Cu^{2+}(aq) + 2OH^-(aq)$	$Cu(OH)_2(s)$
Symbols	1Cu ✓, 2O ✓, 2H ✓	1Cu ✓, 2O ✓, 2H ✓
Charges	2+ and 2− = 0 ✓	0 ✓

△ Fig. 1.65 Copper(II) hydroxide.

END OF EXTENDED

End of topic checklist

Key terms

chemical formula, chemical symbol, diatomic, ionic equation, product, radical, reactant, spectator ions, state symbols

During your study of this topic you should have learned:

○ How to use the symbols of the elements to write the formulae of simple compounds.

○ How to deduce the formula of a simple compound from the numbers of atoms present.

○ How to deduce the formula of a simple compound from a model or a diagrammatic representation.

○ How to construct and use word equations.

○ How to interpret and balance simple symbol equations.

○ EXTENDED How to determine the formula of an ionic compound from the charges on the ions present.

○ EXTENDED How to construct and use symbol equations with state symbols, including ionic equations.

End of topic questions

Note: The marks given for these questions indicate the level of detail required in the answers. In the examination, the number of marks given to questions like these may be different.

1. Work out the chemical formulae of the following compounds:

   a) sodium chloride (1 mark)

   b) magnesium fluoride (1 mark)

   c) aluminium nitride (1 mark)

   d) lithium oxide (1 mark)

   e) carbon(IV) oxide (carbon dioxide). (1 mark)

2. Work out the chemical formulae of the following compounds:

   a) iron(III) oxide (1 mark)

   b) phosphorus(V) chloride (1 mark)

   c) chromium(III) bromide (1 mark)

   d) sulfur(VI) oxide (sulfur trioxide) (1 mark)

   e) sulfur(IV) oxide (sulfur dioxide). (1 mark)

3. Work out the chemical formulae of the following compounds:

   a) potassium carbonate (1 mark)

   b) ammonium chloride (1 mark)

   c) sulfuric acid (1 mark)

   d) magnesium hydroxide (1 mark)

   e) ammonium sulfate. (1 mark)

4. EXTENDED Write symbol equations from the following word equations:

   a) carbon + oxygen → carbon dioxide (1 mark)

   b) iron + oxygen → iron(III) oxide (1 mark)

   c) iron(III) oxide + carbon → iron + carbon dioxide (1 mark)

   d) calcium carbonate + hydrochloric acid → calcium chloride + carbon dioxide + water. (1 mark)

End of topic questions continued

5. EXTENDED Write ionic equations for the following reactions:

   a) calcium ions and carbonate ions form calcium carbonate   **(2 marks)**

   b) iron(III) ions and hydroxide ions form iron(III) hydroxide   **(2 marks)**

   c) silver(I) ions and bromide ions form silver(I) bromide   **(2 marks)**

Modern physical chemistry originated in the nineteenth century. It is not as clearly defined a category as organic chemistry, but it is still a useful description of this branch of science. Physical chemistry focuses on chemical processes at the 'macro level' (where properties can be observed) more than at the 'micro level' (too small to see) of individual atoms, molecules and ions. However, observed physical properties can still be explained in terms of what the atoms, molecules or ions are doing.

In this section you will explore the chemical reactions that can be caused by using electricity, a process known as electrolysis. You will then investigate some chemical reactions that produce significant amounts of heat energy, as well as some strange ones that seem to absorb energy and make everything cooler. The speed or rate of chemical reactions will also be explored, together with chemists' strategies to try to control them. You will learn about redox reactions, which are reactions involving reduction and oxidation, as well as learning about acids, bases and salts. Finally, you will look at some of the simple analytical techniques that can be used to identify ions and gases.

STARTING POINTS

1. How many non-renewable fuels can you name? What products do they form when they burn?

2. Give an example of a very rapid, almost instantaneous, chemical reaction. Now give an example of a very slow one.

3. Explain how you can easily distinguish between an acid and an alkali.

4. What is a catalyst? Name two examples where catalysts are used in everyday life.

5. Acids react with alkalis in neutralisation reactions. What is meant by neutralisation?

SECTION CONTENTS

a) Electricity and chemistry

b) Chemical energetics

c) Rate of reaction

d) Redox reactions

e) Acids, bases and salts

f) Identification of ions and gases

2
Physical chemistry

△ Physical chemistry deals with properties that can be observed.

Electricity and chemistry

△ Fig. 2.1 Industrial electroplating is a form of electrolysis.

INTRODUCTION

Most elements in nature are found combined with other elements as compounds. These compounds must be broken down to obtain the elements that they contain. One of the most efficient and economical ways to break down some compounds is by using electricity in a process called electrolysis. Simple electrolysis experiments can be performed in the laboratory, and electrolysis is also used in large-scale industrial processes to produce important chemicals such as aluminium and chlorine.

This topic deals with the underlying principles of electrolysis as well as some of the experiments that can be performed in the laboratory.

KNOWLEDGE CHECK

✓ Know the different arrangements of the particles in solids, liquids and gases.
✓ Understand the terms 'conductor' and 'insulator'.
✓ Understand the differences between ionic and covalent bonding.

LEARNING OBJECTIVES

✓ Be able to define electrolysis as the breaking down of an ionic compound, either molten or in aqueous solution, by the passage of electricity.
✓ Be able to use the terms *inert electrode*, *electrolyte*, *anode* and *cathode*.
✓ Be able to describe the electrode products and the observations made in the electrolysis of:
  – molten lead(II) bromide
  – concentrated aqueous sodium chloride
  – dilute sulfuric acid
  between inert electrodes (platinum or carbon).
✓ EXTENDED Be able to describe electrolysis in terms of the ions present and the reactions at the electrodes, in terms of gain of electrons by cations and loss of electrons by anions to form atoms.
✓ EXTENDED Be able to predict the products of the electrolysis of a binary (two-element) compound in the molten state.
✓ EXTENDED Be able to describe electrolysis in terms of the ions present and reactions at the electrodes in specific examples.

ELECTROLYTES AND NON-ELECTROLYTES

Compounds that can conduct electricity are called **electrolytes** – they undergo a reaction called **electrolysis**. Experiments can be carried out using a simple electrical cell, as shown in Fig. 2.2.

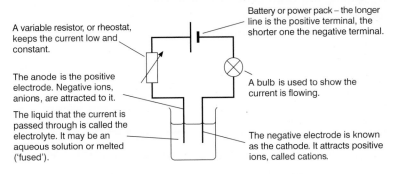

A variable resistor, or rheostat, keeps the current low and constant.

Battery or power pack – the longer line is the positive terminal, the shorter one the negative terminal.

The anode is the positive electrode. Negative ions, anions, are attracted to it.

A bulb is used to show the current is flowing.

The liquid that the current is passed through is called the electrolyte. It may be an aqueous solution or melted ('fused').

The negative electrode is known as the cathode. It attracts positive ions, called cations.

△ Fig. 2.2 A simple electrolysis cell.

When the solution in the beaker is an electrolyte, a complete circuit will form and the bulb will light. The electric current that flows is caused by electrons moving in the electrodes and wires of the circuit, and by ions moving in the solution. If a current does not flow, then the beaker must contain a non-electrolyte. Because of this, a simple circuit like this can be used to distinguish between electrolytes and non-electrolytes.

CONDITIONS FOR ELECTROLYSIS

The substance being electrolysed (the electrolyte) must contain ions, and these ions must be free to move. In other words, the substance must either be molten or dissolved in water. In electrolysis an inert electrode is used. The word 'inert' means 'unreactive' so the electrode will not react with the electrolyte.

A direct current (d.c.) voltage must be used. The **electrode** connected to the positive terminal of the power supply is known as the **anode**. The electrode connected to the negative terminal is known as the **cathode**. The electrical circuit can be drawn as shown in Fig. 2.3.

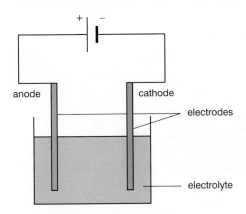

anode
cathode
electrodes
electrolyte

△ Fig. 2.3 A typical electrical circuit used in electrolysis.

During electrolysis electrical charge is transferred as follows:
- negative ions move to the anode and give up electrons
- the electrons travel through the anode and round the circuit in the connecting wires to the cathode
- the electrons reaching the cathode are taken up by positive ions.

QUESTIONS

1. What is meant by the term *electrolysis*?

2. What is the name given to the positive electrode?

3. What two conditions must exist for a substance to be an electrolyte and allow an electric current to pass through it?

SCIENCE LINK **PHYSICS – ELECTRIC CIRCUITS**

- The rules for electric circuits – there must be a complete circuit, there must be an energy source, there must be mobile charge carriers ('charged particles') – apply both to circuits involving electrolysis and to circuits with bulbs and batteries.

- The charge carriers may be different – ions in molten materials or in solutions, compared with electrons in wires – but the measurements of electric current and potential difference are defined in exactly the same way.

- Ion formation through the gain or loss of electrons is another idea that is common to both areas.

ELECTROLYSIS OF MOLTEN LEAD(II) BROMIDE

When an electric current passes through an electrolyte, new substances are formed. The examples below show how you can work out what products will form.

Lead(II) bromide ($PbBr_2$) is ionically bonded and contains Pb^{2+} ions and Br^- ions. When the solid is melted and a voltage is applied, the ions are able to move. The positive lead ions move to the negative electrode (the cathode), and the negative bromide ions move to the positive electrode (the anode). The electrodes are usually made of carbon, which is inert. This means they do not undergo any **chemical change** during the electrolysis. The products of the electrolysis are lead and bromine. Silvery deposits of lead form near the bottom of the dish, and brown bromine vapour near the anode.

At the cathode (negative electrode), the lead ions accept electrons to form lead atoms:

$$Pb^{2+}(l) + 2e^- \rightarrow Pb(l)$$

At the anode (positive electrode), the bromide ions give up electrons to form bromine atoms, and then bromine molecules:

$$2Br^-(l) \rightarrow Br_2(g) + 2e^-$$

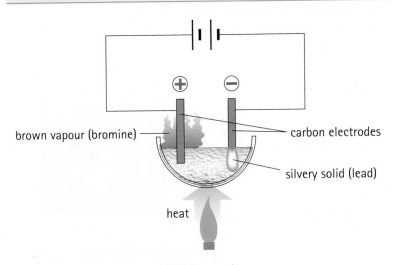

△ Fig. 2.4 Electrolysis of molten lead(II) bromide.

Note: the two equations above are known as half-equations. Unlike normal chemical equations, they do not show the whole chemical change – just the change occurring at an electrode. In the half-equations above, you will see that the numbers of electrons accepted and released are the same. The electric current is produced by this flow of electrons around the external circuit.

ELECTROLYSIS OF SODIUM CHLORIDE SOLUTION

When concentrated sodium chloride solution is electrolysed, hydrogen ions (from the water solvent) form hydrogen molecules at the cathode and chloride ions form chlorine molecules at the anode.

This experiment can be performed using a cell as shown in Fig. 2.2. Again, inert carbon electrodes are used.

When the ionic compound sodium chloride dissolves in water, the sodium and chloride ions separate and are free to move independently. In addition, the water provides a small quantity of hydrogen (H^+) and hydroxide (OH^-) ions:

$$NaCl(aq) \rightarrow Na^+(aq) + Cl^-(aq)$$

$$H_2O(l) \rightarrow H^+(aq) + OH^-(aq)$$

This process is known as **dissociation**. The water breaks up and forms ions. In fact, the ions also combine to form water – the reaction goes both ways: it is a **reversible reaction**. Although there are very few ions present, if they are removed they will be immediately replaced. Therefore, whenever you consider the electrolysis of an aqueous solution you must always include the H^+ and OH^- ions.

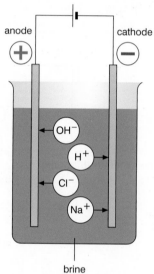

$\triangle$ Fig. 2.5 The electrolysis of sodium chloride solution (brine).

- At the cathode (negative electrode):

two ions, Na^+ and H^+, move to the cathode but only H^+ ions are discharged. The sodium ions remain as ions, but the solution turns alkaline because the loss of hydrogen ions leaves a surplus of hydroxide ions.

$$2H^+(aq) + 2e^- \rightarrow H_2(g)$$

The hydrogen ions accept electrons and form hydrogen molecules.

- At the anode (positive electrode):

two ions, Cl^- and OH^-, move to the anode. Either ion could be discharged depending on the concentration of the solution. If the solution is very dilute, OH^- ions are discharged; if the solution is concentrated, Cl^- ions are discharged.

$$2Cl^-(aq) \rightarrow Cl_2(g) + 2e^-$$

The chloride ions give up electrons and form chlorine molecules.

Bubbling or effervescence is seen at each of the two electrodes, and the products of the electrolysis are hydrogen and oxygen and/or chlorine.

When the sodium chloride solution is concentrated, the main product at the anode is chlorine, which forms as a pale green gas.

When the sodium chloride solution is dilute, the main product at the anode is oxygen, which forms as a colourless gas.

Whatever the concentration of the sodium chloride solution, hydrogen forms as a colourless gas at the cathode.

When dilute sodium chloride solution is electrolysed, the solution becomes increasingly alkaline as sodium hydroxide is formed.

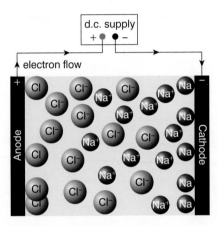

△ Fig. 2.6 Electrolysing molten sodium chloride.

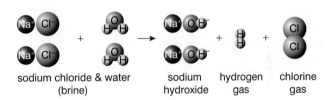

sodium chloride & water (brine) → sodium hydroxide + hydrogen gas + chlorine gas

△ Fig. 2.7 Electrolysing brine. At which electrode is hydrogen formed?

ELECTROLYSIS OF DILUTE SULFURIC ACID

If a d.c. electric current is passed through a solution of dilute sulfuric acid, effervescence will be observed at each inert electrode. The colourless gas oxygen will form at the anode and the colourless gas hydrogen will form at the cathode.

The ions present in dilute sulfuric acid are H^+, OH^- and SO_4^{2-}.

At the cathode, the H^+ ions will be discharged and hydrogen gas will form.

$$2H^+ + 2e^- = H_2$$

At the anode, two ions will be attracted to the anode, SO_4^{2-} and OH^- ions. The OH^- ions will be discharged and oxygen gas will be formed.

$$4OH^- = 2H_2O + O_2 + 4e^-$$

PREDICTING THE PRODUCTS OF ELECTROLYSIS

Predicting the products of the electrolysis of simple molten ionic compounds is relatively straightforward. The metal forms at the cathode and the non-metal forms at the anode. For example, the electrolysis of molten aluminium oxide forms aluminium (at the cathode) and oxygen (at the anode).

REMEMBER

In electrolysis, negative ions give up electrons and usually form molecules (such as Cl_2, Br_2). Positive ions accept electrons and usually form metallic atoms (such as Cu, Al) or hydrogen gas.

END OF EXTENDED

QUESTIONS

1. **a)** What is an *inert* electrode?

   b) Give an example of a substance that is often used as an inert electrode.

2. What products are formed when the following molten solids are electrolysed?

   a) lead(II) chloride

   b) magnesium oxide

   c) aluminium oxide.

End of topic checklist

Key terms

anode, cathode, chemical change, dissociation, electrode, electrolysis, electrolyte, reversible reaction

During your study of this topic you should have learned:

○ That electrolysis is the breaking down of an ionic compound, molten or in aqueous solution, by the passage of electricity.

○ How to use the terms *inert electrode*, *electrolyte*, *anode* and *cathode*.

○ How to describe the electrode products and the observations made in the electrolysis, using inert electrodes of platinum or carbon, of:

- molten lead(II) bromide
- concentrated aqueous sodium chloride
- dilute sulfuric acid.

○ **EXTENDED** How to describe electrolysis in terms of the ions present and reactions at the electrodes, in terms of gain of electrons by cations and loss of electrons by anions to form atoms, in the above examples.

○ How to predict the product of the electrolysis of a binary (two-element) compound in the molten state.

End of topic questions

Note: The marks given for these questions indicate the level of detail required in the answers. In the examination, the number of marks given to questions like these may be different.

1. Explain the following terms:

   a) *electrolysis* (1 mark)

   b) *electrolyte* (1 mark)

   c) *electrode* (1 mark)

   d) *anode* (1 mark)

   e) *cathode.* (1 mark)

2. Zinc bromide, $ZnBr_2$, is an ionic solid. Why does the solid not conduct electricity? (2 marks)

3. Copy and complete the following table, which shows the products formed when molten electrolytes undergo electrolysis. (4 marks)

Electrolyte	Product at the anode	Product at the cathode
Silver bromide		
Lead(II) chloride		
Aluminium oxide		
	Iodine	Magnesium

4. Sodium chloride, NaCl, is ionic. What are the products at the anode and cathode in the electrolysis of molten sodium chloride (2 marks)

Chemical energetics

INTRODUCTION

When chemicals react together, the reactions cause energy changes. This is obvious when a fuel is burned and heat energy is released into the surroundings. Heat changes in other reactions may be less dramatic but they still take place. A knowledge of chemical bonding can really help to understand how these energy changes occur.

△ Fig. 2.8 Fireworks are carefully controlled chemical reactions.

ENERGY CHANGES IN CHEMICAL REACTIONS

In most reactions, energy is transferred to the surroundings and the temperature goes up. These reactions are **exothermic**. Some examples of exothermic reactions are combustion, respiration and neutralisation. In a minority of cases, energy is absorbed from the surroundings as a reaction takes place and the temperature goes down. These reactions are **endothermic**. Some examples of endothermic reactions are photosynthesis and thermal decomposition.

For example, when magnesium ribbon is added to dilute hydrochloric acid, the temperature of the acid increases – the reaction is exothermic. In contrast, when sodium hydrogencarbonate is added to hydrochloric acid, the temperature of the acid decreases – the reaction is endothermic.

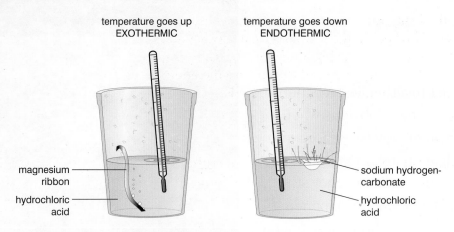

temperature goes up
EXOTHERMIC

temperature goes down
ENDOTHERMIC

magnesium
ribbon

hydrochloric
acid

sodium hydrogen-
carbonate

hydrochloric
acid

△ Fig. 2.9 Measuring energy changes in reactions.

Energy changes in reactions like these can be measured using a polystyrene cup (an insulator) as a calorimeter. If a lid is added to the cup, very little energy is transferred to the air and reasonably accurate results can be obtained.

QUESTIONS

1. What is an *exothermic* reaction?

2. What is an *endothermic* reaction?

3. Why do polystyrene cups make good calorimeters for measuring energy changes in some chemical reactions?

All reactions involving the **combustion** of fuels are exothermic. The energy transferred when a fuel burns can be measured using **calorimetry**, as shown in the diagram.

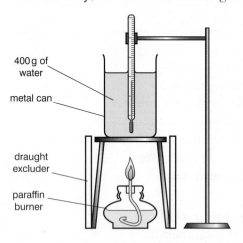

400 g of
water

metal can

draught
excluder

paraffin
burner

◁ Fig. 2.10. Measuring the energy produced by burning a liquid fuel.

The rise in temperature of the water is a measure of the energy transferred to the water. This technique will not give a very accurate answer because much of the energy will be transferred to the surrounding air. Nevertheless, it can be used to compare the energy released by **burning** the same amounts of different fuels.

ENTHALPY CHANGE

The heat energy in chemical reactions is called enthalpy. An **enthalpy change** is given the symbol ΔH. The enthalpy change for a particular reaction is shown at the end of the balanced equation. The units are kJ/mol.

QUESTIONS

1. Measuring energy changes when burning fuels in a liquid burner does not give very accurate results. Why do you think this is?

2. A group of students was comparing the energy released on burning different liquid fuels in spirit burners using apparatus similar to that shown in Fig. 2.10. The results obtained are shown here:

Name of fuel	Amount of fuel burned (g)	Rise in temperature of 200 cm³ of water in a metal can (°C)	Temperature rise of the water per g of fuel burned (°C/g)
Ethanol	1.1	32	
Paraffin	0.9	30	
Pentane	1.5	38	
Octane	0.5	20	

a) How do you think the students worked out how much fuel was burned in each experiment?

b) Complete the last column of the table by working out the temperature rise in each experiment per gram of fuel burned. (Give your answers to the nearest whole number)

  i) Which fuel produced the greatest temperature rise per gram burned?

  ii) If the octane experiment were repeated using 400 cm³ of water in the metal can, approximately what temperature rise would you expect? Explain how you worked out your answer.

3. Another group of students used a glass beaker rather than a metal can in their experiments. Which group of students would you expect to get more accurate results? Explain your answer.

ENERGY PROFILES AND ΔH

Energy level diagrams show the enthalpy difference between the reactants and the products.

In an exothermic reaction, the energy content of the reactants is greater than the energy content of the products. Energy is being lost to the surroundings. ΔH is negative.	In an endothermic reaction, the energy content of the products is greater than the energy content of the reactants. Energy is being absorbed from the surroundings. ΔH is positive.

All ΔH values should have a + or – sign in front of them to show if they are endothermic or exothermic.

Activation energy is the *minimum* amount of energy required for a reaction to occur. Fig. 2.11 shows the activation energy of a reaction.

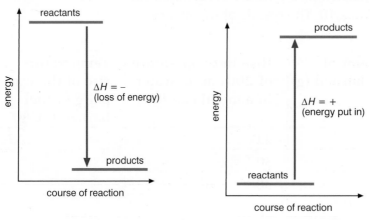

△ Fig. 2.11 Energy level diagrams for exothermic and endothermic reactions.

The energy profile can now be completed as shown. The reaction for this profile is exothermic, with ΔH negative.

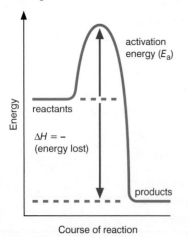

◁ Fig. 2.12 Energy profile.

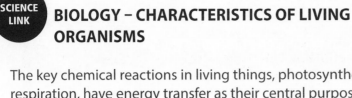

BIOLOGY – CHARACTERISTICS OF LIVING ORGANISMS

- The key chemical reactions in living things, photosynthesis and respiration, have energy transfer as their central purpose.

- In photosynthesis, the energy transfer is to allow the energy to be stored.

- In respiration, energy is released so that it can be used to power further changes.

- The effects of energy transfers (often unwanted) from, for example, combustion, can also have important effects on ecosystems.

PHYSICS – ENERGY RESOURCES

- The storage and transfer of energy play a key part in many physical processes, for example, releasing energy from petrol in order for a car to gain kinetic energy.

- Energy resources on a global scale, for the generating of heat and electricity, is a key feature of modern life.

QUESTIONS

1. The enthalpy change for a particular reaction is positive. Is the reaction endothermic or exothermic?

2. On an energy profile, what is the name given to the minimum amount of energy required for a reaction to occur?

3. EXTENDED Methane burns in excess oxygen to form carbon dioxide and water. When 1 mole of methane is burnt 882 kJ of energy is released. Draw an energy level diagram for this reaction.

4. EXTENDED Draw and label energy level diagrams for each of the reactions shown in the table below:

Reaction	Activation energy (kJ/mol)	Enthalpy change (kJ/mol)
A	120	−90
B	80	+10

Developing investigative skills

Two students used a polystyrene cup to compare the energy changes in two reactions:

- magnesium with hydrochloric acid
- sodium hydrogencarbonate with hydrochloric acid.

They added 50 cm³ of 0.5 M hydrochloric acid to a polystyrene cup and measured its temperature. They then added a known mass of magnesium ribbon to the acid. They stirred the reaction mixture with a glass stirring rod until the reaction was complete and then took the final temperature. They then repeated the procedure using sodium hydrogencarbonate. Their results are shown in the table.

Reaction	Mass of solid used (g)	Initial temperature of the acid (°C)	Final temperature of the acid (°C)	Temperature change (°C)
Magnesium + hydrochloric acid	0.1	22	35	
Sodium hydrogencarbonate + hydrochloric acid	2.5	22	18	

Using and organising techniques, apparatus and materials

❶ What apparatus do you think was used to measure the volumes of hydrochloric acid?

❷ Why did the students stir the reaction mixtures until the reaction was complete and the final temperature was taken?

Observing, measuring and recording

❸ What would you expect to observe when magnesium ribbon is added to dilute hydrochloric acid?

Interpreting observations and data

❹ Work out the temperature change in each reaction.

❺ Work out the temperature change per gram of solid in each reaction.

❻ In each case is the reaction exothermic or endothermic?

❼ Draw a simple energy level diagram to represent the reaction between magnesium and hydrochloric acid.

Evaluating methods

❽ Energy is often lost from the polystyrene cup, making the temperature change lower than it should be. Suggest one way this error could be reduced.

WHERE DOES THE ENERGY COME FROM?

The reaction that occurs when a fuel is burning can be considered to take place in two stages. In the first stage the covalent bonds between the atoms in the fuel molecules and in the oxygen molecules are broken. In the second stage the atoms combine and new covalent bonds are formed. For example, in the combustion of propane:

propane	+	oxygen	→	carbon dioxide	+	water
$C_3H_8(g)$	+	$5O_2(g)$	→	$3CO_2(g)$	+	$4H_2O(l)$

$\Delta H = -2202\,kJ/mol$

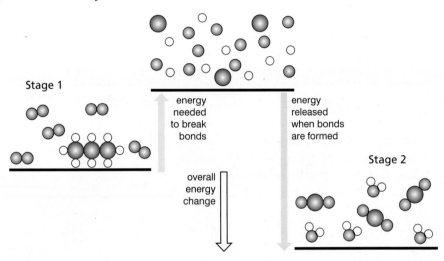

△ Fig. 2.13 Energy changes in an exothermic reaction.

Stage 1: Energy is needed (absorbed from the surroundings) to break the bonds. This process is endothermic.

Stage 2: Energy is released (transferred to the surroundings) as the new bonds form. This process is exothermic.

The overall reaction is exothermic because forming the new bonds releases more energy than is needed initially to break the old bonds. Fig. 2.14 is a simplified energy level diagram showing the exothermic nature of the reaction.

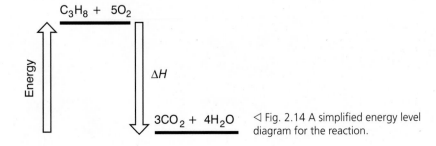

◁ Fig. 2.14 A simplified energy level diagram for the reaction.

The larger the alkane molecule, the more the energy is released on combustion. This is because although more bonds must be broken in the first stage of the reaction, more bonds are formed in the second stage.

Alkane		Molar enthalpy of combustion (kJ/mol)
Methane	CH_4	−882
Ethane	C_2H_6	−1542
Propane	C_3H_8	−2202
Butane	C_4H_{10}	−2877
Pentane	C_5H_{12}	−3487
Hexane	C_6H_{14}	−4141

△ Table 2.1 Molar enthalpy of combustion of alkanes.

REMEMBER

In an exothermic reaction, the energy released on forming new bonds is greater than that needed to break the old bonds.
In an endothermic reaction, more energy is needed to break the old bonds than is released when new bonds are formed. The energy changes in endothermic reactions are usually relatively small.

END OF EXTENDED

QUESTIONS

1. What does the sign of ΔH indicate about a reaction?

2. EXTENDED Is energy needed or released when bonds are broken?

3. EXTENDED In an endothermic reaction is more or less energy needed to break the old bonds than is recovered when new bonds are formed?

HOW COMMON ARE ENDOTHERMIC REACTIONS?

Almost all chemical reactions in which simple compounds or elements react to make new compounds are exothermic. One exception is the formation of nitrogen oxide (NO) from nitrogen and oxygen. Overall, energy is needed to create this compound, with less energy being released on forming bonds than was needed to break the bonds of the reactants. Nitrogen oxide is often formed in lightning storms. The lightning provides enough energy to split the nitrogen and oxygen molecules before the atoms combine to form nitrogen oxide:

$$N_2(g) + O_2(g) \rightarrow 2NO(g) \quad \Delta H \text{ positive}$$

△ Fig. 2.15 These plants are making food by photosynthesis, an endothermic reaction.

Another exception is **photosynthesis**. Plants use energy from sunlight to convert carbon dioxide and water into glucose and oxygen:

$$6CO_2(g) + 6H_2O(l) \rightarrow C_6H_{12}O_6(aq) + 6O_2(g) \quad \Delta H \text{ positive}$$

'Cold packs', which you can buy in some countries, can be used to help you keep cool. Usually you have to bend a pack to break a partition inside and allow two substances to mix. The pack will then stay cold for an hour or longer. However, it may not be an endothermic reaction that is working in the cold pack. Dissolving chemicals like urea or ammonium nitrate in water also cause the temperature of water to fall, but dissolving is a **physical change**, not a chemical change. Whether it is an endothermic reaction or not is the manufacturer's secret.

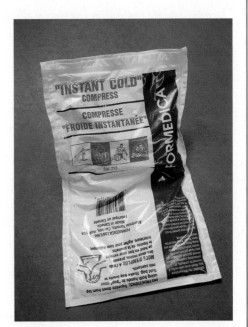

△ Fig. 2.16 A cold pack.

End of topic checklist

Key terms

activation energy, burning, calorimetry, combustion, endothermic, enthalpy change, exothermic, photosynthesis, physical change

During your study of this topic you should have learned:

○ How to describe the meaning of exothermic and endothermic reactions.

○ **EXTENDED** How to describe bond breaking as endothermic and bond forming as exothermic.

○ **EXTENDED** Be able to draw and label energy level diagrams for exothermic and endothermic reactions using data provided.

○ **EXTENDED** How to interpret energy level diagrams showing exothermic and endothermic reactions and the activation energy of a reaction.

End of topic questions

Note: The marks given for these questions indicate the level of detail required in the answers. In the examination, the number of marks given to questions like these may be different.

1. Explain each of the following:

    a) A polystyrene cup is used when measuring energy changes in simple reactions, such as adding magnesium ribbon to an acid. **(2 marks)**

    b) When sodium hydrogencarbonate is added to a solution of an acid the temperature of the acid falls. **(1 mark)**

2. An estimate of the energy produced when a fuel burns can be made by burning the fuel under a container holding water and measuring the temperature rise of the water.

    a) What type of material should the container be made of? Explain your answer. **(2 marks)**

    b) Why does this method give an estimate rather than an accurate value? **(2 marks)**

    c) How can the accuracy of this method be improved? **(2 marks)**

3. EXTENDED Calcium oxide reacts with water as shown in the equation:

$$CaO(s) + H_2O(l) \rightarrow Ca(OH)_2(s)$$

An energy level diagram for this reaction is shown below.

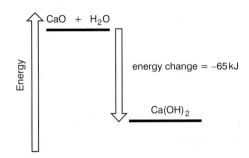

a) What does the energy level diagram tell us about the type of energy change that takes place in this reaction? **(1 mark)**

b) What does the energy level diagram indicate about the amounts of energy required to break original bonds and form new bonds in this reaction? **(1 mark)**

4. EXTENDED Chlorine (Cl_2) and hydrogen (H_2) react together to make hydrogen chloride (HCl). The equation can be written as:

$$H–H + Cl–Cl \rightarrow H–Cl + H–Cl$$

When this reaction occurs, energy is transferred to the surroundings. Explain this in terms of the energy transfer processes taking place when bonds are broken and when bonds are made. **(2 marks)**

Rate of reaction

△ Fig. 2.17 Petrol igniting.

INTRODUCTION

Some chemical reactions take place extremely quickly. For example, when petrol is ignited it combines with oxygen almost instantaneously. Reactions like these have a *high rate*. Other reactions are much slower, for example, when an iron bar rusts in the air; reactions like these have a *low rate*. Chemical reactions can be controlled and made to be quicker or slower. This can be very important in food production, either by slowing down or increasing the rate at which food ripens, or in the chemical industry where the rate of a reaction can be adjusted to an optimum level.

KNOWLEDGE CHECK

✓ Know the arrangement, movement and energy of the particles in the three states of matter: solid, liquid and gas.
✓ Understand how the course of a reaction can be shown in an energy level diagram.
✓ Be able to write and interpret balanced chemical equations.

LEARNING OBJECTIVES

✓ Be able to describe a practical method for investigating the rate of a reaction involving the evolution of a gas.
✓ Be able to interpret data obtained from experiments concerned with rate of reaction.
✓ Be able to describe the effects of concentration, particle size, catalysts and temperature on the rates of reactions.
✓ **EXTENDED** Be able to suggest apparatus, given information, for experiments, including collection of gases and measurement of rates of reaction.
✓ **EXTENDED** Be able to describe and explain the effect of changing concentration in terms of collisions between reacting particles.
✓ **EXTENDED** Be able to explain that an increase in temperature causes an increase in collision rate *and* more of the colliding particles have sufficient energy (activation energy) to react, whereas an increase in concentration only causes an increase in collision rate.

COLLISION THEORY

For a chemical reaction to occur, the reacting particles (atoms, molecules or ions) must collide. The energy involved in the collision must be enough to break the chemical bonds in the reacting particles – or the particles will just bounce off one another.

A collision that has enough energy to result in a chemical reaction is an **effective collision**.

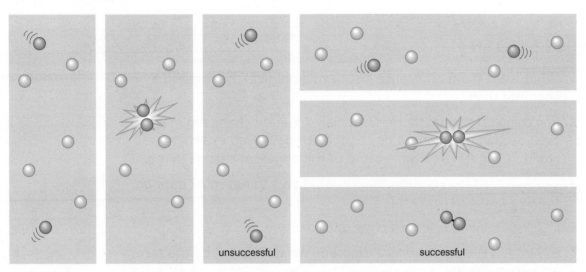

Δ Fig. 2.18 Particles must collide with sufficient energy to make an effective collision.

Some chemical reactions occur extremely quickly (for example, the explosive reaction between petrol and oxygen in a car engine) and some more slowly (for example, iron rusts over days or weeks). This is because they have different activation energies. Activation energy acts as a barrier to a reaction. It is the minimum amount of energy required in a collision for a reaction to occur. As a general rule, the bigger the activation energy, the slower the reaction will be at a particular temperature.

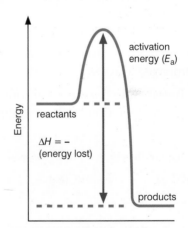

Δ Fig. 2.19 Reaction profile.

REMEMBER

The 'barrier' preventing a reaction from occurring is called the activation energy. If the activation energy of a reaction is low, more of the collisions will be effective and the reaction will proceed quickly. If the activation energy is high, a smaller proportion of collisions will be effective and the reaction will be slow.

END OF EXTENDED

RATE OF A REACTION

A quick reaction takes place in a short time. It has a high **rate of reaction**. As the time taken for a reaction to be completed increases, the rate of the reaction decreases. In other words:

Speed	Rate	Completion time
Quick or fast	High	Short
Slow	Low	Long

△ Table 2.2 Speed, rate and time.

QUESTIONS

1. What is the main difference between a physical change and a chemical change?

2. In a chemical change there is often an apparent change in mass even though mass cannot be created nor destroyed in a chemical reaction. What is a possible cause of this apparent change in mass?

3. EXTENDED In the collision theory, what two things must happen for two particles to react?

4. EXTENDED What is an *effective collision*?

5. EXTENDED Describe, using a diagram, what is meant by the term *activation energy*.

MONITORING THE RATE OF A REACTION

The rate of a reaction changes as the reaction proceeds. There are some easy ways of monitoring this change.

When marble (calcium carbonate) reacts with hydrochloric acid, the following reaction starts straight away:

calcium carbonate + hydrochloric acid → calcium chloride + carbon dioxide + water

$$CaCO_3(s) + 2HCl(aq) \rightarrow CaCl_2(aq) + CO_2(g) + H_2O(l)$$

The reaction can be monitored as it proceeds either by measuring the volume of gas being formed or by measuring the change in mass of the reaction flask.

The volume of gas produced in this reaction can be measured using the apparatus shown in Fig. 2.20. The hydrochloric acid is put into the conical flask, the marble chips are added, the bung is quickly fixed into the neck of the flask and the stopclock is started.

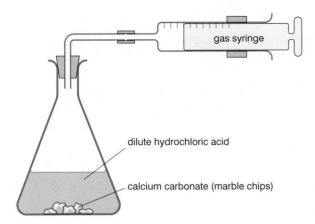

dilute hydrochloric acid

calcium carbonate (marble chips)

◁ Fig. 2.20 Monitoring the rate of a reaction.

The reaction will start immediately, effervescence (bubbling) will occur in the flask as the carbon dioxide gas is produced and the plunger on the syringe will start to move. Measuring the volume of gas in the syringe every 10 seconds will indicate how the total amount of gas produced changes as the reaction proceeds. The change in the rate of the reaction with time can be shown on a graph of the results (see Fig. 2.22).

EXTENDED

To measure the change in mass in the same reaction, the apparatus shown in Fig. 2.21 can be used. The hydrochloric acid is put into the conical flask, the marble chips are added, the cotton wool plug is put in the neck of the flask and the stopclock is started. The mass of the flask and contents is measured as soon as the plug is inserted and then every 10 seconds as the reaction occurs. The mass will decrease as carbon dioxide gas escapes from the flask.

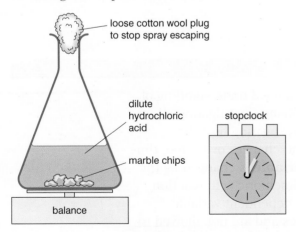

loose cotton wool plug to stop spray escaping

dilute hydrochloric acid

stopclock

marble chips

balance

◁ Fig. 2.21 Measuring the change in mass.

As before, drawing a graph of the results shows the change in the rate of the reaction over time.

Graphs of the results from both experiments have almost identical shapes. The rate of the reaction decreases as the reaction proceeds.

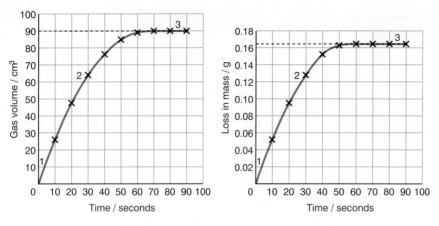

△ Fig. 2.22 Volume of carbon dioxide produced or loss in mass.

Loss in mass during the reaction

The rate of the reaction at any point can be calculated from the gradient of the curve. The shapes of the graphs can be divided into three regions.

1. At this point, the curve is the steepest (has the highest gradient) and the reaction has its highest rate. The maximum number of reacting particles are present and the number of effective collisions per second is at its greatest.

2. The curve is not as steep (has a lower gradient) at this point and the rate of the reaction is lower. Fewer reacting particles are present and so the number of effective collisions per second is lower.

3. The curve is horizontal (gradient is zero) and the reaction is complete. At least one of the reactants has been completely used up and so no further collisions can occur between the two reactants.

END OF EXTENDED

REMEMBER

In experiments like these it is helpful to have a good understanding of the types of **variables** involved. The factor you are investigating is called the **independent variable** – when investigating how the reaction between marble and hydrochloric acid changes over time, time is the independent variable. A **dependent variable** is changed by the independent variable – in the marble and hydrochloric acid reaction, the volume of carbon dioxide produced is the dependent variable. Other variables involved are **control variables** and are not allowed to change to ensure a 'fair test'. So, temperature could be a control variable in the reaction between marble and hydrochloric acid.

In chemical reactions it is very rare that exact (as predicted by the equation) quantities of reactants are used. In the marble and hydrochloric acid reaction all the marble may be used up (it is called the *limiting reactant*) but not all the hydrochloric acid; some is left when the reaction has stopped (it is *in excess*).

1. What piece of apparatus can accurately measure the volume of gas produced in a reaction?

2. EXTENDED On a volume versus time graph, what does a horizontal line show?

3. EXTENDED When comparing two reactions, will the slower or quicker reaction have a steeper volume/time gradient at the beginning?

WHAT CAN CHANGE THE RATE OF A REACTION?

There are four key factors that can change the rate of a reaction:

- concentration (of a solution)
- temperature
- particle size (of a solid)
- a catalyst.

EXTENDED

A simple **collision theory** can be used to explain how these factors affect the rate of a reaction. Two important parts of the theory are:

- The reacting particles must collide with each other.
- There must be sufficient energy in the collision to overcome the activation energy.

END OF EXTENDED

Concentration

Increasing the concentration of a reactant will increase the rate of reaction. When a piece of magnesium ribbon is added to a solution of hydrochloric acid, the following reaction occurs:

magnesium	+	hydrochloric acid	→	magnesium chloride	+	hydrogen
$Mg(s)$	+	$2HCl(aq)$	→	$MgCl_2(aq)$	+	$H_2(g)$

As the magnesium and acid come into contact, there is effervescence ('bubbling') and hydrogen gas is given off. Two experiments were performed using the same length of magnesium ribbon, but different concentrations of acid. In experiment 1 the hydrochloric acid used was $2.0 \, mol/dm^3$, in experiment 2 the acid was $0.5 \, mol/dm^3$. The graph in Fig. 2.23 shows the results of the two experiments.

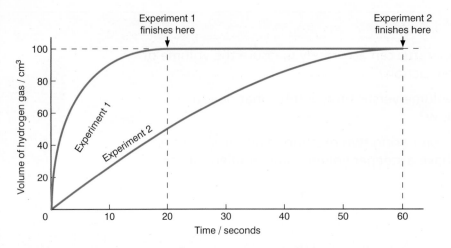

△ Fig. 2.23 Volume of hydrogen produced in the reaction between magnesium and hydrochloric acid.

In experiment 1 the curve is steeper (has a higher gradient) than in experiment 2. In experiment 1 the reaction is complete after 20 seconds, whereas in experiment 2 it takes 60 seconds. The initial rate of the reaction is higher with $2.0 \, mol/dm^3$ hydrochloric acid than with $0.5 \, mol/dm^3$ hydrochloric acid.

EXTENDED

In the $2.0 \, mol/dm^3$ hydrochloric acid solution there are more hydrogen ions in a given volume, a higher concentration of hydrogen ions, and so there will be a lot more effective collisions per second with the surface of the magnesium ribbon than in the $0.5 \, mol/dm^3$ hydrochloric acid.

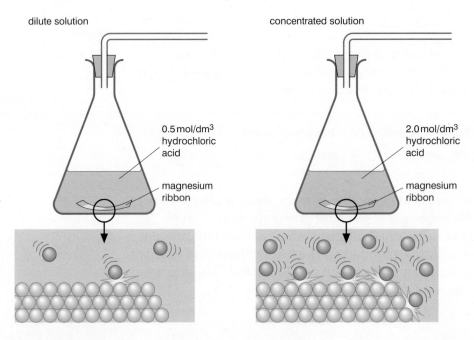

△ Fig. 2.24 Using dilute and concentrated solutions in a reaction.

Developing investigative skills

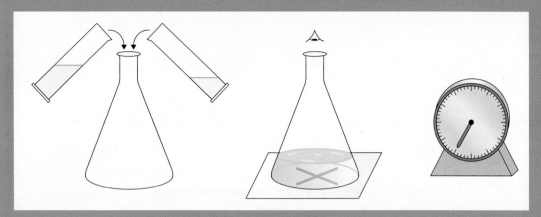

△ Fig. 2.25 Experiment with sodium thiosulfate solution and hydrochloric acid.

A student was investigating the reaction between sodium thiosulfate solution and hydrochloric acid. As the reaction takes place, a precipitate of sulfur forms in the solution and makes it change from colourless (and clear) to pale yellow (and opaque). The time it takes for a certain amount of sulfur to form can be used as a measure of the rate of the reaction.

The student used 1.0 mol/dm³ sodium thiosulfate solution and made up different concentrations of the solution by using the volumes of the solution and water shown in the table that follows.

She then drew a cross in pencil on a piece of paper.

She then added 5 cm³ of dilute hydrochloric acid to the solution in one of the flasks, the stopclock was started, the mixture was quickly stirred or swirled and then the conical flask was put on top of the pencilled cross.

The student looked through the conical flask to the cross and stopped the stopclock as soon as the cross could no longer be seen.

She then repeated the process with the other four solutions. Her results are shown in the table:

Volume of sodium thiosulfate solution (cm³)	50	40	30	20	10
Volume of water (cm³)	0	10	20	30	40
Volume of hydrochloric acid (cm³)	5	5	5	5	5
Time for the cross to be obscured (s)	14	18	23	36	67

Using and organising techniques apparatus and materials

❶ Why was the total volume in the flask always 55 cm³?

❷ Why was the stopclock started when the acid was added and not when the flask was put on the pencilled cross?

❸ What apparatus would you use to measure the volume of sodium thiosulfate solution?

END OF EXTENDED

TEMPERATURE

Increasing the temperature of the reactants will increase the rate of a reaction.

EXTENDED

Warming a substance transfers kinetic energy to its particles. More kinetic energy means that the particles move faster. Because they are moving faster there will be more collisions each second. The increased energy of the collisions also means that the proportion of collisions that are effective will increase. A reaction was carried out at two different temperatures – first at 20 °C and then at 30 °C.

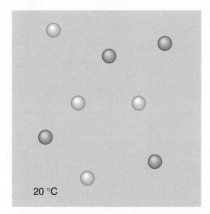
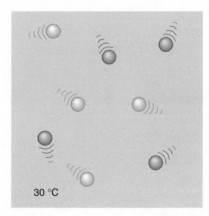

△ Fig. 2.26 Effect of increasing temperature on particles.

Increasing the temperature of the reaction between some marble chips and hydrochloric acid will not increase the final amount of carbon dioxide produced. The same amount of gas will be produced in a shorter time. The rates of the two reactions are different but the final loss in mass is the same.

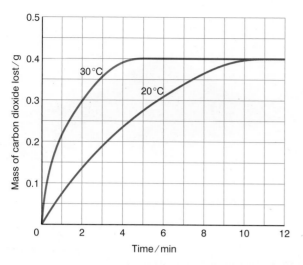

△ Fig. 2.27 The effect of temperature on the reaction between hydrochloric acid and marble chips.

END OF EXTENDED

QUESTIONS

1. What units are used to measure the concentration of solutions?

2. EXTENDED In terms of particles colliding, why does increasing the concentration of a solution increase the rate of reaction?

3. EXTENDED Give two reasons why increasing temperature increases the rate of reaction.

PARTICLE SIZE

Decreasing the particle size (or increasing the **surface area**) of a solid reactant will increase the rate of a reaction.

EXTENDED

A reaction can only take place if the reacting particles collide. This means that the reaction takes place at the surface of a solid. The particles within the solid cannot react until those on the surface have reacted and moved away.

END OF EXTENDED

Powdered calcium carbonate has a smaller particle size (or much larger surface area) than the same mass of marble chips. A lump of coal will burn slowly in the air, whereas coal dust can react explosively. This is a hazard in coal mines where coal dust can react explosively with air. In addition, as well as the danger of explosive mixtures of coal dust and air, the build-up of methane gas can also form an explosive mixture with the air.

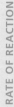

△ Fig. 2.28 Powdered carbon has a much larger surface area than the same mass in larger lumps.

CATALYSTS

A **catalyst** is a substance that alters the rate of a chemical reaction and is chemically unchanged at the end of the reaction. An **enzyme** is a biological catalyst, for example, amylase, which is found in saliva.

Note: Enzymes are involved in the fermentation of glucose. Enzymes are present in yeast and these increase the rate at which glucose is converted into ethanol and carbon dioxide. The reaction rate increases as the yeast multiplies – but as the concentration of ethanol increases, the rate decreases because the ethanol begins to kill or denature the enzymes.

QUESTION

1. What is a *catalyst*?

SCIENCE LINK

BIOLOGY – ENZYMES

- The factors that affect how quickly a chemical reaction happens link directly to the role of enzymes in the maintenance of body processes.

- Describing how the energy of the particles changes at higher temperatures also allows us to explain why an enzyme will not work above a certain temperature.

PHYSICS – SIMPLE KINETIC MODEL

- Explaining why the different factors affect the rate of a chemical reaction uses the same particle model that gives us the simple structure of solids, liquids and gases.

- Thinking about the forces between the particles and the energy involved in the interactions between particles leads to a common explanation in terms of particle speed and kinetic energy.

End of topic checklist

Key terms

catalyst, collision theory, control variable, dependent variable, effective collision, enzyme, independent variable, rate of reaction, surface area, variable

During your study of this topic you should have learned:

○ How to describe a practical method for investigating the rate of a reaction involving the evolution of a gas.

○ How to interpret data obtained from experiments concerned with rate of reaction.

○ How to describe the effect of concentration, particle size, catalysts and temperature on the rate of reactions.

○ EXTENDED The suitable apparatus, given information, for experiments, including collection of gases and measurement of rates of reaction.

○ EXTENDED How to describe and explain the effect of changing concentration on reaction rate in terms of collisions between reacting particles.

○ EXTENDED That an increase in temperature causes an increase in collision rate and more of the colliding particles have sufficient energy (the activation energy) to react.

End of topic questions

Note: The marks given for these questions indicate the level of detail required in the answers. In the examination, the number of marks given to questions like these may be different.

1. This question is about the reaction between magnesium and hydrochloric acid.

   a) Draw and label a diagram of the apparatus that could be used to monitor the rate of the reaction by measuring the volume of hydrogen produced. **(2 marks)**

   b) How will the following changes affect the rate of the reaction?

   i) Using powdered magnesium rather than magnesium ribbon. **(1 mark)**

   ii) Using a less concentrated solution of hydrochloric acid. **(1 mark)**

   iii) Lowering the temperature of the hydrochloric acid. **(1 mark)**

2. EXTENDED For a chemical reaction to occur, the reacting particles must collide. Why don't all collisions between the particles of the reactants lead to a chemical reaction? **(2 marks)**

3. The diagrams below show the activation energies of two different reactions, A and B.

   a) What is the *activation energy* of a reaction? **(1 mark)**

   b) Which reaction is likely to have the higher rate of reaction at a particular temperature? Explain your answer. **(2 marks)**

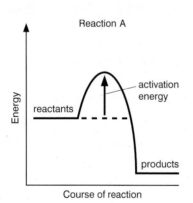

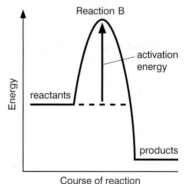

End of topic questions continued

4. EXTENDED Look at the table of results obtained when dilute hydrochloric acid is added to marble chips.

Time (seconds)	0	10	20	30	40	50	60	70	80	90
Volume of gas (cm³)	0	20	36	49	58	65	69	70	70	70

a) What is the name of the gas produced in this reaction? **(1 mark)**

b) Write a balanced equation, including state symbols, for the reaction. **(2 marks)**

c) Draw a graph of volume of gas (*y*-axis) against time (*x*-axis).

   Label it 'Graph 1'. **(3 marks)**

d) Use the results to calculate the volume of gas produced:

   i) in the first 10 seconds **(1 mark)**

   ii) between 10 and 20 seconds **(1 mark)**

   iii) between 20 and 30 seconds **(1 mark)**

   iv) between 80 and 90 seconds. **(1 mark)**

e) Explain why the rate of the reaction changes as the reaction takes place. **(2 marks)**

f) Use the collision theory to explain the change in the rate of reaction. **(2 marks)**

g) The reaction was repeated using the same volume and concentration of hydrochloric acid with the same mass of marble, but as a powder instead of chips. Draw another curve on your graph paper, using the same axes as before (label as 'Graph 2'), to show how the original results will change. **(3 marks)**

h) The reaction was repeated, but this time using the original mass of new marble chips and the same volume of hydrochloric acid, but with the acid only half as concentrated as originally. Draw another curve on your graph paper, using the same axes as before (label as 'Graph 3'), to show how the original results will change. **(3 marks)**

Redox reactions

INTRODUCTION

Oxidation reactions are very familiar in everyday life – with examples such as the rusting of iron and bleaching, which is effective because bleach is a powerful oxidising agent. Whenever anything burns, an oxidation reaction takes place between the fuel and oxygen in the air. Reduction reactions may seem less familiar, but oxidation and reduction go hand in hand – if an element or compound in a chemical reaction is oxidised, then another element or compound in the same reaction must be reduced. So even when a bonfire is burning furiously and using oxygen from the air, reduction is taking place at the same time!

Δ Fig. 2.29 Oxidation and reduction are both taking place in this bonfire.

KNOWLEDGE CHECK

✓ Know about ions and ion charges.
✓ Be able to interpret chemical equations and associated state symbols.

LEARNING OBJECTIVES

✓ Know the definitions of oxidation and reduction in terms of oxygen loss or gain.
✓ Know that oxidation states are used to name ions, for example, iron(II), iron(III), copper(II), manganate(VII).
✓ **EXTENDED** Know that an oxidising agent is a substance that oxidises another substance during a redox reaction.
✓ **EXTENDED** Know that a reducing agent is a substance that reduces another substance during a redox reaction.

OXIDATION, REDUCTION AND REDOX

When oxygen is added to an element or a compound, the process is called **oxidation**:

$$2Cu(s) + O_2(g) \rightarrow 2CuO(s)$$

The copper has been oxidised.

Removing oxygen from a compound is called **reduction**:

CuO(s) + Zn(s) → ZnO(s) + Cu(s)

The copper(II) oxide has been *reduced*.

If we look more carefully at this last reaction, we see the zinc has changed to zinc oxide: that is, it has been oxidised at the same time as the copper(II) oxide has been reduced.

This is one example of reduction and oxidation taking place at the same time, in the same reaction. These are called **redox** reactions.

EXTENDED

In the equation above showing the reaction between copper(II) oxide and zinc, the copper(II) oxide has been reduced by the zinc. Zinc is therefore acting as a reducing agent. The zinc itself has been oxidised by the copper(II) oxide, so the copper(II) oxide is acting as an oxidising agent.

END OF EXTENDED

OXIDATION STATES

When you learned to write chemical formulae, you were introduced to the use of Roman numerals for metals that had more than one ion, for example, iron as Fe^{2+} or Fe^{3+}:

- iron(II) oxide = FeO
- iron(III)oxide = Fe_2O_3

The II and III are called **oxidation states**.

- Fe^{2+} has an oxidation state of +2.
- Fe^{3+} has an oxidation state of +3.
- oxygen has an oxidation state of −2.

You take the ion charge and reverse it, so an ion of 3− has oxidation number −3.

The oxidation state of elements is always 0 (zero).

An oxidation state describes how many electrons an atom loses or gains when it forms a chemical bond.

QUESTIONS

1. Define the term *reduction*.

2. What is the oxidation state of the metal ion in each of the following compounds?

   a) copper(II) oxide

   b) iron(III) chloride

   c) potassium manganate(VII).

End of topic checklist

Key terms

oxidation, oxidation state, redox, reduction

During your study of this topic you should have learned:

○ The definitions of oxidation and reduction in terms of oxygen loss or gain.

○ That oxidation states are used to name ions, for example, iron(II), iron(III), copper(II), manganate(VII).

○ **EXTENDED** That an oxidising agent is a substance that oxidises another substance during a redox reaction.

○ **EXTENDED** That a reducing agent is a substance that reduces another substance during a redox reaction.

End of topic questions

Note: The marks given for these questions indicate the level of detail required in the answers. In the examination, the number of marks given to questions like these may be different.

1. The following equation shows a redox reaction:

$Mg(s) + ZnO(s) \rightarrow MgO(s) + Zn(s)$

a) What has been oxidised in the reaction? (1 mark)

b) What has been reduced in the reaction? (1 mark)

c) EXTENDED What is the oxidising agent? (1 mark)

d) EXTENDED What is the reducing agent? (1 mark)

2. EXTENDED Look at the following equation showing the reaction between lead(II) oxide and hydrogen. Name the oxidising agent in this reaction. Explain your answer.

$2PbO(s) + C(s) \rightarrow 2Pb(s) + CO_2(g)$

Acids, bases and salts

INTRODUCTION

Acids are commonly used in everyday life. Many of them, such as hydrochloric acid and sulfuric acid, are extremely toxic and corrosive. About 20 million tonnes of hydrochloric acid are manufactured worldwide each year. Some of this is used to make important chemicals, such as PVC (polyvinyl chloride) plastic. Alkalis and bases are less common in everyday use, yet about 60 million tonnes of sodium hydroxide are produced worldwide each year and used in the manufacture of paper and **soap**. Sodium hydroxide is harmful and corrosive. Common salt, sodium chloride, is an example of a salt.

△ Fig. 2.30 Sodium hydroxide.

KNOWLEDGE CHECK

✓ Know the names of some common acids, including hydrochloric acid and sulfuric acid.
✓ Know that vegetable dyes can be used as indicators to identify acids and alkalis.
✓ Be able to use state symbols such as (s), (l), (g) and (aq).

LEARNING OBJECTIVES

✓ Be able to describe neutrality and relative acidity and alkalinity in terms of pH measured using universal indicator.
✓ Be able to describe the characteristic properties of acids as reactions with metals, bases, carbonates and the effect on litmus paper.
✓ Be able to describe and explain the importance of controlling acidity in soil.
✓ Be able to describe the preparation, separation and purification of soluble salts.
✓ Be able to demonstrate knowledge and understanding of preparation, separation and purification of salts.
✓ **EXTENDED** Be able to suggest a method of making a given salt from a suitable starting material, given appropriate information.

AQUEOUS SOLUTIONS

When any substance dissolves in water, it forms an aqueous solution, shown by the state symbol (aq). Aqueous solutions can be acidic, alkaline or neutral. A neutral solution is neither acidic nor alkaline.

Indicators are used to tell if a solution is acidic, alkaline or neutral. They can be used either as liquids or in paper form, and they turn different colours with different solutions. There are different indicators that can be used.

The most common indicator is **litmus**. Its colours are shown in the table:

Colour of litmus	Type of solution
Red	Acidic
Blue	Alkaline

△ Table 2.3 Litmus.

Universal indicator can show how strongly acidic or how strongly alkaline a solution is because it has more colours than litmus. Each colour is linked to a number ranging from 0 (most strongly acidic solution) to 14 (most strongly alkaline solution). A neutral substance has a pH of 7. This range is called the **pH scale** and is related to the concentration of hydrogen ions ($H^+(aq)$).

Concentration of hydrogen ions compared to distilled water			Examples of solutions and their respective pH
1/10 000 000		14	Liquid drain cleaner, caustic soda
1/1 000 000		13	Bleaches, oven cleaner
1/100 000		12	Soapy water
1/10 000		11	Household ammonia (11.9)
1/1 000		10	Milk of magnesia (10.5)
1/100		9	Toothpaste (9.9)
1/10		8	Baking soda (8.4), seawater, eggs
0		7	Pure water (7)
10		6	Urine (6) milk (6.6)
100		5	Acid rain (5.6) black coffee (5)
1 000		4	Tomato juice (4.1)
10 000		3	Grapefruit and orange juice, soft drink
100 000		2	Lemon juice (2.3) vinegar (2.9)
1 000 000		1	Hydrochloric acid secreted from the stomach lining (1)
10 000 000		0	Battery acid

△ Fig. 2.31 The pH scale.

WHAT ARE ACIDS?

Acids are substances that contain replaceable hydrogen atoms. These hydrogen atoms are replaced in chemical reactions by metal atoms, forming a compound known as a salt. Acids have pHs in the range 0–7.

Acid name	Acid formula
Hydrochloric acid	HCl
Nitric acid	HNO_3
Sulfuric acid	H_2SO_4
Phosphoric acid	H_3PO_4

△ Table 2.4 Common acids.

The typical reactions of acids include:

- Acid + metal makes a salt and hydrogen gas.
- Acid + carbonate makes a salt, carbon dioxide and water.
- Acid + base makes a salt and water.

QUESTION

1. Two solutions are tested with universal indicator paper. Solution A has a pH of 8 and solution B has a pH of 14. What does this tell you about the two solutions?

THE IMPORTANCE OF CONTROLLING ACIDITY IN SOIL

If soil is too acidic, then it can be neutralised using quicklime. Quicklime is made from limestone, which is quarried from limestone rocks. It is heated in lime kilns at 1200 °C to make calcium oxide or quicklime:

$$1200 °C \uparrow$$

limestone $\rightarrow$ quicklime + carbon dioxide

$$CaCO_3(s) \rightarrow CaO(s) + CO_2(g)$$

When quicklime is added to water, it makes calcium hydroxide, which is an alkali and so can neutralise the acidic soil:

quicklime + water $\rightarrow$ slaked lime

$$CaO(s) + H_2O(l) \rightarrow Ca(OH)_2(s)$$

Different plants grow better in different types of soil. The pH of a soil is an important factor in the growth of different plants – some plants prefer slightly acidic conditions and others slightly alkaline conditions.

Adding fertilisers to soil can also affect the pH and so the soil may have to be treated by adding acids or alkalis. The pH of soil can be measured by taking a small sample, putting it in a test tube with distilled water and adding indicator solution or using indicator paper. The pH can be found from a pH chart.

SCIENCE LINK

BIOLOGY – CHARACTERISTICS OF LIVING THINGS, HUMAN EFFECTS ON ECOSYSTEMS

- A number of plants grow flowers of different colours, depending on whether they are growing in acidic soil or alkaline soil – this property forms the basis of many indicators used in chemistry, such as litmus.

- A knowledge of acids and bases is important in understanding the effects of acid rain and some other environmental effects – it is also important in trying to reduce any damaging consequences.

WHAT ARE SALTS?

Acids contain replaceable hydrogen atoms. When metal atoms take their place, a compound called a **salt** is formed. The names of salts have two parts, as shown in Fig. 2.32.

sodium chloride (NaCl)

the name of the metal that replaced the hydrogen

the part of the salt name showing which acid was used

Δ Fig. 2.32 Salt – sodium chloride.

Table 2.5 shows the four most common acids and their salt names.

Acid	Salt name
Hydrochloric (HCl)	Chloride (Cl^-)
Nitric (HNO_3)	Nitrate (NO_3^-)
Sulfuric (H_2SO_4)	Sulfate (SO_4^{2-})
Phosphoric (H_3PO_4)	Phosphate (PO_4^{3-})

Δ Table 2.5 Common acids and their salt names.

△ Fig. 2.33 Sodium chloride crystals.

△ Fig. 2.34 Copper(II) sulfate crystals.

Salts are ionic compounds. The names of these compounds are created by taking the first part of the name from the metal ion, which is a positive ion (cation), and the second part of the name from the acid, which is a negative ion (anion). For example:

copper(II) sulfate: $Cu^{2+} + SO_4^{2-} \rightarrow CuSO_4$

                cation   anion   salt

Salts are often found in the form of crystals. Crystals of many salts contain **water of crystallisation**, which is responsible for their crystal shape. Water of crystallisation is shown in the chemical formula of a salt. For example:

copper(II) sulfate crystals: $CuSO_4.5H_2O$

iron(II) sulfate crystals: $FeSO_4.7H_2O$

Salts that do not contain water of crystallisation are **anhydrous**.

MAKING SALTS

The solubility of salts in water

Here are the general rules that describe the solubility of common types of salts in water:

- All common sodium, potassium and ammonium salts are soluble.
- All nitrates are soluble.
- Common chlorides, bromides and iodides are soluble – except those of silver chloride, silver bromide and silver iodide.
- Common sulfates are soluble – except those of barium, lead and calcium.
- Common carbonates are insoluble – except those of sodium, potassium and ammonium.

Making soluble salts

1.

acid   +   alkali   →   a salt   +   water

For example:

$$HCl(aq) + NaOH(aq) \rightarrow NaCl(aq) + H_2O(l)$$

2.

acid   +   base   →   a salt   +   water

For example:

$$H_2SO_4(aq) + CuO(s) \rightarrow CuSO_4(aq) + H_2O(l)$$

3.

acid   +   carbonate   →   a salt   +   water   +   carbon dioxide

For example:

$$2HNO_3(aq) + CuCO_3(s) \rightarrow Cu(NO_3)_2(aq) + H_2O(l) + CO_2(g)$$

4.

acid   +   metal   →   a salt   +   hydrogen

For example:

$$2HCl(aq) + Mg(s) \rightarrow MgCl_2(aq) + H_2(g)$$

Here is a shortcut for remembering the four general equations above. Remember the initials of the reactants:

A (acid) + A (alkali)

A (acid) + B (base)

A (acid) + C (carbonate)

A (acid) + M (metal)

The symbol '(aq)' after the formula of the salt shows that it is a soluble salt.

Neutralisation describes the reactions of acids with alkalis and bases. When acids react with alkalis, the reaction is between H^+ ions and OH^- ions to make water, as in:

$$H^+(aq) + OH^-(aq) \rightarrow H_2O(l)$$

Reactions of acids with alkalis are used in the experimental procedure of **titration**, in which solutions react together to give the end-point shown by an indicator. Calculations can then be performed to find the concentration of the acid or the alkali.

In the laboratory

Of the four methods for making soluble salts, shown by the symbol (aq), only one uses two solutions:

1. acid(aq) + alkali(aq) → a salt(aq) + water(l)

The other three methods involve adding a solid(s) to a solution(aq):

2. acid(aq) + base(s) → a salt(aq) + water(l)

3. acid(aq) + carbonate(s) → a salt(aq) + water(l) + carbon dioxide(g)

4. acid(aq) + metal(s) → a salt(aq) + hydrogen(g)

Method 1 involves the titration method. An indicator is used to show when exact quantities of acid and alkali have been mixed. The procedure is then repeated using the same exact volumes of acid and alkali, but without the indicator. The resulting solution is evaporated to the point of crystallisation, then left to cool and the salt to crystallise.

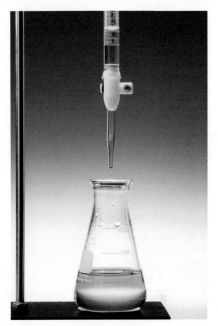

△ Fig. 2.35 Using the neutralisation method for a titration.

EXTENDED

REMEMBER

You should know that because acids in water form $H^+(aq)$ ions and alkalis form $OH^-(aq)$ ions, the neutralisation reaction of acids with alkalis can be summarised as:

$$H^+(aq) + OH^-(aq) \rightarrow H_2O(l)$$

All neutralisation reactions can be represented by this equation, whatever the acid or alkali used.

END OF EXTENDED

The general procedure used for each of the methods 2, 3 and 4 is the same:

• The solid (base, carbonate or metal) is added to the acid with stirring until no more solid will react. Heating may be necessary.

- The mixture is filtered to remove unreacted solid and the solution is collected as the **filtrate** in an evaporating dish.
- The solution is evaporated to the point of crystallisation and is then left to cool and the salt to crystallise.

The process is summarised in Fig. 2.36:

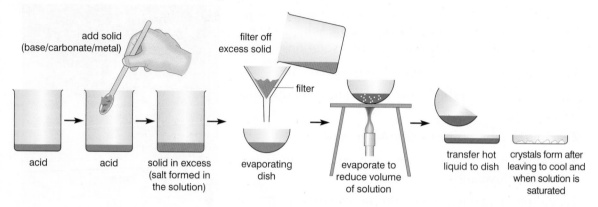

add solid
(base/carbonate/metal)

filter off
excess solid

filter

acid   acid   solid in excess
(salt formed in
the solution)

evaporating
dish

evaporate to
reduce volume
of solution

transfer hot
liquid to dish

crystals form after
leaving to cool and
when solution is
saturated

△ Fig. 2.36 Making soluble salts from solids.

QUESTIONS

1. What is a *salt*?

2. Which acid would you use to make a sample of sodium sulfate?

3. Would you expect potassium chloride to be soluble or insoluble in water? Explain your answer.

4. What is the name of the salt formed when calcium carbonate reacts with nitric acid?

5. What does the word *neutralisation* mean?

SOME INTERESTING FACTS ABOUT ACIDS AND ALKALIS

1. Pure sulfuric acid is a clear, oily, highly corrosive liquid. It was well known to Islamic, Greek and Roman scholars in the ancient world, when it was called 'oil of vitriol'. Although pure sulfuric acid does not occur naturally on Earth because of its attraction for water, dilute sulfuric acid is found in acid rain and in the upper atmosphere of the planet Venus. It has a wide range of industrial uses from making

△ Fig. 2.37 A wasp.

fertilisers, dyes, paper and pharmaceuticals to batteries, steel and iron. Sulfuric acid is not toxic but it is highly reactive with water, in a strongly exothermic reaction. It can cause severe burns. The acid must be stored in glass containers (never plastic or metal) and handled with extreme care.

2. Hydrochloric acid, although classified as toxic and corrosive, is part of the gastric acid in the stomach and is involved in digestion. Excess acid in the stomach can cause indigestion but 'anti-acid' (alkali) medications can be taken to neutralise this.

3. Perhaps the strongest acid is a mixture of nitric acid and hydrochloric acid, known as 'aqua regia' because it reacts with the 'royal' metals. Unlike other acids, it reacts with very unreactive metals such as gold and platinum. However, some metals like titanium and silver are not affected.

4. Formic acid (now called methanoic acid) is in the venom of ant and bee stings. Such stings can be relieved by (or neutralised with) an alkali such as sodium bicarbonate (sodium hydrogencarbonate). Wasp stings, however, contain an alkali and so need to be neutralised by a weak acid such as vinegar.

The differences can be remembered using:

Bee – **B**icarb

Wasp (W looks like two **V**s) – **V**inegar

EXTENDED

Choosing the method for making a salt

It is important to be able to choose a suitable method for making a salt. For example:

1. Making zinc sulfate by using dilute sulfuric acid. There are a number of possibilities for the reaction:

   a) Zinc + dilute sulfuric acid (zinc is above hydrogen in the reactivity series and so will react with the acid)

   b) Zinc oxide + dilute sulfuric acid

   c) Zinc carbonate + dilute sulfuric acid

   The method is practically the same whichever solid is used, however in the case of zinc oxide the acid will need to be warmed to ensure the zinc oxide reacts. Zinc and zinc carbonate will react with the acid at room temperature. The overall method is shown in Fig 2.36.

2. Making copper(II) chloride from dilute hydrochloric acid. In this case there are only two possibilities for the reaction: In this case there are only two possibilities for the reaction:

a) Copper(II) oxide + dilute hydrochloric acid

b) Copper(II) carbonate + dilute hydrochloric acid

Copper does not react with dilute hydrochloric acid as copper is below hydrogen in the reactivity series. As in example 1, if copper(II) oxide is chosen the acid will need to be warmed to ensure the copper(II) oxide reacts. Again, the overall method is shown in Fig 2.36.

END OF EXTENDED

Developing investigative skills

A student wanted to make a sample of copper(II) sulfate crystals, $CuSO_4.5H_2O$. She used the following steps in her method.

She put on eye protection and warmed $50\,cm^3$ of dilute sulfuric acid in a $250\,cm^3$ beaker and then copper(II) oxide was added a spatula at a time, stirring the reaction mixture with a glass rod.

When no more copper(II) oxide would react, she filtered the mixture and collected the copper(II) sulfate solution in an evaporating basin.

She then heated the solution in the evaporating basin until she could see crystals starting to form and then allowed it to cool.

After several hours, blue crystals of copper(II) sulfate had formed. She drained off any remaining liquid and dried the crystals between filter papers.

Using and organising techniques, apparatus and materials

❶ Copper(II) sulfate crystals are hydrated. Explain what this means.

❷ Draw a diagram showing the apparatus the student could have used to filter the mixture.

❸ What is the general name given to a liquid that passes through a filter paper?

❹ How could the student have tested for the crystallisation point while heating the filtrate?

Observing, measuring and recording

❺ What is the colour of copper(II) oxide?

❻ What would be the colour of the solution the student evaporated?

Evaluating methods

❼ While heating the solution in the evaporating dish the student noticed some very pale blue powder around the edges of the evaporating basin. What do you think this powder was and how do you think it had formed?

❽ **EXTENDED** Write a fully balanced equation for the reaction.

End of topic checklist

Key terms

acid, anhydrous, filtrate, indicator, litmus, neutralisation, pH scale, salt, soap, titration, universal indicator, water of crystallisation

During your study of this topic you should have learned:

○ How to describe neutrality and relative acidity and alkalinity in terms of pH measured using universal indicator paper.

○ How to describe the characteristic properties of acids as reactions with metals, bases, carbonates and the effect on litmus paper.

○ How to describe and explain the importance of controlling acidity in soil.

○ How to describe the preparation, separation and purification of salts as prepared by the following reactions:

- acid + metal
- acid + base
- acid + carbonate
- acid + alkali.

○ EXTENDED How to suggest a method of making a given salt from a suitable starting material, given appropriate information.

End of topic questions

Note: The marks given for these questions indicate the level of detail required in the answers. In the examination, the number of marks given to questions like these may be different.

1. **a)** What is an *indicator*? (1 mark)

   b) What is the *pH scale*? (1 mark)

   c) What do the following pH numbers indicate about a solution that has been tested?

   i) pH 6 (1 mark)

   ii) pH 8 (1 mark)

   iii) pH 14. (1 mark)

2. **a)** What is an *acid*? (1 mark)

   b) What is an *alkali*? (1 mark)

   c) What is the name of the process when an acid reacts with an alkali to form water? (1 mark)

3. Calcium chloride can be made from calcium oxide and dilute hydrochloric acid.

   a) What type of chemical is calcium oxide? (1 mark)

   b) What type of chemical is calcium chloride? (1 mark)

   c) Is calcium chloride soluble or insoluble in water? (1 mark)

   d) Describe the different stages in the preparation of calcium chloride crystals. (4 marks)

   e) EXTENDED Write a fully balanced equation, including symbols, for the reaction. (2 marks)

4. EXTENDED Copy and complete the following equations and include state symbols:

   a) $2KOH(aq) + H_2SO_4(aq) \rightarrow$ _____ + _____ (2 marks)

   b) $2HCl(aq) + MgO(s) \rightarrow$ _____ + _____ (2 marks)

   c) $2HNO_3(aq) + BaCO_3(s) \rightarrow$ _____ + _____ + _____ (2 marks)

   d) $2HCl(aq) + Zn(s) \rightarrow$ _____ + _____ (2 marks)

   e) $ZnCl_2(aq) + K_2CO_3(aq) \rightarrow$ _____ + _____ (2 marks)

Identification of ions and gases

INTRODUCTION

It is important to be able to analyse different substances and identify the different elements or components. The techniques used today are fairly sophisticated but many of them are based on simple laboratory tests. With improved understanding of the beneficial and harmful properties of chemical substances, it has become more and more important to identify metals and non-metals in chemical processes and in the environment.

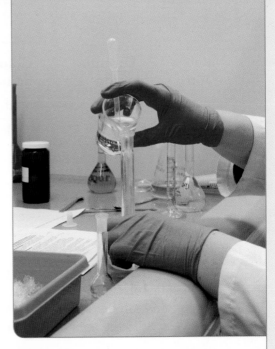

△ Fig. 2.38 A chemist performs a chemical test on a substance in her pharmaceutical laboratory.

KNOWLEDGE CHECK

✓ Understand the nature of the chemical bonding in ionic compounds.
✓ Be familiar with the terms anion and cation.
✓ Know some of the characteristics of Group I and Group VII elements.
✓ Know the order of the common metals that are included in the reactivity series.

LEARNING OBJECTIVES

✓ Be able to describe the tests for the aqueous cations of ammonium, calcium, copper(II), iron(II), iron(III) and zinc using aqueous sodium hydroxide and aqueous ammonia.
✓ Be able to use a flame test to identify lithium, sodium, potassium and copper(II) cations.
✓ Be able to describe the tests for the anions carbonate, chloride, nitrate and sulfate.
✓ Be able to describe the tests for the gases ammonia, carbon dioxide, chlorine, hydrogen and oxygen.

IDENTIFYING METAL IONS (CATIONS)

Ions of metals are **cations** – positive ions – and are found in ionic compounds. There are two ways of identifying metal cations:

- *either* from solids of the compound
- *or* from solutions of the compound.

Flame tests

In a flame test, a piece of nichrome wire is dipped into concentrated hydrochloric acid, then into the solid compound, and then into a blue Bunsen flame. The colour seen in the flame identifies the metal ion in the compound.

Name of ion	Formula of ion	Colour seen in flame
lithium	Li$^+$	bright red
sodium	Na$^+$	golden yellow/orange
potassium	K$^+$	lilac (light purple)
copper	Cu^{2+}	green

△ Table 2.6 Colours of ions in a flame.

Tests on solutions in water

Metal ions are found in ionic compounds, so most of them will dissolve in water to form solutions. These solutions can be tested with sodium hydroxide solution to identify the aqueous cation.

△ Fig. 2.39 The colour of the flame can be used to identify the metal ions present.

Name of ion in solution	Formula	Result
Calcium	Ca^{2+}(aq)	White precipitate formed; remains, even when excess sodium hydroxide solution added
Copper(II)	Cu^{2+}(aq)	Light blue precipitate formed; insoluble in excess sodium hydroxide solution
Iron(II)	Fe^{2+}(aq)	Green precipitate formed; insoluble in excess sodium hydroxide solution; after a few minutes starts to change to reddish-brown colour
Iron(III)	Fe^{3+}(aq)	Reddish-brown precipitate formed; insoluble in excess sodium hydroxide solution
Zinc	Zn^{2+}(aq)	White precipitate formed; dissolves in excess sodium hydroxide solution

△ Table 2.7 Tests for identifying cations by adding sodium hydroxide solution to excess.

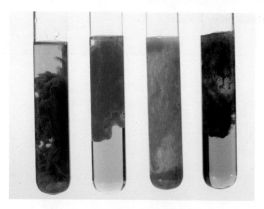

△ Fig. 2.40 Colourful hydroxide precipitates.

Note: Similar results can be obtained if these reactions are performed using ammonia solution instead of sodium hydroxide solution.
However, there is a noticeable difference in the case of copper(II) ions.
At first, as with sodium hydroxide solution, a pale blue precipitate is formed, but then as excess ammonia solution is added the precipitate dissolves to form a royal blue solution.

EXTENDED

The reactions in the table can be represented by ionic equations.
For example:

$$Ca^{2+}(aq) \quad + \quad 2OH^-(aq) \quad \rightarrow \quad Ca(OH)_2(s)$$

white precipitate

$$Cu^{2+}(aq) \quad + \quad 2OH^-(aq) \quad \rightarrow \quad Cu(OH)_2(s)$$

light blue precipitate

$$Fe^{3+}(aq) \quad + \quad 3OH^-(aq) \quad \rightarrow \quad Fe(OH)_3(s)$$

reddish-brown precipitate

END OF EXTENDED

QUESTIONS

1. a) What would you observe if you added sodium hydroxide solution to a solution of calcium chloride?

   b) In what way would the observations be different if zinc chloride solution were used instead of calcium chloride solution?

2. What test can be used to distinguish between Fe^{2+} and Fe^{3+} ions? What is the result of the test with each ion?

IDENTIFYING AMMONIUM IONS, NH$_4^+$

The test for the ammonium ion is shown in Fig. 2.41.

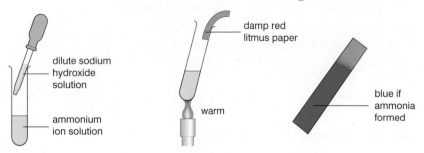

dilute sodium hydroxide solution

ammonium ion solution

damp red litmus paper

warm

blue if ammonia formed

△ Fig. 2.41 Test for the ammonium ion NH$_4^+$.

The suspected ammonium compound is dissolved in water in a test tube and a few drops of dilute sodium hydroxide are added. The mixture is then warmed over a Bunsen burner and some damp red litmus paper (or universal indicator paper) is placed in the mouth of the test tube. A colour change in the indicator to blue (alkaline) shows that an ammonium compound is present.

IDENTIFYING ANIONS

Negative ions (**anions**) can be tested as solids or in solution.

Testing for anions in solids or solutions

The following test for anions in solids applies only to **carbonates**.

Dilute hydrochloric acid is added to the solid, and any gas produced is passed through limewater. If the limewater goes cloudy/milky, the solid contains a carbonate.

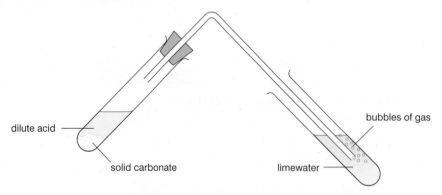

dilute acid

solid carbonate

bubbles of gas

limewater

△ Fig. 2.42 Testing for anions in solids.

This reaction is as follows:

acid + carbonate → a salt + water + carbon dioxide

$$2HCl(aq) + Na_2CO_3(s) \rightarrow 2NaCl(aq) + H_2O(l) + CO_2(g)$$
$$2HCl(aq) + ZnCO_3(s) \rightarrow ZnCl_2(aq) + H_2O(l) + CO_2(g)$$

The reaction between an acid and a carbonate can be represented by the following ionic equation:

$$CO_3^{2-}(s) \quad + \quad 2H^+(aq) \quad \rightarrow \quad CO_2(g) \quad + \quad H_2O(l)$$

Many ionic compounds are soluble in water, and so they form solutions that contain anions.

The tests and results used to identify some other common anions are shown in the table.

Name of ion	Formula	Test	Result
Chloride	$Cl^-(aq)$	Add: 1. Dilute nitric acid 2. Silver nitrate solution	White precipitate (of AgCl)
Sulfate	$SO_4^{2-}(aq)$	Add: 1. Dilute hydrochloric acid 2. Barium chloride solution	White precipitate (of $BaSO_4$)
Nitrate	$NO_3^-(s)$	1. Add sodium hydroxide solution and warm 2. Add aluminium powder 3. Test any gas produced with damp red litmus paper	Red litmus paper goes blue (ammonia gas is produced)

△ Table 2.8 Tests for anions.

◁ Fig. 2.43 Test for the chloride ion – a white precipitate with silver nitrate solution.

This reaction can be represented by an ionic equation:

$$Ag^+(aq) \quad + \quad Cl^-(aq) \quad \rightarrow \quad AgCl(s)$$

white precipitate

QUESTIONS

1. Describe how you would test for an ammonium compound. Give the result of the test.

2. When a carbonate is reacted with dilute hydrochloric acid, a gas is given off.

   a) What is the name of the gas?

   b) What is the test for the gas? Give the result of the test.

3. Sodium hydroxide solution is added to solution X and a reddish-brown precipitate is formed. What metal ion was present in solution X?

4. A mixture of dilute nitric acid and silver nitrate solution is added to solution Y in a test tube. A white precipitate forms. What anion is present in solution Y?

5. EXTENDED Metallic ions in solution can be identified using sodium hydroxide solution.

   Sodium hydroxide is useful because it forms coloured precipitates with many metallic ions although it will form white precipitates with others.

   a) Copy and complete the table with the names of two cations that form white precipitates and three cations that give coloured precipitates.

Name of cation	Colour of precipitate

   b) When testing for sulfate ions, why is it important to add dilute hydrochloric acid before adding barium chloride?

6. A forensic scientist has been provided with a small sample of a blue compound which is suspected to be copper(II) sulfate, and a white compound that is suspected to be sodium carbonate. Devise a series of tests that could be followed to identify the ions. Indicate in your plan the expected results if the samples are to be positively identified.

IDENTIFYING GASES

Many chemical reactions produce a gas as one of the products. Identifying the gas is often a step towards identifying the compound that produced it in the reaction.

Gas	Formula	Test	Result of test
Hydrogen	H_2	Put in a lighted splint (a flame)	'Pop' or 'squeaky pop' heard (flame usually goes out)
Oxygen	O_2	Put in a glowing splint	Splint relights, producing a flame
Carbon dioxide	CO_2	Pass gas through limewater	Limewater goes cloudy/milky
Chlorine	Cl_2	Put in a piece of damp blue litmus paper or universal indicator paper	Paper goes red then white (bleached)
Ammonia	NH_3	Put in a piece of damp red litmus or universal indicator paper	A strong smell is produced and the indicator paper goes blue

△ Table 2.9 Tests for gases.

REMEMBER

Carbon dioxide: cloudiness with limewater is caused by insoluble calcium carbonate. If carbon dioxide continues to be passed through, the cloudiness disappears: $CaCO_3(s)$ is changed to soluble calcium hydrogencarbonate, $Ca(HCO_3)_2(aq)$.

Chlorine: the gas is acidic, but also a bleaching agent.

Ammonia: the only basic gas.

QUESTIONS

1. What is the name of a gas that is alkaline?

2. What is the name of a gas that supports combustion?

3. What is the name of a gas that acts as a bleach?

End of topic checklist

Key terms

anion, carbonate, cation

During your study of this topic you should have learned:

⭕ How to carry out tests for the aqueous cations ammonium, calcium, copper(II), iron(II), iron(III) and zinc using aqueous sodium hydroxide and aqueous ammonia.

⭕ How to use a flame test to identify lithium, sodium, potassium and copper(II).

⭕ How to carry out tests for the anions:

- carbonate by reaction with dilute acid and then limewater
- chloride by reaction under acidic conditions with aqueous silver nitrate
- nitrate by reduction with aluminium
- sulfate by reaction under acidic conditions with aqueous barium ions.

⭕ How to carry out tests for gases:

- ammonia using damp red litmus paper
- carbon dioxide using limewater
- chlorine using damp litmus paper
- hydrogen using a lighted splint
- oxygen using a glowing splint.

End of topic questions

Note: The marks given for these questions indicate the level of detail required in the answers. In the examination, the number of marks given to questions like these may be different.

1. What ions are likely to be present in the compounds X, Y, and Z?

   a) Solution X forms a pale blue precipitate when sodium hydroxide solution is added. **(1 mark)**

   b) Solution Y forms no precipitate when sodium hydroxide solution is added, but produces a strong-smelling, alkaline gas when the mixture is heated. **(1 mark)**

   c) Solution Z forms an orange-brown precipitate when sodium hydroxide solution is added. **(1 mark)**

2. Copy and complete the table about the identification of gases. **(3 marks)**

Gas	Test	Observations
Chlorine	Damp universal indicator paper	
	Bubble through limewater	White precipitate or suspension forms
Hydrogen		Burns with a 'pop'

3. A white powder is labelled 'lithium carbonate'. What test could you do to prove it was a carbonate? **(2 marks)**

4. How would you test a solid to identify the presence of each of the ions shown below?

   a) the sulfate ion, SO_4^{2-} **(3 marks)**

   b) the nitrate ion, NO_3^- **(3 marks)**

5. EXTENDED Write ionic equations for the following reactions:

   a) between copper(II) sulfate and sodium hydroxide solution **(2 marks)**

   b) between sodium carbonate and dilute hydrochloric acid. **(2 marks)**

This section concentrates on a 'branch' of chemistry known as inorganic chemistry. As the title suggests, it focuses on the chemical elements, of which there are over 100. This may seem rather a lot, but the good news is that you will not study all 100 elements! However, because the chemical elements are arranged in a particular pattern, known as the Periodic Table, learning about one element often provides a very good idea of how other elements may behave. So, it should be possible to learn about the chemistry of about 45 elements from studying this section.

The section starts with the Periodic Table. You will learn how the elements are arranged into groups and periods. You will then study a group of metals and a group of non-metals, followed by the transition metals and noble gases. The topic on metals highlights differences in the reactivity of metals and how this influences the methods used to extract them from their ores. A topic on air and water allows a consideration of the environmental impact of living in an industrial world.

STARTING POINTS

1. What is an element – how would you define the term?

2. What does the proton number of an atom tell you about its structure?

3. In terms of electronic structures, what is the difference between a metal and a non-metal?

4. You will be learning about the Periodic Table of elements. Look at the Periodic Table and make a list of the things that you notice about it.

5. You will be learning about the composition of gases in the air. What is the most abundant gas in the air?

6. Make a list of six to eight metals that you have come across. Which metal in your list do you think is the most reactive? Which metal do you think is the least reactive? Explain your choices.

SECTION CONTENTS

a) The Periodic Table

b) Group I elements

c) Group VII elements

d) Transition metals and noble gases

e) Metals

f) Air and water

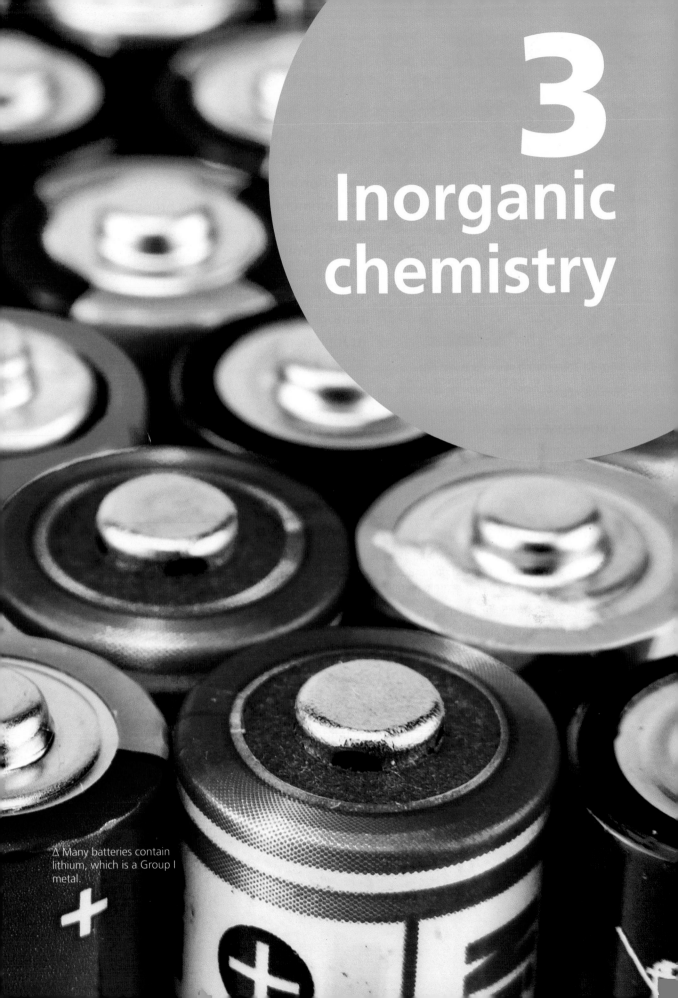

3 Inorganic chemistry

△ Many batteries contain lithium, which is a Group I metal.

The Periodic Table

INTRODUCTION

With over 100 different elements in existence, it's very important to have some way of ordering them. The Periodic Table puts elements with similar properties into columns, with a gradual change in properties moving from left to right along the rows. This topic looks at some of the basic features of the Periodic Table. Later topics will look in more detail at particular groups and arrangements of the elements.

△ Fig. 3.1 This ordering of elements was first published in 1871 by the Russian chemist Dmitri Mendeleev.

KNOWLEDGE CHECK

✓ Understand that all matter is made up of elements.
✓ Know that the proton number of an element gives the number of protons (and electrons) in an atom of the element.
✓ Know that electrons are arranged in shells around the nucleus of the atom.

LEARNING OBJECTIVES

✓ Be able to describe the Periodic Table as a method of classifying elements and recognise its use in predicting properties of elements.
✓ Be able to describe the change from metallic to non-metallic character across a period.

✓ **EXTENDED** Be able to describe the relationship between group number, number of valency electrons and metallic/non-metallic character.

THE ARRANGEMENT OF THE PERIODIC TABLE

As new elements were discovered in the nineteenth century, chemists tried to organise them into patterns based on the similarities in their properties. The English chemist John Newlands was the first to classify elements according to their properties and produced his classification system before Mendeleev produced his. When the structure of the atom was better known, elements were arranged in order of increasing proton number, and then the patterns started to make more sense. (Proton number is the number of protons in an atom.)

HOW ARE ELEMENTS CLASSIFIED IN THE MODERN PERIODIC TABLE?

More than 110 elements have now been identified, and each has its own properties and reactions. In the **Periodic Table**, elements with similar properties and reactions are shown close together.

The Periodic Table arranges the elements in order of increasing proton number. They are then arranged in periods and groups.

Groups	I	II												III	IV	V	VI	VII	VIII or 0
Periods																			
1							H hydrogen 1												He helium 2
2	Li lithium 3	Be beryllium 4												B boron 5	C carbon 6	N nitrogen 7	O oxygen 8	F fluorine 9	Ne neon 10
3	Na sodium 11	Mg magnesium 12		transition metals										Al aluminium 13	Si silicon 14	P phosphorus 15	S sulfur 16	Cl chlorine 17	Ar argon 18
4	K potassium 19	Ca calcium 20	Sc scandium 21	Ti titanium 22	V vanadium 23	Cr chromium 24	Mn manganese 25	Fe iron 26	Co cobalt 27	Ni nickel 28	Cu copper 29	Zn zinc 30		Ga gallium 31	Ge germanium 32	As arsenic 33	Se selenium 34	Br bromine 35	Kr krypton 36
5	Rb rubidium 37	Sr strontium 38	Y yttrium 39	Zr zirconium 40	Nb niobium 41	Mo molybdenum 42	Tc technetium 43	Ru ruthenium 44	Rh rhodium 45	Pd palladium 46	Ag silver 47	Cd cadmium 48		In indium 49	Sn tin 50	Sb antimony 51	Te tellurium 52	I iodine 53	Xe xenon 54
6	Cs caesium 55	Ba barium 56	La lanthanum 57	Hf hafnium 72	Ta tantalum 73	W tungsten 74	Re rhenium 75	Os osmium 76	Ir iridium 77	Pt platinum 78	Au gold 79	Hg mercury 80		Tl thallium 81	Pb lead 82	Bi bismuth 83	Po polonium 84	At astatine 85	Rn radon 86

metal	non metal	transition metal	metalloid

Δ Fig. 3.2 The Periodic Table.

Periods

Horizontal rows of elements are arranged in increasing proton number from left to right. Rows correspond to **periods**, which are numbered from 1 to 7.

Moving across a period, each successive atom of the elements gains one proton and one electron (in the same outer shell/orbit).

You can see how this works in Fig. 3.3.

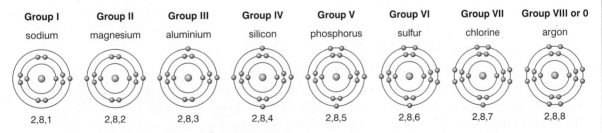

Group I	Group II	Group III	Group IV	Group V	Group VI	Group VII	Group VIII or 0
sodium	magnesium	aluminium	silicon	phosphorus	sulfur	chlorine	argon
2,8,1	2,8,2	2,8,3	2,8,4	2,8,5	2,8,6	2,8,7	2,8,8

Δ Fig. 3.3 Moving across a period shows the atomic structure of each element.

Moving across a period such as Period 3 (sodium to argon), the following trends take place:

- Metals on the left going to non-metals on the right.
- Group I elements are the most reactive metal group, and as you go to the right the reactivity of the groups decreases. Group IV elements are the least reactive.
- Continuing right from Group IV, the reactivity increases until Group VII, the most reactive of the non-metal groups.

Groups

Vertical columns contain elements with the proton number increasing down the column – they are called **groups**. They are numbered from I to VII and 0 (Group 0 is sometimes referred to as Group VIII).

Groups are referred to as 'families' of elements because they have similar characteristics, just like families – the alkali metals (Group I), the alkaline earth metals (Group II) and the halogens (Group VII).

REMEMBER

It is important to understand the relationship between group number, number of outer electrons, and metallic and non-metallic character across periods.

QUESTIONS

1. Find the element calcium in the Periodic Table. Answer these questions about calcium:

  a) What is its proton number?

  b) What information does the proton number give about the structure of a calcium atom?

  c) Which group of the Periodic Table is calcium in?

  d) Which period of the Periodic Table is calcium in?

  e) Is calcium a metal or a non-metal?

2. What is the family name for the Group VII elements?

3. Are the Group VII elements metals or non-metals?

EXTENDED

CHARGES ON IONS AND THE PERIODIC TABLE

We can explain why elements in the same group have similar reactions in terms of the electron structures of their atoms. Elements with the same number of electrons in their outer shells have similar chemical properties. The relationship between the group number and the number of electrons in the outer electron shell is shown in Table 3.1.

Group number	I	II	III	IV	V	VI	VII	VIII or 0
Electrons in the outer electron shell	1	2	3	4	5	6	7	2 or 8

△ Table 3.1 Relationship between group number and number of electrons in outer shell.

The ion formed by an element can be worked out from the element's position in the Periodic Table. The elements in Group IV and Group VIII (or 0) generally do not form ions.

Group number	I	II	III	IV	V	VI	VII	VIII (or 0)
Ion charge	1+	2+	3+	Typically no ions	3–	2–	1–	No ions
Metallic or non-metallic	Metallic			Non-metallic, metalloid and metallic	Non-metallic (except for some metalloids)			

△ Table 3.2 Groups and their ions.

REACTIVITIES OF ELEMENTS

Going from the top to the bottom of a group in the Periodic Table, metals become more reactive, but non-metals become less reactive. As the metal atom gets bigger, the outer electrons get further away from the nucleus and can be removed more easily to form positive ions. So, the larger metal atoms can react more easily with other elements and form compounds.

The reverse is true for a group of non-metal atoms: the smaller the atom, the easier it is to accept electrons and form ions. So, the smaller non-metal atoms react more easily with other elements to form compounds.

Group VIII or 0 elements, known as the noble gases, are very unreactive. They already have full outer electron shells (two electrons for helium and eight electrons for the other noble gases).

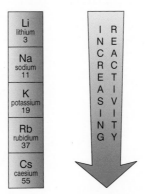

△ Fig. 3.4 The Group VII elements (non-metal) become more reactive further up the group.

△ Fig. 3.5 Group I elements (metals) become more reactive further down the group.

QUESTIONS

1. How many electrons does an aluminium atom have in its outer shell?

2. What ion charge does oxygen have?

3. Which is the most reactive element in Group VII?

4. Which is the most reactive element in Group II?

END OF EXTENDED

SCIENCE IN CONTEXT

THE FIRST PERIODIC TABLE

In 1871 the Russian chemist Dmitri Mendeleev published his work on the Periodic Table. It included the 66 elements that were known at the time. Interestingly, Mendeleev left gaps in his arrangement when the next element in his order did not seem to fit. He predicted that there should be elements in the gaps but that they had yet to be discovered. One such element is gallium (discovered in 1875), which Mendeleev predicted would be between aluminium and indium.

By June 2011 there were 118 known elements, but only 91 of these occurred naturally – the others had been made artificially. Some of these artificial elements can now be detected in small quantities in the environment, for example, the element americium (Am, proton number 95), which is used in smoke detectors.

End of topic checklist

Key terms

group, period, Periodic Table

During your study of this topic you should have learned:

○ How to describe the Periodic Table as a method of classifying elements and its use to predict properties of elements.

○ How to describe the change from metallic to non-metallic character across a period.

○ **EXTENDED** How to describe the relationship between group number, number of valency electrons and metallic/non-metallic character.

End of topic questions

Note: The marks given for these questions indicate the level of detail required in the answers. In the examination, the number of marks given to questions like these may be different.

1. Look at the diagram representing the Periodic Table. The letters stand for elements.

    a) Which element is in Group IV?   (1 mark)

    b) Which element is in the second period?   (1 mark)

    c) Which element is a noble gas?   (1 mark)

    d) Which element is a transition metal?   (1 mark)

    e) Which elements are non-metals?   (1 mark)

    f) Which element is most likely to be a gas?   (1 mark)

2. What are the electron arrangements in the following atoms?

    a) sodium (proton number = 11)   (1 mark)

    b) silicon (proton number = 14)   (1 mark)

    c) fluorine (proton number = 9)   (1 mark)

3. How does the metallic and non-metallic nature of the elements change across Period 3 of the Periodic Table?   (1 mark)

4. Why do elements in the same group have similar chemical properties?   (1 mark)

5. What ions would you expect the following atoms to form?

    a) sodium   (1 mark)

    b) chlorine.   (1 mark)

6. In the Periodic Table, what is the trend in reactivity:

    a) down a group of metals?   (1 mark)

    b) down a group of non-metals?   (1 mark)

7. Explain why the noble gases in Group 0 are very unreactive.   (2 marks)

Group I elements

INTRODUCTION

Metals are positioned on the left-hand side and in the middle of the Periodic Table. Therefore the Group I elements are metals, but rather different from the metals in everyday use. In fact, when you see how the Group I metals react with air and water, it is hard to think how they could be used outside the laboratory. This very high reactivity makes them interesting to study. Our focus is on the first three elements in the group: lithium, sodium and potassium. Rubidium, caesium and francium are not available in schools because they are too reactive.

△ Fig. 3.6 Potassium reacting with water.

REACTIVITY OF GROUP I ELEMENTS

All Group I elements react with water to produce an alkaline solution. This makes them recognisable as a 'family' of elements, often called the **alkali metals.**

These very reactive metals all have only one electron in their outer electron shell. This electron is easily given away when the metal reacts with non-metals. The more electrons a metal atom has to lose in a reaction, the more energy is needed to start the reaction. This is why the Group II elements are less reactive – they have to lose two electrons when they react.

Reactivity increases down the group because as the atom gets bigger the outer electron is further away from the nucleus and so can be removed more easily, as the atoms react to form positive ions.

PROPERTIES OF GROUP I METALS

The properties of Group I metals are as follows:

- Soft to cut.
- Shiny when cut, but quickly tarnish in the air.
- Very low melting points compared with most metals – melting points decrease down the group.
- Very low densities compared with most metals. Lithium, sodium and potassium float on water. Densities increase down the group.
- React very easily with air, water and elements such as chlorine. The alkali metals are so reactive that they are stored in oil to prevent reaction with air and water. Reactivity increases down the group.

△ Fig. 3.7 The freshly cut surface of sodium.

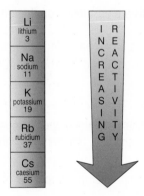

△ Fig. 3.8 Group I elements become more reactive as you go down the group.

QUESTIONS

1. Why are the Group I elements known as the *alkali metals*?

2. How many electrons do the Group I elements atoms have in their outer shell?

3. The Group I metals are unusual metals. Give one property they have that is different to most other metals.

4. **EXTENDED** Predict how the melting point of rubidium will compare to that of sodium. Explain your answer.

5. **EXTENDED** Why is potassium more reactive than lithium?

Reaction	Observations	Equations
Air or oxygen	The metals burn easily and their compounds colour flames: lithium – red sodium – orange/yellow potassium – lilac A white solid oxide is formed.	lithium + oxygen → lithium oxide $4Li(s) + O_2(g) \rightarrow 2Li_2O(s)$ sodium + oxygen → sodium oxide $4Na(s) + O_2(g) \rightarrow 2Na_2O(s)$ potassium + oxygen → potassium oxide $4K(s) + O_2(g) \rightarrow 2K_2O(s)$
Water	The metals react vigorously. They float on the surface, moving around rapidly. With both sodium and potassium, the heat of the reaction melts the metal so it forms a sphere; bubbles of gas are given off and the metal 'disappears'. With the more reactive metals (such as potassium) the hydrogen gas produced burns. The resulting solution is alkaline.	lithium + water → lithium hydroxide + hydrogen $2Li(s) + 2H_2O(l) \rightarrow 2LiOH(aq) + H_2(g)$ sodium + water → sodium hydroxide + hydrogen $2Na(s) + 2H_2O(l) \rightarrow 2NaOH(aq) + H_2(g)$ potassium + water → potassium hydroxide + hydrogen $2K(s) + 2H_2O(l) \rightarrow 2KOH(aq) + H_2(g)$
Chlorine	The metals react easily, burning in the chlorine to form a white solid, the metal chloride.	lithium + chlorine → lithium chloride $2Li(s) + Cl_2(g) \rightarrow 2LiCl(s)$ sodium + chlorine → sodium chloride $2\ Na(s) + Cl_2(g) \rightarrow 2NaCl(s)$ potassium + chlorine → potassium chloride $2K(s) + Cl_2(g) \rightarrow 2KCl(s)$

△ Table 3.3 Reactions of Group I metals.

COMPOUNDS OF THE GROUP I METALS

The compounds of Group I metals are usually colourless crystals or white solids and always have ionic bonding. Most of them are soluble in water. Some examples are sodium chloride (NaCl) and potassium nitrate (KNO_3).

The compounds of the alkali metals are widely used:

- lithium carbonate – as a hardener in glass and ceramics
- lithium hydroxide – removes carbon dioxide in air-conditioning systems
- sodium chloride – table salt
- sodium carbonate – a water softener
- sodium hydroxide – used in paper manufacture
- monosodium glutamate – a flavour enhancer
- sodium sulfite – a preservative
- potassium nitrate – a fertiliser; also used in explosives.

△ Fig. 3.9 Sodium burning in chlorine.

QUESTIONS

1. Sodium burns in oxygen to make sodium oxide. What colour would you expect sodium oxide to be?

2. What gas is produced when potassium reacts with water? What is the name of the solution formed in this reaction?

3. Are the compounds of the Group I metals usually soluble or insoluble in water?

It is important to understand the structure of the Periodic Table and how it relates to the properties and reactions of the elements. These questions link electronic structure with the reactivity trends of the different elements in the Periodic Table.

4. EXTENDED Predict how the melting point of rubidium will compare to that of sodium.

5. EXTENDED These questions are based on Group I of the Periodic Table.

   a) What do you understand by a *group* in the Periodic Table?

   b) Name the first three metals of Group I.

   c) Why do the elements in this group have similar chemical properties?

   d) Group I elements are very reactive. Suggest reasons for this.

SCIENCE IN CONTEXT

FACTS ABOUT THE GROUP I METALS

1. Lithium is found in large quantities (estimated at 230 billion tonnes) in compounds in seawater.

2. Sodium is found in many minerals and is the sixth most abundant element overall in the Earth's crust (amounting to 2.6% by weight).

3. Potassium is also found in many minerals and is the seventh most abundant element in the Earth's crust (amounting to 1.5% by weight).

4. Rubidium was discovered by Bunsen (of Bunsen burner fame) in 1861. It is more abundant than copper, about the same as zinc, and is found in very small quantities in a large number of minerals. Because of this low concentration in mineral deposits, only 2 to 4 tonnes of rubidium are produced each year worldwide.

5. Caesium is more abundant than tin, mercury and silver. However, its very high reactivity makes it very difficult to extract from mineral deposits.

6. Francium was discovered as recently as 1939 as a product of the radioactive decay of an isotope of actinium.

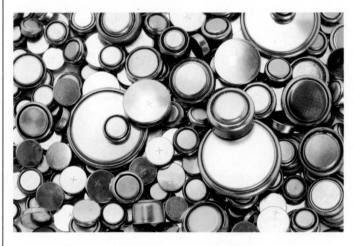

△ Fig. 3.10 Lithium is used in all of these batteries.

End of topic checklist

Key terms

alkali metal

During your study of this topic you should have learned:

○ How to describe lithium, sodium and potassium in Group I as a collection of relatively soft metals showing trends in melting point, density and reaction with water.

○ **EXTENDED** How to predict the properties of other elements in Group I, given data where appropriate.

End of topic questions

Note: The marks given for these questions indicate the level of detail required in the answers. In the examination, the number of marks given to questions like these may be different.

1. This question is about the Group I elements lithium, sodium and potassium.

   a) Which is the most reactive of these elements? **(1 mark)**

   b) Why are the elements stored in oil? **(1 mark)**

   c) Which element is the easiest to cut? **(1 mark)**

   d) Why do the elements tarnish quickly when they are cut? **(1 mark)**

   e) Why does sodium float when added to water? **(1 mark)**

2. Why are the Group I elements known as the 'alkali metals'? **(2 marks)**

3. Write word equations and balanced equations for the following reactions:

   a) lithium and oxygen **(3 marks)**

   b) potassium and water **(3 marks)**

   c) potassium and chlorine. **(3 marks)**

4. **EXTENDED** This question is about rubidium (symbol Rb), which is a less common Group I element.

   a) What state of matter would you expect rubidium to be in at room temperature and pressure? **(1 mark)**

   b) When rubidium is added to water:

   i) Which gas is formed? **(1 mark)**

   ii) What chemical compound would be formed in solution? What result would you predict if universal indicator was added to the solution? **(2 marks)**

   c) Would you expect rubidium to be more or less reactive than potassium? Explain your answer. **(2 marks)**

5. **EXTENDED** Explain why potassium is more reactive than sodium. **(3 marks)**

Group VII elements

INTRODUCTION

Group VII elements are located on the right-hand side of the Periodic Table with the other non-metals. They look very different from each other, so it may seem strange that they are in the same group. However, their chemical properties are very similar, and all of them are highly reactive. This topic focuses on chlorine, bromine and iodine. Fluorine is a highly reactive gas and astatine is a radioactive black solid with a very short half-life (so will exist in only very small quantities).

△ Fig. 3.11 At room temperature and atmospheric pressure, chlorine is a pale green gas, bromine a red–brown liquid and iodine is a black solid.

KNOWLEDGE CHECK

✓ Understand that non-metals are positioned on the right-hand side of the Periodic Table.
✓ Know that the elements in a group have similar electron arrangements.
✓ Be familiar with the terms oxidation and reduction.

LEARNING OBJECTIVES

✓ Be able to describe chlorine, bromine and iodine in Group VII as a collection of diatomic non-metals showing trends in colour and physical state.
✓ **EXTENDED** Be able to describe the reactions of chlorine, bromine and iodine with other halide ions.
✓ **EXTENDED** Be able to predict the properties of other elements in Group VII, given data where appropriate.

REACTIVITY OF GROUP VII ELEMENTS

The Group VII elements are sometimes referred to as the **halogen** elements or halogens.

'Halogen' means 'salt-maker' – halogens react with most metals to make salts.

Halogen atoms have seven electrons in their outermost electron shell, so they need to gain only one electron to obtain a full outer shell. This is what makes them very reactive. They react with metals, gaining an electron and forming a singly charged negative ion.

The reactivity of the elements decreases down the group because as the atoms gets bigger, an eighth electron will be further from the attractive force of the nucleus. This makes it harder for the atom to gain this electron.

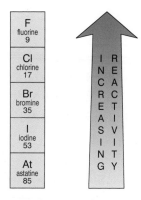

△ Fig. 3.12 Increasing reactivity goes up Group VII.

PROPERTIES OF GROUP VII ELEMENTS

The properties of the Group VII elements are as follows.

- Fluorine is a pale yellow gas; chlorine is a pale green gas; bromine is a red–brown liquid; iodine is a black shiny solid.
- All the atoms have seven electrons in their outermost electron shell.
- All exist as diatomic molecules, that is, each molecule contains two atoms. For example, F_2, Cl_2, Br_2, I_2.
- Halogens react with water and react with metals to form salts. Iodine has very low solubility and little reaction with water.

- They undergo **displacement reactions**.

Reaction	Observations	Equations
Water 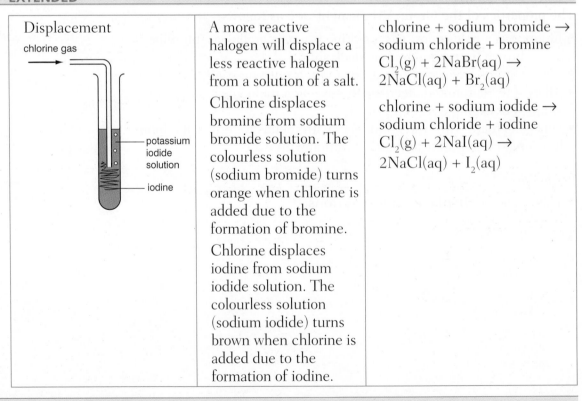	The halogens dissolve in water and also react with it, forming solutions that behave as bleaches. Chlorine solution is pale yellow. Bromine solution is orange. Iodine solution is yellow/brown.	chlorine + water → hydrochloric acid + chloric(I) acid $Cl_2(g) + H_2O(l) → HCl(aq) + HClO(aq)$
Metals	The halogens form salts with all metals. For example, gold leaf will catch fire in chlorine without heating. With a metal such as iron, brown fumes of iron(III) chloride form.	iron + chlorine → iron(III) chloride $2Fe(s) + 3Cl_2(g) → 2FeCl_3(s)$ Fluor*ine* forms salts called fluor*ides* Chlor*ine* forms salts called chlor*ides* Brom*ine* forms salts called brom*ides* Iod*ine* forms salts called iod*ides*

EXTENDED

Displacement	A more reactive halogen will displace a less reactive halogen from a solution of a salt. Chlorine displaces bromine from sodium bromide solution. The colourless solution (sodium bromide) turns orange when chlorine is added due to the formation of bromine. Chlorine displaces iodine from sodium iodide solution. The colourless solution (sodium iodide) turns brown when chlorine is added due to the formation of iodine.	chlorine + sodium bromide → sodium chloride + bromine $Cl_2(g) + 2NaBr(aq) → 2NaCl(aq) + Br_2(aq)$ chlorine + sodium iodide → sodium chloride + iodine $Cl_2(g) + 2NaI(aq) → 2NaCl(aq) + I_2(aq)$

END OF EXTENDED

△ Table 3.4 Properties of the Group VII elements.

QUESTIONS

1. How many electrons do the Group VII element atoms have in their outer shell?

2. Why are the Group VII elements particularly reactive when compared with other non-metals?

3. Chlorine exists as diatomic molecules. Explain what this means.

4. EXTENDED Astatine is an element in Group VII. Predict whether you would expect it to be a solid, liquid or gas at room temperature. Explain your answer.

5. EXTENDED What is meant by a 'displacement reaction' involving the Group VII elements?

EXTENDED

Developing investigative skills

A student was provided with three aqueous solutions containing chlorine, bromine and iodine. She added a few drops of cyclohexane to each solution in separate test tubes and then stirred each with a clean glass rod. Cyclohexane does not mix with water, it floats on top of aqueous solutions. When each cyclohexane layer separated from the solution, she recorded the colours of the cyclohexane layers in the table:

Solution in water	Colour of the cyclohexane layer
chlorine	colourless
bromine	orange
iodine	violet

She cleaned out the tubes and then performed a series of test tube reactions as indicated in the table below. In each case she mixed small quantities of solution A with twice the volume of solution B, added 10 drops of cyclohexane and then stirred the mixture with a clean glass rod. Once the cyclohexane layer had separated, she recorded her results.

Solution A	Solution B	Colour of cyclohexane layer
aqueous chlorine	sodium bromide	orange
aqueous chlorine	sodium iodide	violet
aqueous bromine	sodium chloride	orange
aqueous bromine	sodium iodide	violet
aqueous iodine	sodium chloride	violet
aqueous iodine	sodium bromide	orange

Using and organising techniques, apparatus and materials

❶ Cyclohexane is highly flammable and the chlorine and bromine solutions are both irritants. What precautions should the student have taken when doing this experiment?

Interpreting observations and data

❷ What can you deduce about the relative reactivity of chlorine, bromine and iodine from the *first two* results in the second table?

❸ Write an equation for the reaction indicated by result 4 in the second table.

Evaluating methods

❹ The student made a mistake in recording one of her results. Which one? Explain how you know.

END OF EXTENDED

USES OF HALOGENS

Halogens and their compounds have a wide range of uses:

- fluorides – in toothpaste help prevent tooth decay
- fluorine compounds – making plastics such as Teflon (the non-stick surface on pans)
- chlorofluorocarbons – propellants in aerosols and refrigerants (now being replaced because of their damaging effect on the ozone layer)
- chlorine – purifying water
- chlorine compounds – household bleaches
- hydrochloric acid – widely used in industry
- bromine compounds – making pesticides
- silver bromide – the light-sensitive film coating on photographic film
- iodine solution – an antiseptic.

SCIENCE IN CONTEXT **FLUORINE**

Fluorine is the most reactive non-metal in the Periodic Table. It reacts with most other elements except helium, neon and argon. These reactions are often sudden or explosive. Even radon, a very unreactive noble gas, burns with a bright flame in a jet of fluorine gas. All metals react with fluorine to form fluorides. The reactions of fluorine with Group I metals are explosive.

Early scientists tried to make fluorine from hydrofluoric acid (HF(aq)) but this proved to be highly dangerous, killing or blinding several scientists who attempted it. They became known as the 'fluorine martyrs'. Today fluorine is manufactured by the electrolysis of the mineral fluorite, which is calcium fluoride.

Fluorine is not an element to play with. You will certainly not see it in your laboratory!

QUESTIONS

1. Why is chlorine used in the treatment of drinking water in many countries?

2. Which halogen element has medical uses as an antiseptic?

3. EXTENDED Fluorine is used to make a plastic material with the common name of 'Teflon'. What is Teflon used for?

You should be familiar with the elements of Group VII, the halogens. These are coloured non-metallic elements of varying reactivity. Although they are potentially harmful, their properties make them very useful. Use your knowledge of atomic structure, bonding and reaction types to answer the questions below.

4. EXTENDED Chlorine is a pale green gas obtained by the electrolysis of an aqueous solution of sodium chloride. Chlorine can be used to kill bacteria and is used in the manufacture of bleach.

   a) The electronic structure of a chlorine atom is 2,8,7. Draw simple diagrams to show the arrangement of the outer electrons in a diatomic molecule of Cl_2 and a chloride ion, Cl^-.

   b) Chlorine will displace bromine from a solution of potassium bromide to form bromine and potassium chloride. Explain why this reaction takes place and describe what you would observe if chlorine water was added to a solution of potassium bromide in a test tube.

5. EXTENDED Fluorine is a pale yellow gas and is the most reactive of the chemical elements. It is so reactive that glass, metals and even water burn with a bright flame in a jet of fluorine gas. Fluorides, however, are often added to toothpaste and, controversially, to some water supplies to prevent dental decay.

   a) Give a reason why fluorine is so reactive.

   b) Potassium fluoride is a compound that may be found in toothpaste. Explain why fluorine cannot be displaced from this compound using either chlorine or iodine.

End of topic checklist

Key terms

displacement reaction, halogens

During your study of this topic you should have learned:

○ How to describe chlorine, bromine and iodine in Group VII as a collection of diatomic non-metals showing trends in colour and density.

○ EXTENDED How to describe the reactions of chlorine, bromine and iodine with other halide ions.

○ EXTENDED How to predict the properties of other elements in Group VII, given data, where appropriate.

End of topic questions

Note: The marks given for these questions indicate the level of detail required in the answers. In the examination, the number of marks given to questions like these may be different.

1. This question is about the Group VII elements: chlorine, bromine and iodine.

   a) Which is the most reactive of these elements? (1 mark)

   b) Which of the elements exists as a liquid at room temperature and pressure? (1 mark)

   c) Which of the elements exists as a solid at room temperature and pressure? (1 mark)

   d) What is the appearance of bromine? (1 mark)

2. Explain the following statements:

   a) The Group VII elements are the most reactive non-metals. (2 marks)

   b) The most reactive halogen is at the top of its group. (2 marks)

3. Write word and balanced equations for the following reactions:

   a) sodium and chlorine (3 marks)

   b) magnesium and bromine (3 marks)

   c) hydrogen and fluorine. (3 marks)

4. EXTENDED Aqueous bromine reacts with sodium iodide solution.

   a) What type of chemical reaction is this? (1 mark)

   b) Write a balanced equation for the reaction. (2 marks)

Transition metals and noble gases

△ Fig. 3.13 This incandescent light bulb contains unreactive argon instead of air.

INTRODUCTION

There are two other important families of elements. The first are the transition elements, a 'block' of metals – including more 'everyday' metals than Group I. The second is the noble gases (Group VIII or 0), a group of elements of interest because of their uses rather than their chemical reactions.

KNOWLEDGE CHECK

✓ Understand that metals are positioned on the left side and the middle of the Periodic Table.
✓ Understand that non-metals are positioned on the right side of the Periodic Table.
✓ Know that elements in a group have similar electron arrangements.

LEARNING OBJECTIVES

✓ Be able to describe the transition elements as a collection of metals with high densities, high melting points and forming coloured compounds.
✓ Know that transition elements and their compounds are often used as catalysts.
✓ Be able to describe the noble gases as being unreactive, monatomic gases and explain this in terms of electronic structure.
✓ Be able to describe the uses of the noble gases in providing an inert atmosphere, such as argon in lamps and helium in balloons.

TRANSITION ELEMENTS

The **transition metals** are grouped in the centre of the Periodic Table and include iron, copper, zinc and chromium.

All the transition metals have more than one electron in their outer electron shell. They are much less reactive than Group I and Group II metals and so are more 'everyday' metals. They have much higher melting points and densities. They react much more slowly with water and with oxygen.

They are widely used as construction metals (particularly iron through steel).

One of the typical properties of transition metals and their compounds is their ability to act as **catalysts** and speed up the rate of a chemical reaction by providing an alternative pathway with a lower activation energy, for example, vanadium(V) oxide in the Contact process and iron in the Haber process.

Property	Group I metal	Transition metal
Melting point	Low	High
Density	Low	High
Colour of compounds	White	Mainly coloured
Reactions with water/air	Vigorous	Slow or no reaction
Reactions with an acid	Violent (dangerous)	Slow or no reaction

△ Table 3.5 Properties of the Group I metals and the transition metals.

The compounds of the transition metals are usually coloured. Copper compounds are usually blue or green; iron compounds tend to be either green or brown. When sodium hydroxide solution is added to a solution of a transition metal compound, a precipitate of the metal hydroxide is formed. The colour of the precipitate helps to identify the metal. For example:

copper(II) sulfate + sodium hydroxide → copper(II) hydroxide + sodium sulfate

$CuSO_4(aq)$ + $2NaOH(aq)$ → $Cu(OH)_2(s)$ + $Na_2SO_4(aq)$

EXTENDED

This can be written as an ionic equation:

$Cu^{2+}(aq) + 2OH^-(aq) \rightarrow Cu(OH)_2(s)$

END OF EXTENDED

Colour of metal hydroxide	Likely metal present
Blue	Copper(II) Cu^{2+}
Green	Nickel(II) Ni^{2+}
Green turning to brown	Iron(II) Fe^{2+}
Orange/brown	Iron(III) Fe^{3+}

△ Table 3.6 Transition metal hydroxides and their colours.

QUESTIONS

1. Would you expect a reaction to happen between copper and water?

2. a) Write a fully balanced equation for the reaction between iron(II) sulfate solution and sodium hydroxide solution.

   b) What will be the colour of the precipitate formed?

THE NOBLE GASES

This is actually a group of *very* unreactive non-metals. They used to be called the inert gases as it was thought that they didn't react with anything. But scientists later managed to produce fluorine compounds of some of the **noble gases**. As far as your school laboratory work is concerned, however, they are completely unreactive.

Name	Symbol
Helium	He
Neon	Ne
Argon	Ar
Krypton	Kr
Xenon	Xe
Radon	Rn

△ Table 3.7 The noble gases.

The unreactivity of the noble gases can be explained in terms of their electronic structures. The atoms all have complete outer electron shells or eight electrons in their outer shell. They don't need to lose electrons (as metals do), or gain electrons (as most non-metals do).

Similarities of the noble gases

- Full outer electron shells
- Very unreactive
- Gases
- Exist as single atoms – they are **monatomic** (He, Ne, Ar, Kr, Xe, Rn)

How are the noble gases used?

- Helium – in balloons
- Neon – in red tube lights
- Argon – in lamps and light bulbs

End of topic checklist

Key terms

catalyst, monatomic, noble gas, transition metal

During your study of this topic you should have learned:

◯ How to describe the transition elements as a collection of metals with high densities, high melting points and forming coloured compounds.

◯ That transition elements and their compounds are often used as catalysts.

◯ How to describe the noble gases as being unreactive, monatomic gases and explain this in terms of electronic structure.

◯ How to describe the uses of the noble gases in providing an inert atmosphere, such as argon in lamps and helium in balloons.

End of topic questions

Note: The marks given for these questions indicate the level of detail required in the answers. In the examination, the number of marks given to questions like these may be different.

1. This question is about the transition metals.

   a) Give two differences in the physical properties of the transition metals compared with the alkali metals. (2 marks)

   b) Transition metals are used as catalysts. What is a *catalyst*? (1 mark)

   c) Suggest why the alkali metals are more reactive than the transition metals. (2 marks)

2. Look at the table of observations.

Compound tested	Colour of compound	Effect of adding sodium hydroxide solution to a solution of the compound
A	White	No change
B	Blue	Blue precipitate formed
C	White	White precipitate formed

   a) Which of the compounds, A, B or C, contains a transition metal? Explain your answer. (1 mark)

   b) Which transition metal do you think it is? (1 mark)

   c) Compound B is a metal sulfate. Write a balanced equation for the reaction between a solution of this transition metal compound and sodium hydroxide solution. (2 marks)

3. Explain why the noble gases are so unreactive. (2 marks)

4. The noble gases are *monatomic*. What does this mean? (1 mark)

5. Although the noble gases are generally very unreactive, reactions do occur with very reactive elements such as fluorine. Which of the noble gases are more likely to react – helium at the top of the group or xenon near the bottom of the group? (1 mark)

Metals

INTRODUCTION

Metals are very important in our everyday lives and many have very similar physical properties. Some metals are highly reactive, such as the Group I metals on the left-hand side of the Periodic Table. Other metals are much less reactive, such as the transition metals in the middle of the Periodic Table. Knowing the order of the reactivity of metals can help chemists make very accurate predictions about how the metals will react with different substances and also what individual metals can be used for.

△ Fig. 3.14 What sort of properties should the metals used in the construction of this helicopter have?

PROPERTIES OF METALS

Most metals have similar physical properties.

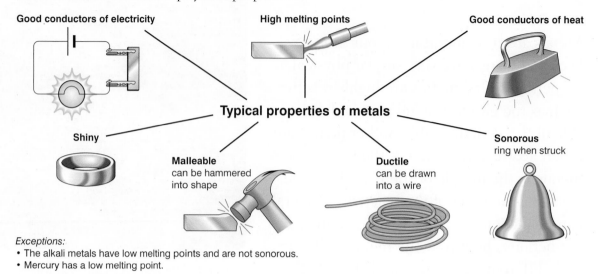

Exceptions:
- The alkali metals have low melting points and are not sonorous.
- Mercury has a low melting point.

△ Fig. 3.15 Properties of metals.

Alloys

An **alloy** is a mixture of a metal with one or more other elements.

Common examples are:

Alloy	Constituents
Brass	Copper (70%), zinc (30%)
Bronze	Copper (90%), tin (10%)
Steel	Iron and small amounts of carbon
Solder	Tin (50%), lead (50%)

△ Table 3.8 Common alloys.

The reason for producing alloys is to 'improve' the properties of a metal. Table 3.9 shows some examples.

Alloy	Property improved
Steel	Hardness/tensile strength
Bronze	Hardness
Solder	Lower melting point
Cupronickel	Cheaper than silver (used for coins)
Stainless steel	Resistance to corrosion
Brass	Easier to shape and stamp into shape

△ Table 3.9 Alloys and their properties.

△ Fig. 3.16 Alloys are used to make coins.

The structure of alloys

The structure of pure metallic elements is usually shown as in Fig. 3.17.

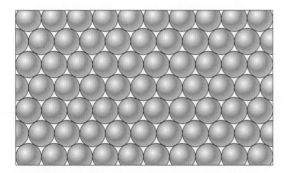

△ Fig. 3.17 Particles in a solid.

This is a simplified picture but, surprisingly, such a structure is very weak. If there is the slightest difference between the planes of atoms, the metal will break at that point.

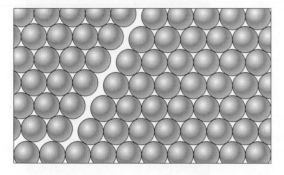

△ Fig. 3.18 The gaps show a weak point of a metal.

The more irregular (jumbled-up) the metal atoms are, the stronger the metal is. This is why alloy structures are stronger: because of the elements added.

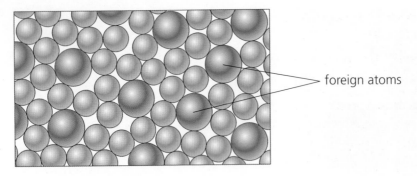

△ Fig. 3.19 Atoms in an alloy.

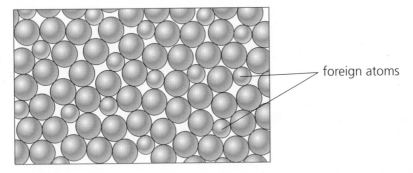

△ Fig. 3.20 Even smaller atoms make the metal stronger.

Steel that is heated to red heat and then plunged into cold water is made harder by the process of 'jumbling up' the metal atoms. Further heat treatment is used to increase the strength and toughness of the alloy.

END OF EXTENDED

 SCIENCE IN CONTEXT FACTS ABOUT METALS

1. The use of metals can be traced back to about 7000 years ago. An archaeological site in Serbia has evidence of the extraction of copper about that time, and gold artefacts dating to about 1000 years later have been found at a burial site at Varna in Bulgaria. In the period up to 2700 years ago, seven metals were known and used. These so-called 'metals of antiquity' were gold, copper, silver, lead, tin, iron and mercury. Of these gold, silver, copper, iron and mercury were found in their native state – that is, as pure elements. (Iron as a pure metal is found only in meteors.)

2. Mercury is the only liquid metal at normal room temperature and pressure.

△ Fig. 3.21 The Burj Khalifa building in Dubai.

3. The most reactive metals are found in Group I of the Periodic Table and include sodium and potassium.

4. Many of the metals used in construction are found in the middle of the Periodic Table and are called the transition metals.

5. Most of the metallic structures around us are not made from pure metals but from alloys. Alloys are mixtures of metals or occasionally mixtures of metals and non-metals: for example, steel is an alloy of iron and carbon. 39 000 tonnes of steel were used in building the Burj Khalifa building in Dubai (the world's tallest skyscraper at 828 metres). Alloys allow the properties of a metal to be modified for a particular purpose. For instance, aluminium is useful in building aircraft because it has a low density, but alloying it with other metals increases its strength. Bronze, an alloy of copper and tin, was the first alloy invented.

6. Some elements have the properties of both metals and non-metals. They are called metalloids. One of the most common metalloids is silicon.

QUESTIONS

1. Metals are usually *malleable.* What does this mean?

2. What is an alloy?

3. Which alloy is used to make many coins?

4. Explain why an alloy, of aluminium is likely to be stronger than pure aluminium.

SCIENCE LINK

BIOLOGY – PLANT NUTRITION

• Metal ions are important for the healthy growth of plants, for example, magnesium ions for making chlorophyll.

PHYSICS – ELECTRIC CIRCUITS

• The metallic bonding structure allows a number of electrons to be 'free' of particular atoms – metals are good conductors.

• Having electrons to conduct electricity is the basis of electric circuits, where the property of electric charge allows us to give a large number of electrons a general 'drift' in one direction – leading to an overall energy transfer.

• Metals are also good thermal conductors since the electrons are also able to transfer thermal energy – in addition to the energy transferred by the ions in the structure which can transfer energy by vibration as in insulators.

REACTIVITY SERIES

The Periodic Table is a way of ordering the chemical elements that highlights their similar and their different properties. The **reactivity series** is another way of classifying elements, this time in order of their reactivity to help explain or predict their reactions. This has many practical applications, such as being able to predict how metals can be extracted from their ores and how the negative effects of the chemical process of rusting can be reduced.

The more reactive metals react with oxygen to form oxides:

calcium   +   oxygen   →   calcium oxide

$2Ca(s)$ + $O_2(g)$ → $2CaO(s)$

Less reactive metals, such as gold, do not react with oxygen.

△ Fig. 3.22 Sodium, magnesium, gold – these are all metals, but they have very different reactivities.

Elements can be arranged in order of their reactivity. The more reactive a metal is, the easier it is to form compounds and the harder it is to break those compounds down. We can predict how metals might react by looking at the reactivity series (Fig. 3.23).

The most reactive metals react with water at room temperature. For example, potassium, sodium and lithium in Group I and calcium in Group II react rapidly with water:

sodium   +   water   →   sodium hydroxide   +   hydrogen

$2Na(s)$ + $2H_2O(l)$ → $2NaOH(aq)$ + $H_2(g)$

most reactive

potassium

sodium

calcium

magnesium

aluminum

(carbon)

zinc

iron

lead

(hydrogen)

copper

silver

gold

REACTIVITY INCREASES

least reactive

△ Fig. 3.23 The reactivity series shows elements, mainly metals, in order of decreasing reactivity.

The less reactive metals such as magnesium and iron react with steam:

magnesium	+	steam	→	magnesium oxide	+	hydrogen
$Mg(s)$	+	$H_2O(g)$	→	$MgO(s)$	+	$H_2(g)$

Some of the mid-reactivity series metals produce hydrogen when they react with dilute acids. So, for example, magnesium, aluminium, zinc and iron all release hydrogen when they react with dilute hydrochloric acid:

zinc	+	hydrochloric acid	→	zinc chloride	+	hydrogen
$Zn(s)$	+	$2HCl(aq)$	→	$ZnCl_2(aq)$	+	$H_2(g)$

The metals below hydrogen in the reactivity series do not react to form hydrogen with water or dilute acids.

Another use of the reactivity series is to predict how metals can be extracted from their ores. The elements below carbon in the reactivity series can be obtained by heating their oxides with carbon:

zinc oxide	+	carbon	→	zinc	+	carbon dioxide
$2ZnO(s)$	+	$C(s)$	→	$2Zn(s)$	+	$CO_2(g)$

copper(II) oxide	+	carbon	→	copper	+	carbon dioxide
$2CuO(s)$	+	$C(s)$	→	$2Cu(s)$	+	$CO_2(g)$

This type of reaction is called a **displacement reaction**. A more reactive element, such as carbon, 'pushes' (or displaces) a less reactive metal, such as copper, out of its compound. In this reaction the copper(II) oxide has lost oxygen and been reduced. The carbon has gained oxygen and been oxidised.

EXTENDED

The position of a metal in the reactivity series depends on how easily it forms ions. More reactive metals will form ions more readily than less reactive metals.

Any element higher up the reactivity series can displace an element lower down the series.

For example, magnesium is higher up the reactivity series than copper. So, if magnesium powder is heated with copper(II) oxide, then copper and magnesium oxide are produced:

magnesium	+	copper(II) oxide	$\rightarrow$	magnesium oxide	+	copper
$Mg(s)$	+	$CuO(s)$	$\rightarrow$	$MgO(s)$	+	$Cu(s)$

This reaction is an example of a redox reaction. The magnesium has been oxidised to magnesium oxide and the copper(II) oxide has been reduced to copper. Because the magnesium is responsible for the reduction of the copper(II) oxide, it is acting as a reducing agent. Similarly, the copper(II) oxide is responsible for the **oxidation** of the magnesium, so it is acting as an oxidising agent. In a redox reaction, the reducing agent is always oxidised and the oxidising agent is always reduced.

What will happen if copper is heated with magnesium oxide? Nothing happens, because copper is lower in the reactivity series than magnesium.

Using displacement reactions to establish a reactivity series

Displacement reactions of metals and their compounds in aqueous solution can be used to work out the order in the reactivity series.

In the same way that a more reactive element can push a less reactive element out of a compound, a more reactive metal ion in aqueous solution can displace a less reactive one.

For example, if you add zinc to copper(II) sulfate solution, the zinc displaces the copper because zinc is more reactive than copper. When the experiment is carried out, the blue colour of the copper(II) ion will fade as copper is produced and zinc ions are made:

zinc	+	copper(II) sulfate solution	$\rightarrow$	zinc sulfate solution	+	copper
$Zn(s)$	+	$Cu^{2+}(aq) + SO_4^{2-}(aq)$	$\rightarrow$	$Zn^{2+}(aq) + SO_4^{2-}(aq)$	+	$Cu(s)$

To build up a whole reactivity series, a set of reactions can be tried to see if metals can displace other metal ions. By following the general rule that a more reactive metal can displace a less reactive metal it is possible to establish the reactivity series.

For example, you may have seen the reaction of copper wire with silver nitrate solution. As the reaction proceeds, a shiny grey precipitate appears (this is silver) and the solution begins to turn blue as $Cu(II)$ ions are produced from the copper.

copper	+	silver nitrate	$\rightarrow$	copper(II) nitrate	+	silver
$Cu(s)$	+	$2AgNO_3(aq)$	$\rightarrow$	$Cu(NO_3)_2(aq)$	+	$2Ag(s)$

This shows that silver can be displaced by copper, and so silver is below copper in the reactivity series.

END OF EXTENDED

QUESTIONS

1. Will copper react with dilute hydrochloric acid to produce hydrogen? Explain your answer.

2. EXTENDED Write a balanced equation for the reaction of potassium with water.

3. Can carbon displace magnesium from magnesium oxide? Explain your answer.

4. EXTENDED Write the balanced equation for the reaction between magnesium and lead(II) oxide.

Developing investigative skills

A student was asked to carry out some possible displacement reactions. She was given samples of four metals A, B, C and D and a solution of each of their metal nitrates. She set up a series of test tube reactions as summarised in the table:

Solution	Metal A	Metal B	Metal C	Metal D
Metal A nitrate, $A(NO_3)_2$ (aq)		Yes	Yes	No
Metal B nitrate, $B(NO_3)_2$ (aq)	No		No	No
Metal C nitrate, $C(NO_3)_2$ (aq)	No	Yes		No
Metal D nitrate, $D(NO_3)_2$ (aq)	10	11	12	

She decided that she would need 12 test tubes. In each test tube she put a 1 cm depth of one of the solutions and then added a small piece of one of the metals. She left the tubes for 10 minutes and then examined the solution and the piece of metal to see if any reaction was evident. She then recorded a 'yes' if a displacement reaction had taken place and a 'no' where no reaction was evident. She didn't have time to record her results for the metal D nitrate solution (tubes 10, 11 and 12).

Using and organising techniques, apparatus and materials

❶ Why didn't the student set up the tubes represented by the white rectangles?

❷ Even though the student didn't record her results for metal D nitrate solution, explain why she would still be able to put the metals in order of reactivity.

Interpreting observations and data

❸ Use the results to put the four metals in order of reactivity. Start with the most reactive metal.

❹ Complete the results you would expect for the three reactions 10, 11 and 12.

❺ Write a balanced equation for the displacement reaction between metal B and metal C nitrate solution. (Use the symbols B and C for the two metals).

❻ Metal D nitrate solution was blue and metal D was a shiny orange colour. Suggest a name for metal D.

EXTRACTION OF METALS

Metals are found in the form of **ores** containing **minerals** mixed with unwanted rock. In almost all cases, the mineral is a compound of the metal, not the pure metal. One exception is gold, which can exist naturally in a pure state.

Extracting a metal from its ore usually involves two steps:

1. The mineral is physically separated from unwanted rock.

2. The mineral is broken down chemically to obtain the metal.

Reactivity of metals

The chemical method chosen to break down a mineral depends on the reactivity of the metal. The more reactive a metal is, the harder it is to break down its compounds. The more reactive metals are obtained from their minerals by the process of electrolysis. For example, aluminium is obtained from its ore, bauxite, by electrolysis.

The less reactive metals can be obtained by heating their oxides with carbon. This method will only work for metals below carbon in the reactivity series. It involves the **reduction** of a metal oxide to the metal.

Metal		Extraction method
Potassium	}	The most reactive metals are obtained using electrolysis
Sodium		
Calcium		
Magnesium		
Aluminium		
(Carbon)		
Zinc	}	These metals are below carbon in the reactivity series and so can be obtained by heating their oxides with carbon
Iron		
Tin		
Lead		
Copper		
Silver	}	The least reactive metals are found as pure elements
Gold		

△ Table 3.10 Methods for extracting different metals.

EXTENDED

Extracting iron

Iron is produced on a very large scale by reduction using carbon. The reaction takes place in a huge furnace called a blast furnace.

Three important raw materials are put in the top of the furnace: iron ore (iron(III) oxide), coke (the source of carbon for the reduction) and limestone, to remove the impurities as slag. Iron ore is also known as hematite.

△ Fig. 3.24 Coke (nearly pure carbon).

△ Fig. 3.25 Iron ore (hematite).

△ Fig. 3.26 Limestone.

△ Fig. 3.27 Molten iron.

△ Fig. 3.28 Slag.

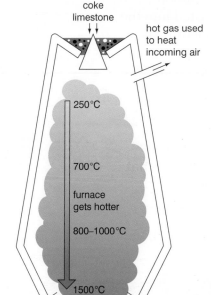

△ Fig. 3.29 How iron is extracted in a blast furnace.

1. Crushed iron ore, coke and limestone are fed into the top of the blast furnace

2. Hot air is blasted up the furnace from the bottom

3. Oxygen from the air reacts with coke to form carbon dioxide:
$$C(s) + O_2(g) \longrightarrow CO_2(g)$$

4. Carbon dioxide reacts with more coke to form carbon monoxide:
$$CO_2(g) + C(s) \longrightarrow 2CO(g)$$

5. Carbon monoxide is a reducing agent. Iron(III) oxide is reduced to iron:
$$\overbrace{Fe_2O_3(s) + 3CO(g) \longrightarrow 2Fe(l) + 3CO_2(g)}^{\text{reduction = loss of oxygen}}$$

6. Dense molten iron runs to the bottom of the furnace and is run off. There are many impurities in iron ore. The limestone helps to remove these as shown in processes 7 and 8.

7. Limestone is broken down by heat to calcium oxide:
$$CaCO_3(s) \longrightarrow CaO(s) + CO_2(g)$$

8. Calcium oxide reacts with impurities like sand (silicon dioxide) to form a liquid called 'slag':
$$\underset{\text{impurity}}{CaO(s) + SiO_2(s)} \longrightarrow \underset{\text{slag}}{CaSiO_3(l)}$$
The liquid slag runs to the bottom of the furnace and is tapped off.

The overall reaction is:

iron(III) oxide + carbon → iron + carbon dioxide

$2Fe_2O_3(s)$ + $3C(s)$ → $4Fe(s)$ + $3CO_2(g)$

The reduction happens in three stages.

Stage 1 – The coke (carbon) reacts with oxygen 'blasted' into the furnace:

carbon + oxygen → carbon dioxide

$C(s)$ + $O_2(g)$ → $CO_2(g)$

Stage 2 – The carbon dioxide is reduced by unreacted coke to form carbon monoxide:

carbon dioxide + carbon → carbon monoxide

$CO_2(g)$ + $C(s)$ → $2CO(g)$

Stage 3 – The iron(III) oxide is reduced by the carbon monoxide to iron:

iron(III) oxide + carbon monoxide → iron + carbon dioxide

$Fe_2O_3(s)$ + $3CO(g)$ → $2Fe(s)$ + $3CO_2(g)$

REMEMBER

In a blast furnace the iron(III) oxide is reduced to iron by carbon monoxide, formed when the carbon reacts with the air blasted into the furnace. In the reduction of iron(III) oxide, the carbon monoxide is oxidised to carbon dioxide.

QUESTIONS

1. What solid raw materials are used in the blast furnace?

2. The iron ore used in the blast furnace is usually *hematite*. What is the name of the main compound present in the ore?

3. What gases will escape from the top of the blast furnace?

4. Write a balanced equation to show the reduction of iron(III) oxide by carbon.

END OF EXTENDED

RECYCLING

Extracting metals from their ores is an expensive process and so recycling metal objects can be economically worthwhile as well as environmentally more efficient. Recycling of metals essentially involves melting the metal and then using the molten metal to form a new object. Steel and aluminium are metals which are often recycled. One potential problem is separating the different types of metal. For example, a motor car as well as being made of steel may include some parts made of aluminium (as well as a range of other materials). Separating the two different metals needs to be done before they can be recycled. In some cases recycling could prove more expensive than extracting the metal form its ore. However, the supplies of metal ores are limited and will eventually be used up. Recycling metals such as steel and aluminium will therefore become increasingly essential.

SCIENCE IN CONTEXT

THE EXTRACTION OF METALS

The reactivity of a metal determines how it can be extracted from ores from the Earth's crust. It also explains why some metals have been used for thousands of years while others have only been used much more recently.

The most unreactive metals can be found in their 'native' state, which is as the pure metal and not combined with other elements. Examples of such metals include gold and silver – metals that have been used for thousands of years. It is estimated that gold was first discovered in about 3000BCE.

△ Fig. 3.30 This gold shoulder cape from North Wales is nearly 4000 years old and still in good condition.

Metals below carbon in the reactivity series can be extracted by heating their ores with carbon. Examples include lead and iron. It is possible that lead was discovered by accident when the silvery element was seen in the ashes of a wood fire that had been made above a deposit of lead ore. It is estimated that lead was first discovered in about 2000BCE.

The most reactive metals, all those above carbon in the reactivity series, have to be extracted from their minerals by electrolysis. This process is a much more recent development and explains why these metals were not used until relatively recently. Aluminium was first extracted in 1825.

End of topic checklist

Key terms

alloy, displacement reaction, mineral, ore, oxidation, reactivity series, reduction

During your study of this topic you should have learned:

○ How to describe the general physical properties of metals as solids with high melting and boiling points, malleable and good conductors of heat and electricity.

○ How to explain, in terms of their properties, why metals are often used in the form of alloys.

○ How to identify representations of an alloy from a diagram of its structure.

○ About the order of reactivity of metals – potassium, sodium, calcium, magnesium, zinc, iron, (hydrogen) and copper – by reference to the reactions, if any, of the metals with water or steam, or dilute hydrochloric acid.

○ About which metal oxides can be reduced by carbon.

○ **EXTENDED** How to describe the reactivity series as being related to the tendency of a metal to form its positive ion, illustrated by its reaction, if any, with the aqueous ions or the oxides of the other listed metals.

○ How to deduce an order of reactivity from a given set of experimental results.

○ **EXTENDED** How to describe the ease of obtaining metals from their ores by relating the metals to their positions in the reactivity series.

○ **EXTENDED** How to describe the essential reactions in the extraction of iron from hematite.

End of topic questions

Note: The marks given for these questions indicate the level of detail required in the answers. In the examination, the number of marks given to questions like these may be different.

1. Arrange the following metals in order of reactivity, starting with the most reactive:

calcium, copper, magnesium, sodium, zinc. **(2 marks)**

2. This question is about four metals represented by the letters Q, X, Y and Z. A series of displacement reactions was carried out and the results are shown below:

Reaction 1: Q oxide + Y → Y oxide + Q

Reaction 2: X oxide + Z → Z oxide + X

Reaction 3: Q oxide + Z→ no change

a) Arrange the metals in order of reactivity starting with the most reactive. **(2 marks)**

b) In reaction 1:

i) Which substance has been oxidised? **(1 mark)**

ii) Which substance has been reduced? **(1 mark)**

iii) Which substance is the oxidising agent? **(1 mark)**

iv) Which substance is the reducing agent? **(1 mark)**

3. The least reactive metals, such as gold and silver, are found in their native state. What do you understand by this? **(1 mark)**

4. **EXTENDED** Iron is extracted from iron ore (iron(III) oxide) in a blast furnace by heating with coke (carbon).

a) Write a balanced equation, including state symbols, for the overall reaction. **(2 marks)**

b) Is the iron(III) oxide oxidised or reduced in this reaction? Explain your answer. **(1 mark)**

c) Why is limestone also added to the blast furnace? **(2 marks)**

End of topic questions continued

5. EXTENDED Zinc can be extracted from zinc oxide by heating with carbon.

a) Write the balanced equation, including state symbols, for this reaction. **(2 marks)**

b) Zinc could also be extracted by the electrolysis of molten zinc oxide. Suggest why heating with carbon is the preferred method of extraction. **(2 marks)**

6. EXTENDED Copper(II) sulfate solution reacts with zinc as shown below:

$CuSO_4(aq) + Zn(s) \rightarrow ZnSO_4(aq) + Cu(s)$

a) What type of chemical reaction is this? **(1 mark)**

b) What can be deduced about the relative reactivities of copper and zinc? **(1 mark)**

Air and water

INTRODUCTION

Clean air is precious. It provides the oxygen that all living things need to survive, and the carbon dioxide that plants need when they photosynthesise. Nitrogen in the air is also very important for healthy plant growth, but not all plants can make use of nitrogen in this form.

Unfortunately, not all the air we breathe is clean. It may contain a number of pollutants that can be harmful to living things and the environment. It is important to understand how these pollutants are produced and how they can be prevented from contaminating the air.

Water vapour is also present in the air. With oxygen, this causes rusting, a process that can be very destructive.

△ Fig. 3.31 The smog over the Forbidden City in Beijing is so thick that it obscures the view from Feng Shui Hill.

KNOWLEDGE CHECK

✓ Know that oxygen is present in the air and forms oxides when substances burn in it.
✓ Know that oxides of non-metals are acidic.
✓ Know that acids react with carbonates to make salts.
✓ Know that salts can be anhydrous or hydrated.

LEARNING OBJECTIVES

✓ Be able to describe chemical tests for identifying the presence of water using cobalt(II) chloride and copper(II) sulfate.
✓ Be able to describe in outline the treatment of the water supply in terms of filtration and chlorination.
✓ Know that clean air is approximately 78% nitrogen and 21% oxygen with the remainder made up of a mixture of noble gases, water vapour and carbon dioxide.
✓ Know that the common pollutants in the air are carbon monoxide, sulfur dioxide and oxides of nitrogen.
✓ Know the adverse effects of the common pollutants on buildings and health.
✓ Be able to state the conditions required for the rusting of iron (presence of oxygen and water).
✓ Be able to describe methods of rust protection including using paint and other coatings to exclude oxygen.
✓ Know that carbon dioxide is formed from the complete combustion of carbon-containing substances, as a product of respiration, as a product of the reaction between an acid and a carbonate and as a product of the thermal decomposition of calcium carbonate.

✓ Know that carbon dioxide and methane are greenhouse gases.

✓ EXTENDED Be able to explain that increased concentrations of greenhouse gases cause an enhanced greenhouse effect, which may contribute to climate change.

✓ Be able to describe the formation of carbon dioxide from the complete combustion of carbon-containing substances, respiration, the reactions between an acid and a carbonate, and the thermal decomposition of a carbonate.

A CHEMICAL TEST FOR WATER

The test for water is to add it to anhydrous copper(II) sulfate solid. If the liquid contains water, the powder will turn from white to blue as hydrated copper(II) sulfate forms.

△ Fig. 3.32 Chemical test for water.

The equation for the reaction is:

anhydrous copper(II) sulfate + water → hydrated copper(II) sulfate

$$CuSO_4(s) + 5H_2O(l) \rightarrow CuSO_4.5H_2O(s)$$

The presence of water can also be detected using anhydrous cobalt(II) chloride. The pink anhydrous cobalt(II) chloride turns to blue hydrated cobalt(II) chloride. A convenient way of performing the test is to use cobalt(II) chloride paper:

anhydrous cobalt(II) chloride + water → hydrated cobalt(II) chloride

$$CoCl_2(s) + 6H_2O(l) \rightarrow CoCl_2.6H_2O(s)$$

Neither of these tests shows the water is pure – only that the liquid has water in it.

THE WATER CYCLE

The recirculation of water that takes place all over the Earth is called the **water cycle**.

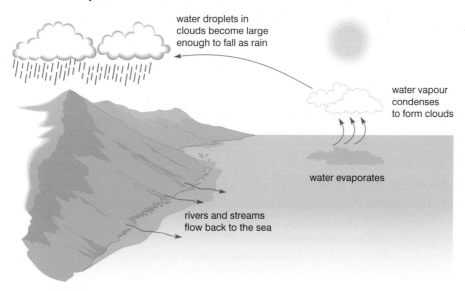

water droplets in clouds become large enough to fall as rain

water vapour condenses to form clouds

water evaporates

rivers and streams flow back to the sea

△ Fig. 3.33 The water cycle.

The pattern of rainfall over the planet determines where there are deserts, rainforests and areas of land that can or cannot be used for growing plants.

Some scientists think that **global warming** is responsible for climate changes that are affecting both where rain falls and how much there is of it. This could be causing both increased risks of flooding in some regions and droughts in others.

Water is essential for life on Earth, and the demand for drinking water is increasing as the world's population grows. Two-thirds of the water is used in homes for washing, cleaning, cooking and in toilets. The rest is used by industry. Most industrial processes use water either as a raw material or for cooling. For example, it takes 200 000 litres of water to make 1 tonne of steel.

Water stored in reservoirs must be purified to produce drinkable tap water.

Fig. 3.34 Water pipes discharging in Thailand.

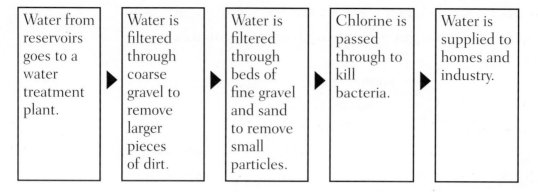

| Water from reservoirs goes to a water treatment plant. | ▶ | Water is filtered through coarse gravel to remove larger pieces of dirt. | ▶ | Water is filtered through beds of fine gravel and sand to remove small particles. | ▶ | Chlorine is passed through to kill bacteria. | ▶ | Water is supplied to homes and industry. |

In addition, tap water in certain areas is treated with sodium fluoride (NaF) to combat tooth decay.

In certain parts of the world supplies of water are very limited. There is insufficient clean water to drink and not enough water to irrigate and support the growth of crops. Water that is not purified can often contain harmful bacteria and so is not safe to drink. In the absence of clean water people have little choice but to drink contaminated water and risk illness or death.

THE COMPOSITION OF CLEAN AIR

Air is a mixture of gases that has remained fairly constant for the last 200 million years. The amount of water vapour varies around the world. For example, air above a desert area has a low proportion of water vapour.

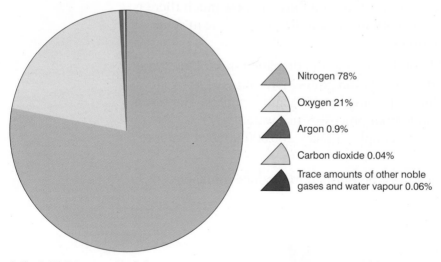

Nitrogen 78%

Oxygen 21%

Argon 0.9%

Carbon dioxide 0.04%

Trace amounts of other noble gases and water vapour 0.06%

△ Fig. 3.35 Components of air.

QUESTIONS

1. What does *anhydrous* mean?

2. What colour change would you observe if water is added to anhydrous cobalt(II) chloride?

3. In the purification of water there are two important stages.

    a) In the first stage, the water is filtered twice. What is used as the filter in each case?

    b) In the second stage, bacteria are killed. What chemical is used to kill the bacteria?

4. This question is about the composition of the air.

    a) What is the percentage of nitrogen in clean air?

    b) What is the percentage of carbon dioxide in clean air?

POLLUTANTS IN THE AIR

Pollutants in the air come from a variety of sources. Some come from burning waste and some from power stations burning coal or gas. Industry produces pollutants as well.

The most common pollutants in the air are:

- carbon monoxide – from the incomplete **combustion** of hydrocarbons (petrol/coal/gas/diesel)
- sulfur dioxide – from burning fossil fuels such as petrol and coal which contain sulfur compounds
- oxides of nitrogen – from burning fossil fuels (petrol/diesel/coal).

Δ Fig. 3.36 Cycling is encouraged in Amsterdam to reduce air pollution.

HOW IS THE ATMOSPHERE CHANGING, AND WHY?

The greenhouse effect

Carbon dioxide, methane and CFCs (chlorofluorocarbons) are known as **greenhouse gases**. The levels of these gases in the atmosphere are increasing due to the burning of **fossil fuels**, pollution from farm animals and the use of CFCs in aerosols and refrigerators.

Short-wave radiation from the Sun warms the ground, and the warm Earth gives off heat as long-wave radiation. Much of this radiation is stopped from escaping from the Earth by the greenhouse gases. This is known as the **greenhouse effect**.

The greenhouse effect is responsible for keeping the Earth warmer than it would otherwise be. This is normal – and important for life on Earth. However, most scientists think that increasing levels of greenhouse gases are stopping even more heat escaping and that the Earth is slowly warming up. This is known as global warming. If global warming continues the Earth's climate may change, polar ice may melt and sea levels may rise flooding low-lying areas – some of them highly populated.

The Earth's average temperature is gradually increasing, but nobody knows for certain if the greenhouse effect is responsible. It may be that the recent rise in global temperatures is part of a natural cycle – there have been ice ages and intermediate warm periods all through history. Many people are concerned, however, that it is not part of a natural cycle and they say we should act now to reduce emissions of these greenhouse gases.

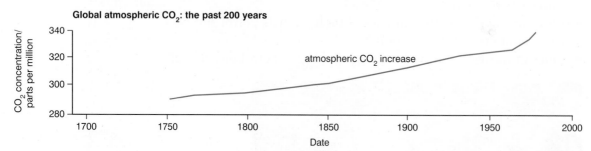

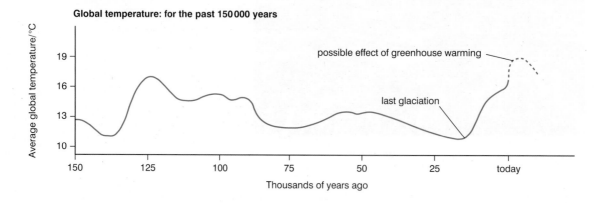

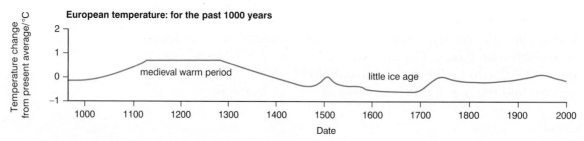

Δ Fig. 3.37 How atmospheric carbon dioxide and temperature have varied.

SCIENCE IN CONTEXT   **SOME INTERESTING FACTS ABOUT METHANE**

1. Methane makes up about 97% of natural gas.

2. It is formed by the decay of plant matter where there is no oxygen (anaerobic decay).

3. Biogas contains 40–70% methane. Biodigesters convert organic wastes into a nutrient-rich liquid fertiliser and biogas, a **renewable** source of electrical and heat energy. These are widely used in non-industrialised countries, particularly India, Nepal and Vietnam. Biodigesters can help families by providing a cheap source of fuel, reducing environmental pollution from the run-off from animal pens, and reducing diseases caused by the use of untreated manure as fertiliser. However, biodigesters only work efficiently in hot countries; they are not as effective at low temperatures.

Δ Fig. 3.38 A commercial biodigester.

4. Methane is one of the greenhouse gases, thought by some scientists to be responsible for global warming. It has almost 25 times the effect of the same volume of carbon dioxide.

5. Ruminant animals such as cattle and sheep produce methane. It has been estimated that a cow can produce as much as 200 litres of methane per day. So could cattle be one of the causes of global warming?

QUESTION

1. a) Methane is a greenhouse gas. Name two sources of methane.

b) Name another greenhouse gas.

CARBON DIOXIDE

Carbon dioxide is an important gas. It is formed as a product in the complete combustion of carbon-containing substances:

$$C(s) + O_2(g) \rightarrow CO_2(g)$$

Carbon dioxide is also formed in respiration. It can be made in the laboratory by the reaction of dilute hydrochloric acid and calcium carbonate in the form of marble chips.

▷ Fig. 3.39 Charcoal is mainly carbon. When it burns it gives off carbon dioxide.

calcium carbonate	+ hydrochloric acid	→ carbon dioxide	+ water	+ calcium chloride
$CaCO_3(s)$	+ $2HCl(aq)$	→ $CO_2(g)$	+ $H_2O(l)$	+ $CaCl_2(aq)$

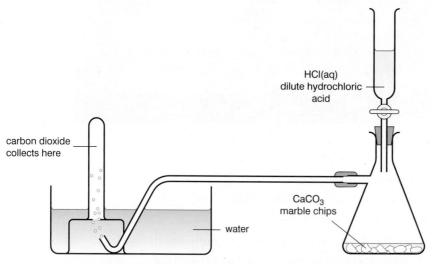

Δ Fig. 3.40 The laboratory preparation of carbon dioxide gas.

If the gas is bubbled through limewater (calcium hydroxide solution), a white precipitate forms. This is used as a laboratory test for carbon dioxide.

INORGANIC CHEMISTRY

Carbon dioxide can also be prepared by the **thermal decomposition** of certain metal carbonates. Copper(II) carbonate and zinc carbonate are examples:

copper(II) carbonate	$\rightarrow$	copper(II) oxide	+	carbon dioxide
$CuCO_3(s)$	$\rightarrow$	$CuO(s)$	+	$CO_2(g)$
green		black		

zinc carbonate	$\rightarrow$	zinc oxide	+	carbon dioxide
$ZnCO_3(s)$	$\rightarrow$	$ZnO(s)$	+	$CO_2(g)$
white		white (yellow when hot)		

METHODS OF PREVENTING RUSTING

Over time, the oxygen and water in the atmosphere affects metals. If they react together the metal is corroded. The corrosion of iron is called **rusting**.

In the presence of water, the following chemical reaction takes place:

$$4Fe(s) + 3O_2(g) \rightarrow 2Fe_2O_3(s)$$

In fact, rust is hydrated iron(III) oxide, $Fe_2O_3 . xH_2O$. The 'x' can vary depending on the conditions.

Here are two methods of preventing rusting:

- Stopping oxygen and water reaching the iron, for example, oiling/greasing, as with bicycle chains; painting, as with car bodies.
- Alloying – iron is mixed with other metals to produce alloys such as stainless steel that do not rust.

Developing investigative skills

A student set up the apparatus as shown in Fig. 3.41 with the long tube turned upside-down in a trough of water. Previously some iron filings had been sprinkled into the tube, and many of these had stuck to the inside of the tube. With the same levels of water in the tube and the trough, the student recorded the volume of air in the tube (100 cm³).

After a few days he returned to the apparatus, equalised the water levels as before and took a second reading of the volume of air in the tube (85 cm³). He then worked out how much of the air had been replaced by water.

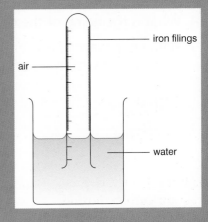

△ Fig. 3.41 Apparatus for experiment.

QUESTIONS

1. Carbon dioxide is formed in the complete combustion of carbon. What product might form if carbon is burned in a limited supply of air?

2. In the laboratory, carbon dioxide can be prepared by the reaction of an acid with a metal carbonate.

   a) Write a word equation for the reaction between copper(II) carbonate and dilute hydrochloric acid.

   b) EXTENDED Write a balanced equation for the reaction in part a).

3. Carbon dioxide can also be prepared by the action of heat on a metal carbonate.

   a) Write a word equation for the action of heat on calcium carbonate.

   b) EXTENDED Write a balanced equation for the reaction in part a).

4. What is the name of the chemical compound present in rust?

5. Name two ways of preventing water and oxygen from getting into contact with the surface of an iron object.

BIOLOGY – HUMAN INFLUENCES ON ECOSYSTEMS

- Pollutant gases, for example, from the burning of fuels, affect ecosystems over a range of distances and time scales.

- Water quality can be affected by a number of factors, including some linked to chemical changes.

PHYSICS – ENERGY TRANSFERS, THERMAL PHYSICS

- The range of sources used to supply energy on a large scale have an impact on the air and water, in particular the production of carbon dioxide gas from the burning of carbon-containing fuels.

- The management of the energy sources is an issue that involves all aspects of the sciences.

- Large scale heating of the atmosphere by the Sun leads to convection currents which drive weather systems.

End of topic checklist

Key terms

combustion, fossil fuel, global warming, greenhouse effect, greenhouse gas, renewable energy, rusting, thermal decomposition, water cycle

During your study of this topic you should have learned:

◯ How to describe chemical tests for identifying the presence of water using cobalt(II) chloride and copper(II) sulfate.

◯ How to describe in outline the treatment of the water supply in terms of filtration and chlorination.

◯ That clean air is approximately 78% nitrogen and 21% oxygen with the remainder made up of a mixture of noble gases, water vapour and carbon dioxide.

◯ That the common pollutants in the air are sulfur dioxide and oxides of nitrogen.

◯ About the adverse effects of the common pollutants on buildings and health.

◯ About the conditions required for the rusting of iron (presence of oxygen and water).

◯ How to describe methods of rust protection, including using paint and other coatings to exclude oxygen.

◯ How to describe the formation of carbon dioxide from:

- the complete combustion of carbon-containing substances
- the process of respiration
- the reactions between an acid and a carbonate
- the thermal decomposition of a carbonate.

◯ That carbon dioxide and methane are greenhouse gases and may contribute to climate change.

◯ EXTENDED How to explain how increased concentrations of greenhouse gases cause an enhanced greenhouse effect, which may contribute to climate change.

End of topic questions

Note: The marks given for these questions indicate the level of detail required in the answers. In the examination, the number of marks given to questions like these may be different.

1. This question is about the composition of a sample of clean air.

   a) What is the proportion of oxygen? (1 mark)

   b) What is the proportion of carbon dioxide? (1 mark)

2. **a)** What could you use to detect the presence of water? (1 mark)

   b) What would you observe if water was present? (2 marks)

3. This question is about the greenhouse effect.

   a) What is the *greenhouse effect*? (2 marks)

   b) Name two greenhouse gases. (2 marks)

   c) Apart from an increase in greenhouse gases, what else could be causing global warming? (1 mark)

4. Carbon dioxide can be prepared using the reaction between calcium carbonate and dilute hydrochloric acid.

   a) How can the gas be collected in this reaction? (1 mark)

   b) EXTENDED Write a balanced equation for the reaction. (2 marks)

5. Carbon dioxide can be made by the thermal decomposition of copper(II) carbonate.

   a) What does *thermal decomposition* mean? (2 marks)

   b) EXTENDED Write a balanced equation for this reaction. (2 marks)

Organic chemistry is distinct from other branches of chemistry, such as inorganic and physical chemistry. It may be described as the chemistry of living processes (often referred to as biochemistry) but extends beyond that. Organic chemistry focuses almost entirely on the chemistry of covalently bonded carbon molecules. As well as life processes, it includes the chemistry of other types of compounds including plastics, petrochemicals, drugs and paint.

Early chemists never imagined that complex chemicals of living processes could ever be manufactured in a laboratory, but they were wrong. Today, medical drugs can be made and then their structures modified to achieve improvements in their effectiveness.

An understanding of organic chemistry begins with knowledge of the structure of a carbon atom and how it can combine with other carbon atoms by forming covalent bonds. In this section you will be introduced to two of the 'families' or series of organic compounds.

STARTING POINTS

1. Where is carbon in the Periodic Table of elements? What can you work out about carbon from its position?

2. What is the atomic structure of carbon? How are its electrons arranged?

3. How does carbon form covalent bonds? Show the bonding in methane (CH_4), the simplest of organic molecules.

4. You will be learning about a series of organic compounds which are hydrocarbons. What do you think a hydrocarbon is?

5. You will be learning about methane. Where can methane be found and what it is used for?

SECTION CONTENTS

a) Fuels

b) Alkanes

c) Alkenes

4
Organic chemistry

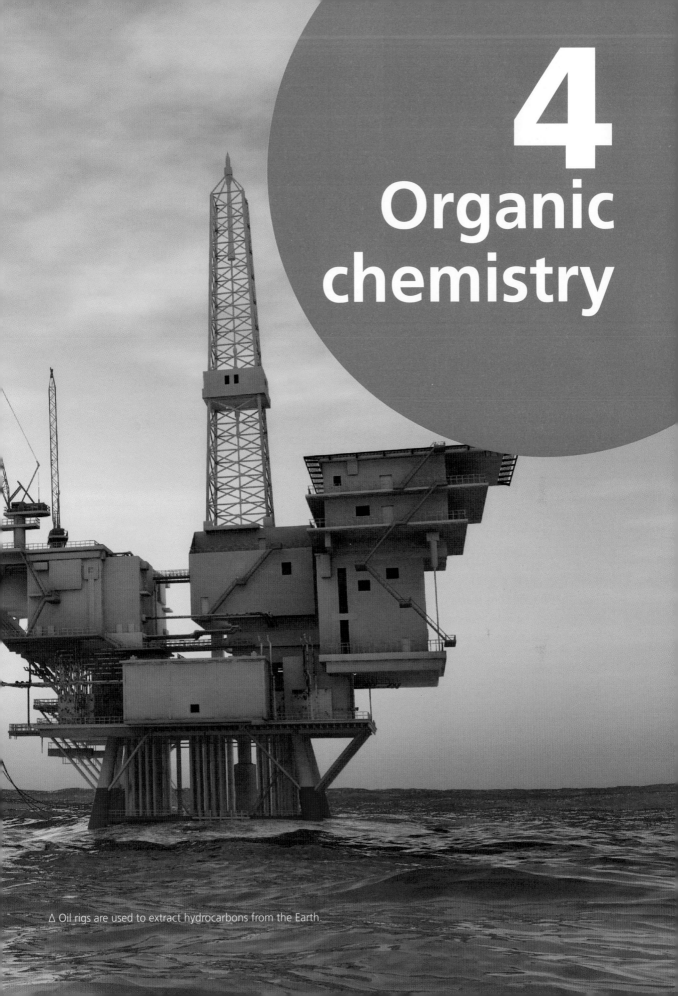

△ Oil rigs are used to extract hydrocarbons from the Earth.

Fuels

INTRODUCTION

The most common fuels used today are either fossil fuels or are made from fossil fuels. There are problems associated with using fossil fuels – burning them produces a number of polluting gases and releases carbon dioxide, a greenhouse gas. Nevertheless, fossil fuels are a very important source of energy.

△ Fig. 4.1 Crude oil contains a mixture of hydrocarbons.

WHAT ARE FOSSIL FUELS?

Petroleum (crude oil), natural gas (mainly methane) and coal are **fossil fuels**.

Crude oil was formed millions of years ago from the remains of animals and plants that were pressed together under layers of rock. It is usually found deep underground, trapped between layers of rock that it can't seep through (impermeable rock). Natural gas is often trapped in pockets above crude oil.

The supply of fossil fuels is limited – having taken millions of years to form, these fuels will eventually run out. They are called finite or **non-renewable** fuels. This makes them extremely valuable resources that must be used efficiently.

△ Fig. 4.2 Fractional distillation takes place in oil refineries, like this one in the Netherlands.

Fossil fuels contain many useful chemicals (known as **fractions**) and these must be separated so that they are not wasted.

FRACTIONAL DISTILLATION

The chemicals in petroleum are separated into useful fractions by a process known as **fractional distillation**.

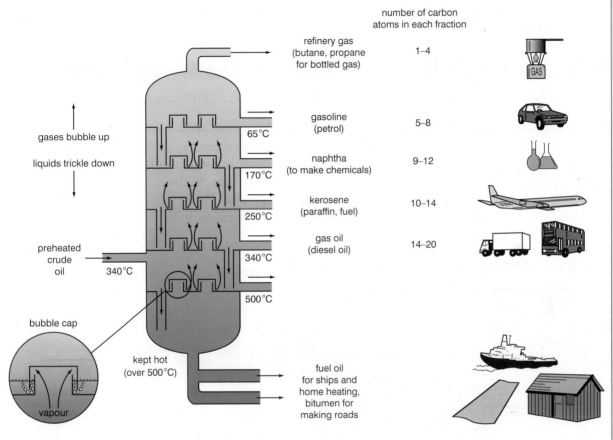

number of carbon atoms in each fraction

refinery gas (butane, propane for bottled gas) — 1–4

gasoline (petrol) — 5–8

naphtha (to make chemicals) — 9–12

kerosene (paraffin, fuel) — 10–14

gas oil (diesel oil) — 14–20

65°C
170°C
250°C
340°C
500°C

gases bubble up

liquids trickle down

preheated crude oil — 340°C

bubble cap

kept hot (over 500°C)

vapour

fuel oil for ships and home heating, bitumen for making roads

△ Fig. 4.3 A fractionating column converts crude oil into many useful fractions.

The crude oil is heated in a furnace and passed into the bottom of a fractionating column. It gives off a mixture of vapours that rise up the column, and the different fractions condense out at different heights. The fractions that come off near the top are light-coloured, runny liquids. Those removed near the bottom of the column are dark and sticky. Thick liquids that are not runny, such as those at the bottom of the fractionating column, are described as **viscous**.

How does fractional distillation work?

The components of petroleum separate because they have different boiling points. A simple particle model explains why their boiling points differ. Petroleum is a mixture of **hydrocarbon** molecules, which contain only carbon and hydrogen. The molecules are chemically bonded in similar ways with strong covalent bonds but contain different numbers of carbon atoms.

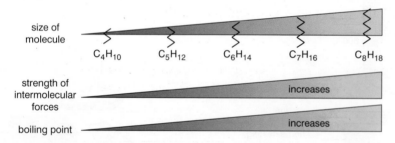

heptane

$$H-\underset{\underset{H}{|}}{\overset{\overset{H}{|}}{C}}-\underset{\underset{H}{|}}{\overset{\overset{H}{|}}{C}}-\underset{\underset{H}{|}}{\overset{\overset{H}{|}}{C}}-\underset{\underset{H}{|}}{\overset{\overset{H}{|}}{C}}-\underset{\underset{H}{|}}{\overset{\overset{H}{|}}{C}}-\underset{\underset{H}{|}}{\overset{\overset{H}{|}}{C}}-\underset{\underset{H}{|}}{\overset{\overset{H}{|}}{C}}-H$$

octane

$$H-\underset{\underset{H}{|}}{\overset{\overset{H}{|}}{C}}-\underset{\underset{H}{|}}{\overset{\overset{H}{|}}{C}}-\underset{\underset{H}{|}}{\overset{\overset{H}{|}}{C}}-\underset{\underset{H}{|}}{\overset{\overset{H}{|}}{C}}-\underset{\underset{H}{|}}{\overset{\overset{H}{|}}{C}}-\underset{\underset{H}{|}}{\overset{\overset{H}{|}}{C}}-\underset{\underset{H}{|}}{\overset{\overset{H}{|}}{C}}-\underset{\underset{H}{|}}{\overset{\overset{H}{|}}{C}}-H$$

△ Fig. 4.4 Octane has one more carbon atom and two more hydrogen atoms than heptane. Their formulae differ by CH_2.

REMEMBER

The longer the molecule, the stronger the attractive force between the molecules.

The weak attractive forces between the molecules must be broken for the hydrocarbon to boil. The longer a hydrocarbon molecule is, the stronger the intermolecular forces between the molecules. The stronger these forces of attraction, the higher the boiling point because more energy is needed to overcome the forces.

size of molecule

C_4H_{10} C_5H_{12} C_6H_{14} C_7H_{16} C_8H_{18}

strength of intermolecular forces — increases

boiling point — increases

△ Fig. 4.5 How the properties of hydrocarbons change as molecules get longer.

The smaller-molecule hydrocarbons are more **volatile** – they form a vapour easily. For example, we can smell petrol (with molecules containing between 5 and 10 carbon atoms) much more easily than we can smell engine oil (with molecules containing between 14 and 20 carbon atoms) because petrol is more volatile.

Another difference between the fractions is how easily they burn and how smoky their flames are.

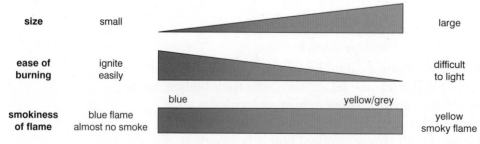

size	small	large
ease of burning	ignite easily	difficult to light
smokiness of flame	blue flame almost no smoke	yellow smoky flame

blue → yellow/grey

△ Fig. 4.6 How different hydrocarbons burn.

QUESTIONS

1. Petroleum is a 'non-renewable' fuel. What does this mean?

2. When drilling for oil, there is often excess gas to be burned off. What is this gas? Where does it come from?

3. One of the oil fractions obtained from the fractional distillation of crude oil is light-coloured and runny.

   Is this fraction more likely to have a small chain of carbon atoms or a long chain?

4. Another of the oil fractions obtained from the fractional distillation of petroleum burns with a very sooty yellow flame.

   Is this fraction more likely to have a small chain of carbon atoms or a long chain?

5. Some fractions obtained from petroleum are very 'volatile'. What does this mean?

CRACKING THE OIL FRACTIONS

The composition of petroleum varies in different parts of the world. Table 4.1 shows the composition of a sample of petroleum from the Middle East after fractional distillation.

Fraction (in order of increasing boiling point)	Typical percentage produced by fractional distillation
Liquefied petroleum gases (LPG)	3
Gasoline	13
Naphtha	9
Kerosene	12
Diesel	14
Heavy oils and bitumen	49

△ Table 4.1 Oil fractions.

Small molecules are much more useful than the larger molecules. Larger molecules can be broken down into smaller ones by **catalytic cracking**.

Cracking requires a high temperature of between 600 to 700 °C and a catalyst of silica or alumina.

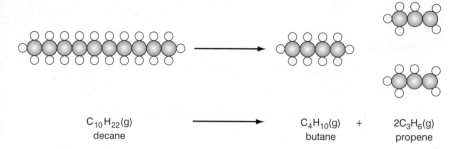

C$_{10}$H$_{22}$(g)   C$_4$H$_{10}$(g)   +   2C$_3$H$_6$(g)
decane   butane   propene

△ Fig. 4.7 The decane molecule (C$_{10}$H$_{22}$) is converted into the smaller molecules butane (C$_4$H$_{10}$) and propene (C$_3$H$_6$) in cracking.

The butane and propene formed in this example have different types of structures.

END OF EXTENDED

REMEMBER

Propene belongs to a family of hydrocarbons called **alkenes**.

Alkenes are much more reactive (and hence more useful) than hydrocarbons such as decane (an alkane).

EXTENDED

Developing Investigative Skills

A group of students set up an experiment to see if they could 'crack' some liquid paraffin. They soaked some mineral wool in liquid paraffin and assembled the apparatus as shown in Fig. 4.8. They then heated the pottery pieces very strongly, occasionally letting the flame heat the mineral wool. Bubbles of gas started to collect in the test tube. After a few minutes they had collected three test tubes full of gas and so they stopped heating. Almost immediately, water from the trough started to travel back up the delivery tube towards the boiling tube.

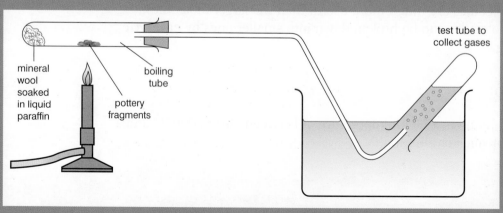

mineral wool soaked in liquid paraffin
boiling tube
pottery fragments
test tube to collect gases

△ Fig. 4.8 Incorrectly set up apparatus for experiment.

Evaluating methods

❶ The gas or gases produced in this reaction can be collected by displacement of water. What property of gas(es) does this demonstrate?

❷ Why did the water start to travel back up the delivery tube when heating was stopped?

Using and organising techniques, apparatus and materials

❸ What are the hazards involved in this experiment? What safety precautions would minimise them?

❹ The first test tube of gas collected did not burn, but the second one did. Explain this difference.

❺ The third test tube of gas decolourised bromine water. What does this suggest about the gas present?

Interpreting observations and data

❻ One of the students suggested that one of the two products was ethene (C_2H_4). Assuming that liquid paraffin has the formula $C_{14}H_{30}$, write an equation for the cracking of the liquid paraffin used in this experiment.

END OF EXTENDED

QUESTIONS

1. The cracking of hydrocarbons often produces ethene. To which homologous series does ethene belong?

2. Why is cracking needed in addition to the fractional distillation of crude oil?

3. EXTENDED What conditions are needed for the cracking of oil fractions?

SCIENCE IN CONTEXT

THE FOSSIL FUEL DILEMMA

There is widespread agreement that supplies of the non-renewable fossil fuels – oil, gas and coal – will eventually run out. However, it is not easy to estimate exactly when they will run out. Many different factors need to be considered, including how much of each deposit is left in the Earth, how fast we are using each fossil fuel at the moment, whether or not countries that have supplies will sell to those that don't, and how this is likely to change in the future. If we start switching to alternative fuel sources that are renewable, the reserves that we currently have will last longer.

Current estimates suggest that crude oil (petroleum) will run out between 2025 and 2070. The estimate for natural gas is similar, with 2060 a possible date.

The situation with coal is very different. Most coal deposits have not yet been tapped, and the decline of the coal mining industry in countries such as the UK means that many coal seams are lying undisturbed. If we carry on using coal at the same rate as we do today, there could be enough to last well over a thousand years. However, as other fossil fuels run out, particularly oil, the use of coal may increase, reducing that timespan considerably.

So, should we increase our efforts to develop renewable forms of energy such as wind and solar energy; should we put greater emphasis on nuclear power; or should we plan to make much greater use of coal? Perhaps we should do all three? Solving this dilemma is likely to depend as much on political decisions as scientific ones. What would you recommend?

△ Fig. 4.9 A coal-fired power station.

End of topic checklist

Key terms

alkene, catalytic cracking, fraction, fractional distillation, fossil fuel, hydrocarbon, non-renewable, viscous, volatile

During your study of this topic you should have learned:

○ About the fossil fuels coal, natural gas and petroleum (crude oil) that produce carbon dioxide on combustion.

○ That methane is the main constituent of natural gas.

○ How to describe petroleum as a mixture of hydrocarbons and its separation into useful fractions by fractional distillation.

○ EXTENDED How to describe the properties of molecules within a fraction.

○ About the uses of the following fractions obtained from petroleum:

- refinery gas for bottled gas for heating and cooking
- gasoline fraction for fuel (petrol) in cars
- naphtha fraction for making chemicals
- diesel/gas oil for fuel in diesel engines
- bitumen for making roads.

End of topic questions

Note: The marks given for these questions indicate the level of detail required in the answers. In the examination, the number of marks given to questions like these may be different.

1. a) How was crude oil (petroleum) formed? **(2 marks)**

    b) Why is crude oil called a 'non-renewable' fuel? **(1 mark)**

2. The diagram shows a column used to separate the components present in petroleum.

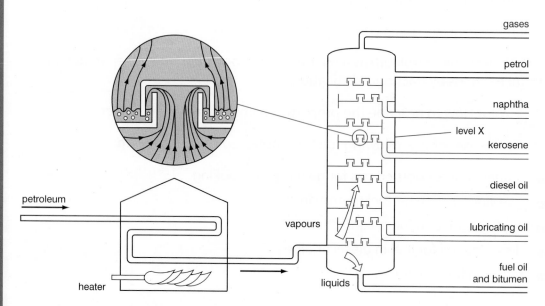

    a) Name the process used to separate petroleum into fractions. **(1 mark)**

    b) What happens to the boiling point of the mixture as it goes up the column? **(1 mark)**

    c) The mixture of vapours arrives at level X. What now happens to the various parts of the mixture? **(2 marks)**

3. The cracking of decane molecules is shown by the equation $C_{10}H_{22} \rightarrow Y + C_2H_4$.

    a) Decane is a *hydrocarbon*. What is a hydrocarbon? **(1 mark)**

    b) EXTENDED What reaction conditions are needed for cracking? **(2 marks)**

    c) Write down the molecular formula for hydrocarbon Y. **(1 mark)**

    d) What 'family' does hydrocarbon Y belong to? **(1 mark)**

    e) Why is the cracking of petroleum fractions so important? **(2 marks)**

4. Petrol is a hydrocarbon with a formula of C_8H_{18}.

   a) What are the products formed when petrol burns in a plentiful supply
   of air? **(2 marks)**

   b) EXTENDED Write a balanced equation, including state symbols, for the reaction
   when petrol burns in a plentiful supply of air. **(2 marks)**

Alkanes

INTRODUCTION

Alkanes are the simplest family or homologous series of organic molecules. The first alkane, methane, is the major component of natural gas, a common fossil fuel. Other alkanes are obtained from petroleum and are widely used as fuels.

△ Fig. 4.10 Natural gas is being extracted from beneath the ocean floor.

KNOWLEDGE CHECK

✓ Understand the nature of covalent bonds.
✓ Know the typical physical properties of compounds that exist as simple molecules.
✓ Understand that combustion involves burning in oxygen or air.

LEARNING OBJECTIVES

✓ EXTENDED Be able to describe a homologous series as a 'family' of similar compounds with similar properties due to the presence of the same functional group.
✓ Be able to describe alkanes as saturated hydrocarbons whose molecules contain only single bonds.
✓ Be able to describe the properties of alkanes (as shown by methane) as being generally unreactive, except in terms of burning.
✓ Be able to describe the complete combustion of hydrocarbons to give carbon dioxide and water.

SCIENCE LINK

BIOLOGY – BIOLOGICAL MOLECULES

- The ability of carbon atoms to form long chains is the basis for many of the compounds that make up living things – leading to the title of 'organic chemistry'.

- Studying hydrocarbons, such as alkanes and alkenes, helps to show the patterns and ideas that will help understand the properties of many biological substances.

- The idea of joining smaller units together to create larger molecules (polymerisation) links to the idea of forming starch molecules by joining together glucose molecules.

WHAT ARE ALKANES?

Alkanes are **hydrocarbons**, which are molecules that contain only carbon and hydrogen atoms. They are made up of carbon atoms linked together by only single covalent bonds and are known as **saturated** hydrocarbons.

Many alkanes are obtained from crude oil by fractional distillation. The smallest alkanes are used extensively as fuels. Apart from burning, however, they are remarkably unreactive.

Alkane	Molecular formula	Displayed formula	Boiling point (°C)	State at room temperature and pressure
Methane	CH_4		−162	Gas
Ethane	C_2H_6		−89	Gas
Propane	C_3H_8		−42	Gas
Butane	C_4H_{10}		0	Gas
Pentane	C_5H_{12}		36	Liquid

△ Table 4.2 Alkanes and their molecular structure.

HOMOLOGOUS SERIES

Alkanes form a **homologous series**. Members of a homologous series have similar chemical properties.

They contain the same **functional group** (the part of the molecule that is responsible for the similar chemical properties) – in this case, the functional group is a C–H single bond.

△ Fig. 4.11 Formula 1 cars use specially blended mixtures of alkane hydrocarbons.

The general characteristics of an homologous series include:

- They have the same general formula. For alkanes this is C_nH_{2n+2}.
- They have similar chemical properties.
- They show a gradual change in physical properties, such as melting point and boiling point.
- They differ from the previous member of the series by $-CH_2-$.

END OF EXTENDED

THE PROPERTIES OF ALKANES

The properties of alkanes are given in Table 4.3.

	Properties of alkanes
General formula	C_nH_{2n+2}
Description	Saturated (no double C═C bond)
Combustion	Burn in oxygen to form CO_2 and H_2O (CO if limited supply of oxygen)
Reactivity	Low
Chemical test	None
Uses	Fuels

△ Table 4.3 Properties of alkanes.

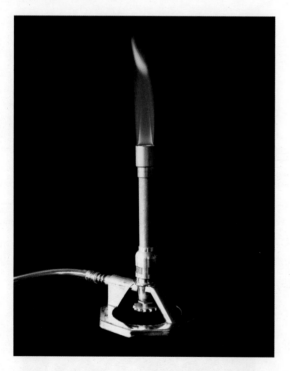

△ Fig. 4.12 Methane is burning in the oxygen in the air to form carbon dioxide and water.

QUESTIONS

1. Alkanes are saturated hydrocarbons.

  a) What is meant by the word *saturated*?

  b) What is meant by the word *hydrocarbon*?

2. a) What is the chemical formula for the alkane with 15 carbon atoms?

b) What products would you expect to be formed if this alkane were burned in a plentiful supply of oxygen?

Combustion of alkanes

In a plentiful supply of air, alkanes burn to form carbon dioxide and water. A blue flame, as produced by a Bunsen burner, indicates complete combustion:

methane	+	oxygen	→	carbon dioxide	+	water
$CH_4(g)$	+	$2O_2(g)$	→	$CO_2(g)$	+	$2H_2O(l)$

SCIENCE IN CONTEXT

When the oxygen supply is limited, as it is when a Bunsen burner burns with a yellow flame, incomplete combustion occurs:

methane	+	oxygen	→	carbon	+	water
$CH_4(g)$	+	$O_2(g)$	→	$C(s)$	+	$2H_2O(l)$

The incomplete combustion of hydrocarbons such as methane can be very dangerous. It can produce carbon monoxide, which is extremely poisonous.

methane	+	oxygen	→	carbon monoxide	+	water
$2CH_4(g)$	+	$3O_2(g)$	→	$2CO(g)$	+	$4H_2O(l)$

Carbon monoxide is difficult to detect without special equipment, because it has no colour or smell. Gas and oil heaters or boilers must be serviced regularly. This is to ensure that jets do not become blocked and limit the air supply, or that exhaust flues don't become blocked and allow small quantities of carbon monoxide to enter the room. The flame in such a boiler or heater should always be blue in colour.

△ Fig. 4.13 The gas in this cooker is burning completely.

QUESTIONS

1. What products are formed when propane gas burns completely?

2. What conditions are essential for an alkane to undergo complete combustion?

End of topic checklist

Key terms

alkane, functional group, homologous series, hydrocarbon, saturated hydrocarbons

During your study of this topic you should have learned:

○ How to describe a homologous series as a 'family' of similar compounds with the same general formula and similar chemical properties.

○ How to describe alkanes as saturated hydrocarbons whose molecules contain only single covalent bonds.

○ How to describe the properties of alkanes (as shown by methane) as being generally unreactive, except in terms of burning.

○ How to describe the complete combustion of hydrocarbons to give carbon dioxide and water.

End of topic questions

Note: The marks given for these questions indicate the level of detail required in the answers. In the examination, the number of marks given to questions like these may be different.

1. What is a *homologous series*? (1 mark)

2. What is the molecular formula for an alkane with 10 carbon atoms? (1 mark)

3. Is the compound with the formula $C_{15}H_{30}$ a member of the alkane series? Explain your answer. (1 mark)

4. Ethane burns in oxygen.

   a) What is the molecular formula of ethane? (1 mark)

   b) Name the products formed when ethane burns in excess oxygen. (2 marks)

   c) What colour flame would indicate that the ethane was burning in excess oxygen? (1 mark)

   d) EXTENDED Write a balanced equation for the reaction. (2 marks)

Alkenes

INTRODUCTION

Alkenes are hydrocarbons and burn in air in the same way that alkanes do. However, alkenes are much more reactive due to the C=C double bond they contain. This makes them very useful starting materials for a number of important industrial processes, including the manufacture of synthetic polymers and margarine.

△ Fig. 4.14 The plastic objects above are made from alkenes.

WHAT ARE ALKENES?

Alkenes are another homologous series, so they have similar chemical properties and physical properties that change gradually from one member to the next.

Alkenes are often formed by the catalytic cracking of larger hydrocarbons (refer to the previous topic on fuels). Hydrogen is also formed in this process. Alkenes contain one or more carbon-to-carbon double bonds. Hydrocarbons with at least one double bond are known as **unsaturated** hydrocarbons. Alkenes burn well and are reactive in other ways also. Their reactivity is due to the carbon-to-carbon double bond.

Alkene	Molecular formula	Structural formula	Boiling point (°C)	State at room temperature and pressure
Ethene	C_2H_4	ethene 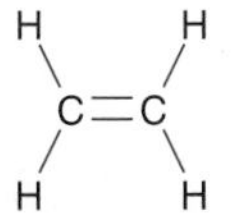 $\quad$ (H₂C=CH₂ structure shown)	−104	Gas
Propene	C_3H_6	propene 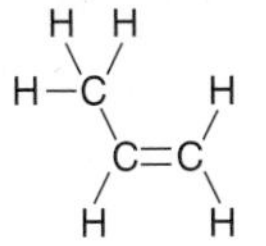 (structure shown) or (structure shown) 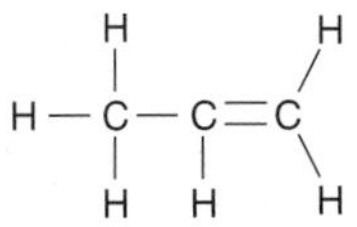	−48	Gas
Butene	C_4H_8	butene 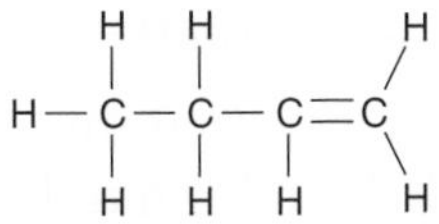 $\quad$ H—C—C—C=C (structure shown)	−6	Gas
Pentene	C_5H_{10}	pentene 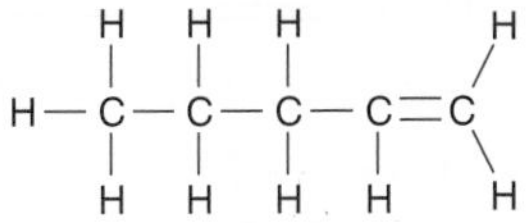 $\quad$ H—C—C—C—C=C (structure shown)	30	Liquid

△ Table 4.4 Structure and state of alkenes.

A simple test to distinguish alkenes from alkanes, or an unsaturated hydrocarbon from a saturated one, is to add bromine water to the hydrocarbon. Alkanes do not react with bromine water, so the colour does not change. An alkene does react with the bromine, and the bromine water loses its colour.

Ethene can also form poly(ethene) or polythene in an addition reaction, a process known as **addition polymerisation**. Ethene is the monomer and reacts with other ethene monomers to form an addition **polymer**, poly(ethene).

Alkenes can be used to make polymers, which are very large molecules made up of many identical smaller molecules called **monomers**. Alkenes are able to react with themselves. They join together into long chains, like adding beads to a necklace. When the monomers add together like this, the material produced is called an addition polymer. Poly(ethene) or polythene is made this way.

By changing the atoms or groups of atoms attached to the carbon-to-carbon double bond, a range of different polymers can be made.

The double bond within the alkene molecule breaks to form a single covalent bond to a carbon atom in an adjacent molecule. This process is repeated rapidly as the molecules link together.

many small molecules

ethene

catalyst and heat

poly(ethene)

one large molecule

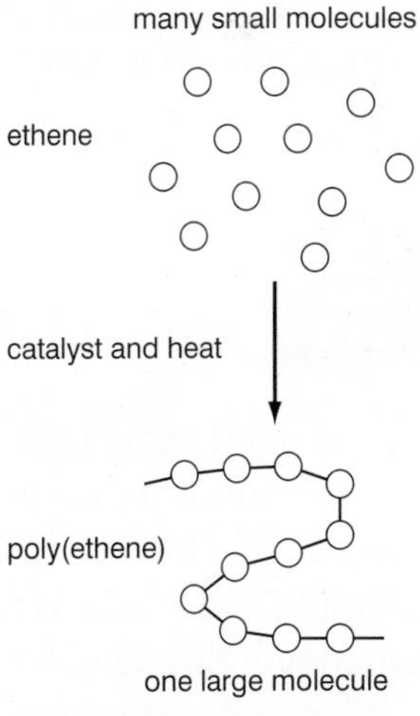

Fig. 4.15 Ethene molecules link together to produce a long polymer chain of poly(ethene).

Δ Fig. 4.16 Poly(ethene) from ethene.

Δ Fig. 4.17 Poly(chloroethene) from chloroethene.

	Properties of alkenes
General formula	C_nH_{2n}
Description	Unsaturated (contain a double C=C bond)
Combustion	Burn in oxygen to form CO_2 and H_2O (CO if limited supply of oxygen)
Reactivity	High (because of the double C=C bonds); undergo addition reactions
Chemical test	Turn bromine water from orange to colourless (an addition reaction)
Uses	Making polymers (addition reactions) such as polyethene

Δ Table 4.5 Properties of alkenes.

QUESTIONS

1. Ethene is an unsaturated hydrocarbon. What does *unsaturated* mean?

2. Name a large-scale use of ethene.

SCIENCE IN CONTEXT

SATURATED AND UNSATURATED FATS

We all need some fat in our diet because it helps the body to absorb certain nutrients. Fat is also a source of energy and provides essential fatty acids. However, it is best to keep the amount of fat we eat at sensible levels and to eat unsaturated fats rather than saturated fats whenever possible. A diet high in saturated fat can cause the level of cholesterol in the blood to build up over time. Raised cholesterol levels increase the risk of heart disease.

Foods high in saturated fat include:

- fatty cuts of meat
- meat products and pies
- butter
- cheese, especially hard cheese
- cream and ice cream
- biscuits and cakes.

Unsaturated fat is found in:

- oily fish such as salmon, tuna and mackerel
- avocados
- nuts and seeds
- sunflower and olive oils.

So, having a carbon-to-carbon double bond does make a difference.

Δ Fig. 4.18 Margarine is made from olive oil, whose unsaturated molecules have been saturated with hydrogen.

End of topic checklist

Key terms

addition polymer, alkene, monomer, polymer, polymerisation, unsaturated

During your study of this topic you should have learned:

○ Describe alkenes as unsaturated hydrocarbons whose molecules contain one double covalent bond.

○ That cracking is a reaction that produces alkenes.

○ EXTENDED How smaller alkanes, alkenes and hydrogen are formed by cracking of larger alkane molecules and know the conditions required for cracking.

○ How to distinguish between saturated and unsaturated hydrocarbons:

- from their molecular structures
- by reaction with aqueous bromine.

○ How to describe the formation of poly(ethene) as an example of addition polymerisation of monomer units.

End of topic questions

Note: The marks given for these questions indicate the level of detail required in the answers. In the examination, the number of marks given to questions like these may be different.

1. Ethene burns in oxygen.

   a) Name the products formed when there is a plentiful supply of oxygen. **(2 marks)**

   b) Write a balanced equation for the burning of ethene in a plentiful supply of oxygen. **(2 marks)**

2. **a)** Draw structural formulae for butane and butene. **(2 marks)**

   b) Which hydrocarbon is unsaturated? **(1 mark)**

   c) Which substance could you use to distinguish between butane and butene? **(1 mark)**

3. Propene gas is bubbled through some bromine water.

   a) Describe the colour change that would occur. **(2 marks)**

   b) EXTENDED Write a balanced equation for this reaction. **(2 marks)**

This section covers concepts that will be important throughout your course. First, you will look at how to measure quantities such as length and time. Then you will look at speed and acceleration before considering mass, weight and density. You will then consider forces and their different effects.

STARTING POINTS

1. What would you use to measure: a) the width of this book; b) the length of the school playing field; c) the amount of milk needed to make a dessert?

2. How could you find the time taken to: a) finish your physics homework; b) run 100 metres?

3. What do we mean when we say a car is travelling at 30 kilometres per hour?

4. If an object is stationary, what must be true about the forces acting on the object?

5. In physics, what do we mean when we say an object is accelerating?

6. How are mass and density related?

CONTENTS

a) Length and time

b) Motion

c) Mass and weight

d) Density

e) Effects of forces

f) Pressure

1
Motion

△ In this topic you will learn about the forces at work on this parachutist.

Length and time

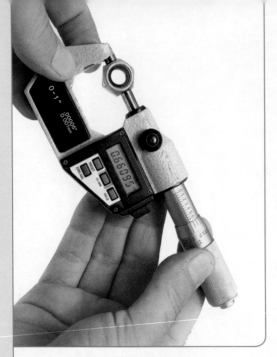

△ Fig. 1.1 Using a micrometer.

INTRODUCTION

Making measurements is very important in physics. Without numerical measurements, physicists would have to rely on descriptions, which could lead to inaccurate comparisons. Imagine trying to build a house if the only descriptions were 'big' and 'small'.

You also need to make sure that you are consistent in your use of units. For example, the Mars Climate Orbiter mission failed in 1999 because not all of the scientists were using the same units.

KNOWLEDGE CHECK

✓ Know how to use a ruler to measure lengths to the nearest millimetre.
✓ Know how to use a stopwatch to measure time to the nearest second.
✓ Know how to use a measuring cylinder to measure volume.

LEARNING OBJECTIVES

✓ Be able to use and describe the use of rulers and measuring cylinders to calculate a length or volume.
✓ Be able to use and describe the use of clocks and devices for measuring an interval of time.
✓ Be able to obtain an average value for a small distance and for a short interval of time by measuring multiples (including the period of a pendulum).

MAKING MEASUREMENTS

When making measurements, physicists use different instruments, such as rulers to measure lengths, measuring cylinders to measure volume and clocks to measure time.

A physicist always takes care to make the measurements as accurate as possible. If she is using a ruler, she will place the ruler along the object to be measured, and read off the scale the positions of the beginning and the end of the object. The length is the difference between these two readings. When the ruler is nearer to her eye than the object being measured, the reading will appear to change as she moves her eye. The correct reading is obtained when her eye is directly above the point being measured.

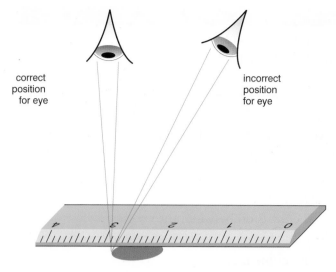

△ Fig. 1.2 Making accurate measurements.

To improve accuracy further, she may take several readings and use the average of these readings as a better result.

△ Fig. 1.3 The Maglev train runs for 30 km between Shanghai and Pudong Airport, and completes the journey in 7 minutes, reaching a top speed of 430 km/h. The train uses magnets to hover 10 mm above the track. The track must be placed within a few millimetres of the planned route, requiring great accuracy in all measurements.

To use a measuring cylinder, she will first make sure that the cylinder is standing on a level table. Then she will make sure that her eye is at the same level as the liquid inside the cylinder. The surface of most liquids will bend up or down near the walls of the measuring cylinder. This bent shape is known as a meniscus. However, most of the surface is flat, and measurements are made to this flat surface.

Warning: Some measuring cylinders have unusual scales and one division may represent an unexpected quantity, perhaps $2\,cm^3$ or $0.5\,cm^3$. Check carefully.

In this book, volumes will usually be measured in cm^3 (or perhaps in m^3). In other places, such as on some measuring cylinders, you will see the millilitre.

A volume of 1 ml is the same as a volume of $1\,cm^3$.

$1000\,cm^3 = 1000\,ml = 1\,l$ (or $1\,dm^3$ to avoid confusion between the number 1 and the letter l)

For measuring large volumes, we also use the cubic metre.

$1\,m^3 = 1000\,dm^3 = 1\,000\,000\,cm^3$

Times are measured by using a stopwatch or stopclock.

Hand-operated stopwatches have an accuracy that is limited by the delay between your eye seeing the moment to start, your brain issuing the command to start the watch and your finger pressing the start button. The total delay is typically around 0.2 s. This delay is known as your 'reaction time', and it increases the danger of some tasks, such as driving a car.

When measuring time accurately is critical, such as in athletics, the clock has to be started and stopped automatically by the athlete breaking a light beam that shines across the track.

If you are measuring the time of an oscillation, such as the swing of a pendulum, it is very easy to improve the accuracy of the measurement by timing a number of swings, perhaps 10 or 20.

It is important to count correctly. Let the swing go, count zero and start the stopwatch as the pendulum crosses a mark at the bottom of the swing (we call this the **fiducial mark**). The next time the pendulum crosses the fiducial mark going in the same direction count one, and so on. In this way the count will be correct.

After measuring the time for 20 swings, say, divide the total time by 20 to give the period of one oscillation of the pendulum.

End of topic checklist

Key term

fiducial mark

During your study of this topic you should have learned:

○ How to use and describe the use of rulers and measuring cylinders to calculate length and volume.

○ How to use and describe the use of clocks and devices for measuring time.

○ How to measure, and describe how to measure, a short interval of time (including the period of a pendulum).

End of topic questions

Note: The marks given for these questions indicate the level of detail required in the answers. In the examination, the number of marks given for questions like these may be different.

1. Rulers that are 30 cm long are often made of wood or plastic that is thicker in the middle and thinner along the edges where the scale is printed. Explain why the user is less likely to make an error if the ruler is thinner at the edge, and suggest reasons why the ruler is thicker in the middle. **(3 marks)**

2. A plastic measuring cylinder is filled with water to the 100 cm³ mark. A student measures the column of water in the cylinder with a ruler and finds that it is 20 cm high.

   a) The student pours 10 cm³ of the water out of the cylinder. How high will the column of water be now? **(2 marks)**

   b) The student then refills the cylinder back to the 100 cm³ mark by holding it under a dripping tap. She finds that it takes 180 drops of water to do this. What is the volume of one of these drops? **(3 marks)**

   c) What is the cross-sectional area of the cylinder? (Hint: The volume of a cylinder is given by the equation: volume = cross-sectional area × length.) **(3 marks)**

   d) From your answer to part **c)**, what is the internal diameter of the measuring cylinder? **(3 marks)**

3. A student tries to measure the period of a pendulum that is already swinging left and right. At the moment when the pendulum is fully to the left, she counts 'one' and starts a stopwatch. She counts successive swings each time that the pendulum returns to the left. When she counts 'ten' she stops the stopwatch, and sees that it reads 12.0 s.

   a) What was her mistake? **(2 marks)**

   b) What is the period of swing of this pendulum? **(3 marks)**

   c) In this particular experiment, explain the likely effect of her reaction time on her answer. **(3 marks)**

Motion

INTRODUCTION

To study almost anything about the world around us or in outer space, we will need to describe where things are, where they were and where we expect them to go. It is even better if we are able to measure these things. Only when we have an organised system for doing this will we be able to look for the patterns in the way things move – the laws of motion – before going a step further and suggesting *why* things move as they do – using ideas about forces.

△ Fig. 1.4 You can use a stopwatch to measure the time taken to run a certain distance.

Think about being a passenger in a car travelling at 90 kilometres per hour. This, of course, means that the car (if it kept travelling at this speed for 1 hour) would travel 90 km. During 1 second, the car travels 25 metres, so its speed can also be described as 25 metres per second. Scientists prefer to measure time in seconds and distance in metres. So they prefer to measure speed in metres per second, usually written as m/s.

KNOWLEDGE CHECK

✓ Know how to measure distances and times accurately.
✓ Know how to calculate the areas of a rectangle and a triangle.
✓ Know how to plot a graph given particular points.
✓ Know how to substitute values into a given formula.

LEARNING OBJECTIVES

✓ Define speed and calculate average speed from total distance/total time.
✓ Be able to plot and interpret a speed–time graph or a distance–time graph.
✓ Recognise from the shape of a speed–time graph when a body is at rest, moving with constant speed or moving with changing speed.
✓ **EXTENDED** Be able to calculate the area under a speed–time graph to work out the distance travelled for motion with constant acceleration.
✓ Demonstrate understanding that acceleration and deceleration are related to changing speed, including qualitative analysis of the gradient of a speed–time graph.
✓ **EXTENDED** Define and calculate acceleration using change of speed/time.
✓ **EXTENDED** Calculate acceleration from the gradient of a speed–time graph.
✓ **EXTENDED** Recognise linear motion for which the acceleration is constant.
✓ **EXTENDED** Recognise motion for which the acceleration is not constant.

CALCULATING SPEED

The **speed** of an object can be calculated using the following formula:

$$\text{speed} = \frac{\text{distance}}{\text{time}}$$

$$v = \frac{s}{t}$$

where: v = speed in m/s,

s = distance in m, and

t = time in s.

Most objects speed up and slow down as they travel. An object's **average speed** can be calculated by dividing the total distance travelled by the total time taken.

REMEMBER

Make sure you can explain *why* this is an average speed. You need to talk about the speed not being constant throughout, perhaps giving specific examples of where it changed. For example, you might consider a journey from home to school. You know how long the journey takes and the distance between home and school. From these, you can work out the average speed using the formula. However, you know that, in any journey, you do not travel at the same speed at all times. You may have to stop to cross the road, or at a road junction. You may be able to travel faster on straight sections of the journey or round corners.

WORKED EXAMPLES

1. Calculate the average speed of a motor car that travels 500 metres in 20 seconds.

Write down the formula:	$v = s / t$
Substitute the values for s and t:	$v = 500 / 20$
Work out the answer and write down the units:	$v = 25$ m/s

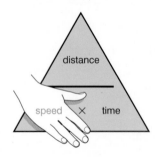

◁ Fig. 1.5 Cover speed to find that speed = distance/time.

2. A horse canters at an average speed of 5 m/s for 2 minutes. Calculate the distance it travels.

Write down the formula in terms of s:	$s = v \times t$
Substitute the values for v and t:	$s = 5 \times 2 \times 60$
Work out the answer and write down the units:	$s = 600$ m

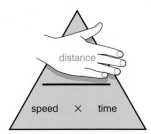

◁ Fig. 1.6 Cover distance to find that distance = speed × time.

QUESTIONS

1. A journey to school is 10 km. It takes 15 minutes in a car. What is the average speed of the car?

2. How far does a bicycle travelling at 1.5 m/s travel in 15 s?

3. A person walks at 0.5 m/s and travels a distance of 1500 m. How long does this take?

USING GRAPHS TO STUDY MOTION

Journeys can be summarised using graphs. The simplest type is a **distance–time graph**, where the distance travelled is plotted against the time of the journey.

At the beginning of any measurement of motion, time is usually given as 0 s and the position of the object as 0 m. If the object is not moving, then time increases but distance does not. This gives a horizontal line. If the object is travelling at a steady speed, then both time and distance increase steadily, which gives a straight line. If the speed is varying, then the line will not be straight. You can calculate the speed of the object by finding the **gradient** of the line on a distance–time graph. In Fig. 1.7, which shows a bicycle journey, the graph slopes when the bicycle is moving. The slope gets steeper when the bicycle goes faster. The slope is straight (has a constant gradient) when the bicycle's speed is constant. The cyclist falls off at about 150 metres from the start. After this, the graph is horizontal because the bicycle is not moving.

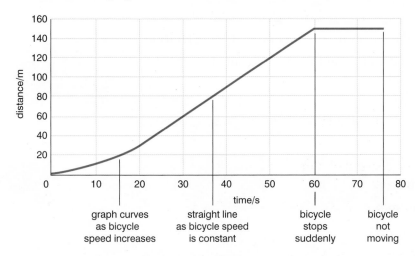

△ Fig. 1.7 A distance–time graph for a bicycle journey.

QUESTIONS

1. How can you tell from a distance–time graph whether the object was moving away from you or towards you?

2. Very often we use sketch graphs to illustrate motion. Describe the main differences between a sketch graph and a graph.

3. Sketch a distance–time graph for a bicycle travelling downhill.

WHAT IS ACCELERATION?

The speedometer of a car displays 50 km/h and then a few seconds later it displays 70 km/h, so the car is accelerating. When a car is slowing down, this is called negative acceleration, or deceleration. **Acceleration** is a change in speed. On a distance–time graph, acceleration is shown by a smooth curve.

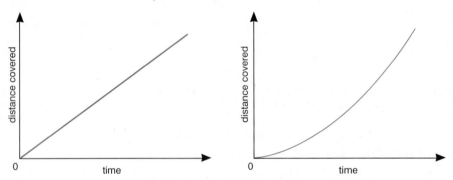

△ Fig. 1.8 Steady speed is shown by a straight line. Acceleration is shown by a smooth curve of increasing gradient.

Imagine that the car is initially travelling at 15 m/s, and that 1 second later it has reached 17 m/s, and that its speed increases by 2 m/s each second after that. Each second its speed increases by 2 metres per second. We can say that its speed is increasing at '2 metres per second *per second*'. This can be written, much more conveniently, as an acceleration of 2 m/s².

EXTENDED

How much an object's speed *changes* in one second is its acceleration.

Acceleration can be calculated using the following formula:

$$\text{acceleration} = \frac{\text{change in speed}}{\text{time taken}}$$

$$a = \frac{(v - u)}{t}$$

where: a = acceleration

v = final speed in m/s

u = starting speed in m/s

t = time in s

Make sure that you are clear what the word 'acceleration' means in physics. It does *not necessarily* mean 'gets faster'. Neither does it measure *how much* the speed changes.

Acceleration measures *how quickly* the speed changes, that is, the *rate of change* of speed.

WORKED EXAMPLE

Calculate the acceleration of a car that travels from 0 m/s to 28 m/s in 10 seconds.

Write down the formula:	$a = (v - u) / t$
Substitute the values for v, u and t:	$a = 28 - 0 / 10$
Work out the answer and write down the units:	$a = 2.8 \text{ m/s}^2$

QUESTIONS

1. As a stone falls, it accelerates from 0 m/s to 20 m/s in 2 seconds. Calculate its acceleration and state the unit.

2. A racing car slows down from 45 m/s to 0 m/s in 3 seconds. Calculate its acceleration and state the unit.

END OF EXTENDED

USING SPEED–TIME GRAPHS

A speed–time graph provides information on speed, acceleration and distance travelled. Steady speed is shown by a horizontal line. Steady acceleration is shown by a straight line sloping up. You can calculate the acceleration of the object from the gradient of a speed–time graph.

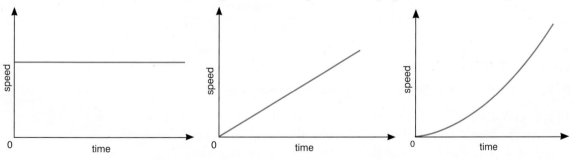

△ Fig. 1.9 Steady speed is shown by a horizontal line. Steady acceleration is shown by a straight line sloping up. Acceleration that is not constant is shown by a curved line.

In the left-hand graph, the object is already moving when the graph begins. In the right-hand graph, the object starts with a speed of zero, and the line therefore starts from the origin.

Note that the object may not move to begin with. In this case the line will start by going along the *x*-axis, showing that the speed stays at zero for a while.

QUESTION

1. An athlete and a fun runner complete a 400 m race. The athlete takes 50 s and the fun runner takes 64 s.

   a) Calculate the average speed for each runner.

   b) Sketch a speed–time graph for the two runners.

EXTENDED

Finding distance from a speed–time graph

The area under a speed–time graph gives the distance travelled because distance = speed × time. Always make sure that the units are consistent, so when the speed is in km/h you must use time in hours too.

The graph in Fig. 1.10 shows how the speed of a car varies as it travels between two sets of traffic lights. The graph can be divided into three regions.

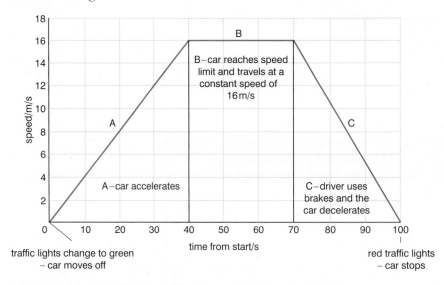

△ Fig. 1.10 The speed of a car travelling between two sets of traffic lights.

In region A, the car has constant acceleration (the line has a constant positive gradient). The gradient is 16/40 so the acceleration is 16/40 m/s² in region A. The distance travelled by the car can be calculated as follows:

average speed in region A = (16 + 0) / 2 = 8 m/s

time = 40 s

so, distance = $v \times t$ = 8 × 40 = 320 m

This can also be calculated from the area under the line:

½ base × height = ½ × 40 × 16 = 320 m

In region B, the car is travelling at a *constant speed* (the line has a gradient of zero). The distance travelled by the car can be calculated:

speed in region B = 16 m/s

time = 30 s

so, distance = $v \times t$ = 16 × 30 = 480 m

This can also be calculated from the area under the line (base × height = 30 × 16 = 480 m).

In region C, the car is *decelerating at a constant rate* (the line has a constant negative gradient). The distance travelled by the car can be calculated:

average speed in region C =

(16 + 0) / 2 = 8 m/s

time = 30 s

so, distance = $v \times t$ = 8 × 30 = 240 m

This can also be calculated from the area under the line:

½ base × height = ½ × 30 × 16 = 240 m

Total distance travelled in 100 s = 320 + 480 + 240 = 1040 m

QUESTIONS

1. Explain how to calculate the acceleration from a speed–time graph.

2. Explain how to calculate the distance travelled from a speed–time graph.

3. Two runners complete a 400 m race. Athlete A takes 50 s and athlete B takes 64 s.

   a) Calculate the average speed of each runner.

   b) Sketch a speed–time graph for each runner.

4. How could you show from the graph that the runners both cover 400 m?

5. From the graph below, calculate the distance travelled between 2 and 12 seconds.

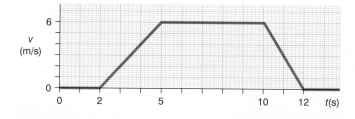

Speed–time graphs can tell us much information about the speed, acceleration and distance of an object when it is travelling.

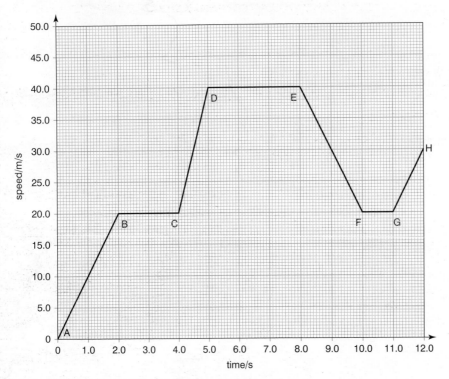

Use the speed–time graph to answer the questions below:

1. What is the speed of the object at 1 s?

2. What can you say about the speed between points B and C?

3. Between which points is there the greatest acceleration?

4. What is happening to the object between points E and F?

5. What is the distance travelled between points D and E?

6. Calculate the acceleration of the object between points A and B.

In Fig. 1.10, the acceleration and deceleration were constant and the lines in regions A and C were straight. This is very often not the case. You will probably have noticed that a car can accelerate much more quickly when it is travelling at 30 km/h than when it is travelling at 120 km/h.

A people-carrying space rocket does exactly the opposite. If you watch one being launched you can see that it has a small acceleration as it leaves the ground. As it is burning several tonnes of fuel per second, it quickly becomes less massive and its acceleration increases.

END OF EXTENDED

Developing investigative skills

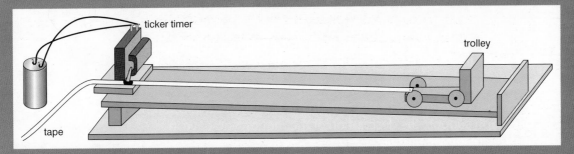

△ Fig. 1.11 Apparatus needed for the investigation.

A student investigated the motion of a trolley rolling down a ramp. To measure the distance travelled by the trolley at different times, she used a ticker timer and ticker tape. A ticker timer has a moving arm that bounces up and down 50 times each second. When the arm moves down and hits the tape it makes a small dot on the tape.

The student attached the tape to the trolley and released the trolley to roll down the ramp.

After that she divided the tape into strips with five dots in each strip – at 50 dots per second this meant that a five-dot strip had taken 0.1 s.

She measured the length of each five-dot strip with a ruler. Her results are shown in the table.

Time/s	Distance from start/cm	Distance covered in the last 0.1 s/cm	Average speed for last 0.1 s/cm/s
0.0		0.0	
0.1		1.8	
0.2		3.4	
0.3		5.2	
0.4		6.0	
0.5		7.7	
0.6		11.1	
0.7		9.9	
0.8		11.9	
0.9		12.5	
1.0		14.0	

Using and organising techniques, apparatus and materials

❶ Suggest how using this method might change the motion of the trolley as it rolls down the ramp.

❷ How else could the student measure the position of the trolley every 0.1 s?

Observing, measuring and recording

❸ Copy the table and complete the second column, showing the total distance travelled up to that time.

❹ Draw a distance–time graph using the data in the first two columns. Use your graph to describe the motion of the trolley.

❺ Use the equation speed = distance/time to complete the final column.

❻ Draw a speed–time graph using the data in the first and fourth columns. Does this graph support the description of the motion you gave in question 4? Explain your answer.

Handling experimental observations and data

❼ The student thought she had made a mistake in measuring the strips. Is there any evidence for this on either of the graphs?

Planning and evaluating investigations

❽ Would repeating the experiment make the data more reliable? Justify your answer.

End of topic checklist

Key terms

acceleration, average speed, distance–time graph, gradient, speed

During your study of this topic you should have learned:

○ How to define speed and calculate speed from total distance–total time.

○ How to plot and interpret a speed–time graph or a distance–time graph.

○ How to recognise from the shape of a speed–time graph when a body is:

- at rest
- moving with constant speed
- moving with changing speed.

○ EXTENDED How to work out the distance travelled from a speed–time graph by calculating the area under the speed–time graph for motion with constant acceleration.

○ That acceleration is related to changing speed and how to analyse speed–time graphs.

○ EXTENDED How to define and calculate acceleration using change of speed/time.

○ EXTENDED How to calculate acceleration from the gradient of a speed–time graph.

○ EXTENDED How to recognise linear motion for which the acceleration is constant and calculate the acceleration.

○ EXTENDED How to recognise motion when the acceleration is not constant.

End of topic questions

Note: The marks given for these questions indicate the level of detail required in the answers. In the examination, the number of marks given for questions like these may be different.

1. A student's journey to school takes 10 minutes and is 3.6 kilometres. What is his average speed in km/min? **(1 mark)**

2. A runner runs 400 metres in 1 minute 20 seconds. What is her speed in m/s? **(1 mark)**

3. EXTENDED A train moves away from a station along a straight track, increasing its speed from 0 to 20 m/s in 16 s. What is its acceleration in m/s^2? **(1 mark)**

4. EXTENDED A rally car accelerates from 100 km/h to 150 km/h in 5 s. What is its acceleration in:

   a) km/h per second **(1 mark)**

   b) m/s^2? **(1 mark)**

5. **a)** On a distance–time graph, what does a horizontal line indicate? **(2 marks)**

   b) A car is travelling at constant speed. What shape would the corresponding distance–time graph have? **(2 marks)**

6. EXTENDED **a)** Define acceleration. **(2 marks)**

   b) Explain how to calculate the distance travelled from a speed–time graph. **(3 marks)**

7. A student cycles to his friend's house. In the first part of his journey, he rides 200 m from his house to a road junction in 20 s. After waiting for 10 s to cross the road, he cycles for 20 s at 8 m/s to reach his friend's house.

   a) What is his average speed for the first part of the journey? **(3 marks)**

   b) How far is it from the road junction to his friend's house? **(2 marks)**

   c) What is his average speed for the whole journey? **(2 marks)**

8. The graph shows a distance–time graph for a car journey.

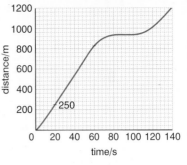

a) What does the graph tell us about the speed of the car between 20 and 60 seconds? **(2 marks)**

b) How far did the car travel between 20 and 60 seconds? **(3 marks)**

c) Calculate the speed of the car between 20 and 60 seconds. **(3 marks)**

d) What happened to the car between 80 and 100 seconds? **(2 marks)**

9. EXTENDED Look at the speed–time graph for a toy tractor.

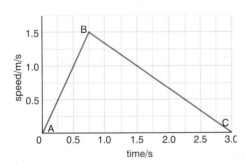

a) Calculate the total distance travelled by the tractor from A to C. **(3 marks)**

b) Calculate the acceleration of the tractor between A and B. **(1 mark)**

Mass and weight

INTRODUCTION

Scientists use the words 'mass' and 'weight' with special meanings. By the mass of an object we mean how much material is present in it. Weight is the force on the object due to gravity. It is measured in newtons. The weight of an object depends on its mass and gravitational field strength. Any mass near the Earth has weight due to the Earth's gravitational pull.

△ Fig. 1.12 An apple being weighed using a newton meter.

MASS AND WEIGHT

Scientists use the words **mass** and **weight** with special meanings. The 'mass' of an object means how much material is present in it. It is measured in kilograms (kg).

Weight is the force on the object due to gravity. It is measured in **newtons** (N). The weight of an object depends on its mass and **gravitational field strength**. Any mass near the Earth has weight due to the Earth's gravitational pull.

Weight is calculated using the equation:

weight = mass × gravitational field strength

$W = mg$

QUESTIONS

1. What is the difference between mass and weight?

2. What is the weight of someone whose mass is 60 kg?

3. What is the mass of someone whose weight is 500 N?

THE EARTH'S GRAVITATIONAL FIELD

Scientists often use the word 'field'. We say that there is a 'gravitational field' around the Earth, and that any object that enters this field will be attracted to the Earth.

The value of the gravitational field strength on Earth is 9.8 N/kg, though we usually round it up to 10 N/kg to make the calculations easier. A gravitational force of 10 N acts on an object of mass 1 kg on the Earth's surface.

Note that gravity does not stop suddenly as you leave the Earth. Satellites go around the Earth and do not escape, because the Earth is still pulling them, even if less strongly than before the satellites were launched. The Earth is even pulling the Moon, and this is why it orbits the Earth once every month. And the Earth goes around the Sun because the Sun's gravity is pulling the Earth.

If you could stand on the Moon you would feel the gravity of the Moon pulling you down. Your mass would be the same as on Earth, but your weight would be less. This is because the gravitational field strength on the Moon is about one-sixth of that on the Earth, and so the force of attraction of an object to the Moon is about one-sixth of that on the Earth. The gravitational field strength on the Moon is 1.6 N/kg, so a force of 1.6 N is needed to support a 1 kg mass.

EXTENDED

Weight is the effect of a gravitational field on a mass.

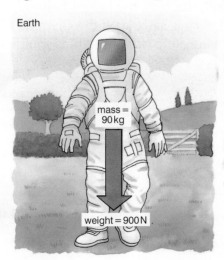
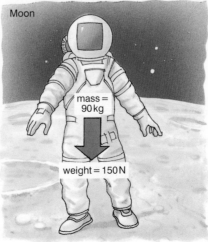

Earth — mass = 90 kg — weight = 900 N

Moon — mass = 90 kg — weight = 150 N

△ Fig. 1.13 Though your mass remains the same, your weight is greater on Earth than it would be on the Moon.

END OF EXTENDED

End of topic checklist

Key terms

gravitational field strength, mass, newton, weight

During your study of this topic you should have learned:

◯ That mass is the amount of matter in a body.

◯ That weight is a force.

◯ How mass and weight are different.

◯ Recall and use the equation $W = mg$.

◯ EXTENDED How to describe, and use the concept of, weight as the effect of a gravitational field on a mass.

End of topic questions

Note: The marks given for these questions indicate the level of detail required in the answers. In the examination, the number of marks given for questions like these may be different.

1. a) Explain the difference between 'mass' and 'weight'. **(2 marks)**

b) Explain why your weight would change if you stood on the surface of different planets. **(3 marks)**

2. EXTENDED The height that you can jump has an inverse relationship to the gravitational field strength. When the field strength doubles, the height halves. The gravitational field strength on the surface of Mars is 3.8 N/kg. If the Olympic Games were held on Mars in a large dome to provide air to breathe, what would happen to the records for:

a) weightlifting (weight in N) **(2 marks)**

b) high jump (height) **(2 marks)**

c) pole vault (height) **(2 marks)**

d) throwing the javelin (distance) **(2 marks)**

e) the 100 m race (time)? **(2 marks)**

In each case, explain if the record will increase, stay similar or decrease.

Density

INTRODUCTION

Which is heavier, a tonne of feathers or a tonne of iron?

That is a trick question, of course – they have the same weight (a tonne). But there would be a noticeable difference if you loaded each one onto a truck: the feathers would take up more space, and we use the idea of density to help explain this difference.

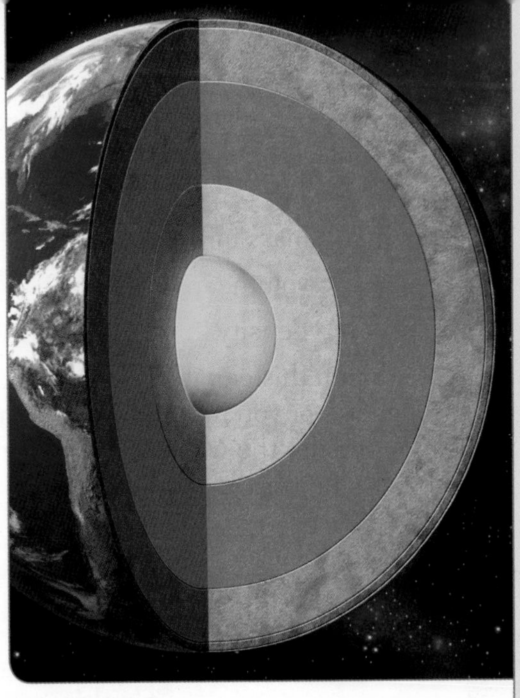

Δ Fig. 1.14 The density of the different layers of the Earth varies.

Density is a useful measure that gives an insight into other areas of physics. For example, density helps explain convection currents, which can lead on to the movement of the continents on the surface of the Earth and the very structure of the Earth itself.

KNOWLEDGE CHECK

✓ Know how to calculate areas of regular shapes, such as squares and rectangles.

LEARNING OBJECTIVES

✓ Recall and use $\rho = m/V$.
✓ Be able to describe an experiment to determine the density of a liquid and of a regularly shaped solid and make the necessary calculation.
✓ EXTENDED Be able to describe the determination of the density of an irregularly shaped solid by the method of displacement and make the necessary calculation.

WHAT IS DENSITY?

You must have noticed that the weight of objects can vary greatly. A plastic teaspoon weighs less than a metal one, and a gold ring weighs twice as much as a silver one, even if the objects are exactly the same size.

The **density** of a material is a measure of how 'squashed up' it is, and a dense object contains more mass than a light object of the same size. The density of a material is defined as the mass per unit **volume**.

The density of a material is calculated using this formula:

$$\rho = m / V$$

where: m = mass in g or kg

V = volume in cm³ or m³

ρ = density in g/cm³ or kg/m³

Note that in this equation you must use g and cm throughout, or you must use kg and m. Also note that if you measure the weight in N you must convert it into mass in g or kg.

THE DENSITY OF A REGULARLY SHAPED OBJECT

△ Fig. 1.15 Gold is one of the densest metals. A block of gold the size of a one-litre carton of milk would have a mass of almost 20 kg.

	g/cubic centimetre	g/cubic metre
Vacuum	0	0
Helium gas	0.000 17	0.17
Air	0.001 24	1.24
Oil (petroleum)	0.88	880
Water	1.0	1000
Sea water	1.03	1030
Plastic	0.9–1.6	900–1600
Wood	0.5–1.3	500–1300
Magnesium	1.74	1740
Aluminium	2.7	2700
Titanium	4.5	4500
Steel	7.8	7800
Mercury (liquid)	13.6	13 600
Silver	10.5	10 500
Gold	19.3	19 300

△ Table 1.1 Some useful densities.

WORKED EXAMPLE

A brick has the dimensions 20 cm × 9 cm × 6.5 cm.

Weight of brick = 22.2 N

What is the density of the brick?

Mass of brick, m $= W/g$

$= 22.2/10 \, \text{kg}$

$= 2.22 \, \text{kg}$

$= 2220 \, \text{g}$

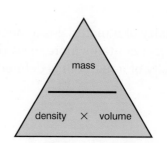

△ Fig. 1.16 The equation triangle for mass, density and volume.

(Remember that 1 kg = 1000 g)

Volume of brick, V $= 20 \times 9 \times 6.5 \, \text{cm}^3$

                    $= 1170 \, \text{cm}^3$

Density of brick, ρ $= \text{mass/volume}$

                    $= 2200/1170 \, \text{g/cm}^3$

                    $= 1.90 \, \text{g/cm}^3$

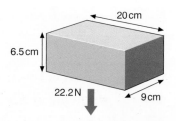

△ Fig. 1.17 Finding the density of a brick.

Developing investigative skills

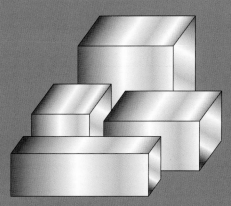

△ Fig. 1.18 Regular shapes for use in the investigation.

A student is finding the density of some different materials. The samples she has are all regular shapes.

The student has a ruler, marked in mm, and an electronic balance that measures to the nearest 0.1 g. The student finds the mass and the volume of each sample. Her data is shown in the table.

Sample	Mass/g	Volume/cm³	Density/?
Aluminium	97.2	36	
Brass	302.4	36	
Copper	321.5	36	
Iron	282.6	36	

Using and organising techniques, apparatus and materials

❶ Describe how the student should use the ruler to find the volume of each sample.

❷ If the student checks that the balance reads zero before she puts the sample on, will this improve the accuracy or the precision of the experiment? Explain your answer.

Observing, measuring and recording

❸ Copy the table and use the equation density = mass / volume to complete it. Include the units at the top of the 'density' column.

❹ How many significant figures should you give your values of density? Explain your answer.

Planning and evaluating investigations

❺ How could the method be changed to find the density of objects with an irregular shape?

MEASURING THE DENSITY OF A LIQUID

The density of a liquid can be measured using an instrument called a **hydrometer**. The hydrometer measures the ratio of the density of the liquid to the density of water and is usually made of glass. It has a cylindrical stem and a bulb, which contains mercury or lead shot to make it float upright. The liquid that is being tested is poured into a measuring cylinder and the hydrometer is lowered into the liquid until it floats. There is a scale on the stem of the hydrometer and the point on the scale at which the surface of the liquid touches the stem of the hydrometer is noted. The scale usually allows the density to be read directly. The type of scale used depends on what the hydrometer is used for.

DID YOU KNOW?

The first hydrometer is credited to the Greek scholar Hypatia, and was probably made sometime in the late fourth or early fifth century. An early description of such a device appears in a letter from Synesius of Cyrene, who asked Hypatia to make one for him.

QUESTIONS

1. A small rectangular block of steel measures 2 cm by 4 cm by 5 cm and has a mass of 312 g. Calculate:

   a) its volume

   b) its density.

2. Why is bread usually less dense than a root vegetable such as a potato or carrot?

DENSITY IN ACTION

In January 2009, an aircraft took off from LaGuardia Airport, New York, bound for Charlotte Douglas International Airport in North Carolina. About three minutes into the flight, it struck a flock of Canada geese, which resulted in a complete loss of thrust from both engines. The pilot realised that they could not safely reach any airfield, so he decided to ditch the aircraft on the Hudson River – in the middle of New York City! The aircraft ditched safely about three minutes after losing power. The pilot said later that he had chosen the ditching location to be as close as possible to boats to maximise the chance of rescue.

△ Fig. 1.19 The ditched aircraft floating in the Hudson River.

Immediately after the aircraft ditched, the crew began to evacuate the passengers. A panicking passenger opened a rear door, which could not be resealed. This made the aircraft fill with water more quickly than it would otherwise have done. All 155 passengers and crew safely evacuated the aircraft, which was almost completely intact but partially submerged and slowly sinking, and nearby ferries and other watercraft quickly rescued them all.

The successful outcome was due to all involved knowing what they had to do and doing it, and to the fact that the density of the aircraft allowed it to stay afloat long enough for the evacuation to take place.

MEASURING THE DENSITY OF AN IRREGULAR OBJECT

This method involves submerging an object in a liquid and measuring the volume of the liquid that is displaced. It only works when the object is denser than the liquid used so that it sinks, although the method can be modified to measure a less dense object by attaching a 'sinker' to the object to hold it beneath the surface. It does not work when the object absorbs the liquid, or if it is damaged by the liquid.

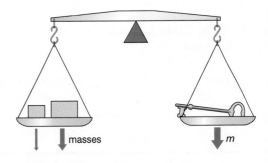

△ Fig. 1.20 Using a balance to find the mass of an object.

1. A balance is used to weigh the object in question, as shown in Fig. 1.20, and find its mass, m.

2. A measuring cylinder is chosen that is wide and deep enough to hold the object. A narrower cylinder will give a more accurate answer than a wider one. Liquid is added to fill the cylinder to a deep enough level so that the object will be completely submerged. The volume of liquid, V_1 is then measured (see Fig. 1.21). The exact amount of liquid that you use is not at all critical. Water is the liquid normally used.

Volume V_1 Volume V_2

△ Fig. 1.21 Measuring the volume of an object.

3. The object is lowered into the liquid (without splashing) and the new reading V_2 is measured (Fig. 1.21). This is the volume of the object and the liquid. The volume of the object is therefore $V_2 - V_1$.

4. The density of the object can now be calculated from the mass and the volume.

WORKED EXAMPLE

The mass of a small metal object, like the one in Fig. 1.22, is found to be 90 g.

A measuring cylinder is filled with water to the 82 cm³ mark. The object is lowered into the measuring cylinder and the water rises to the 91 cm³ mark.

What metal is the object made of?

Volume of the object: $V = 91 - 82$

$= 9 \, \text{cm}^3$

Write down the formula: $\rho = m/v$

Substitute the values for m and V: $\rho = 90/9$

Work out the answer and write down the units: $\rho = 10 \, \text{g/cm}^3$

So, from Table 1.1, what metal could the object be made of?

An experiment of this type is never perfectly accurate, so the density that you measure will never be exactly the same as the values given in tables.

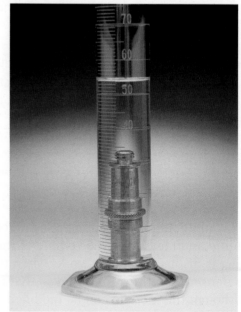

△ Fig. 1.22 Measuring the volume of an irregular metal object by displacing water.

END OF EXTENDED

End of topic checklist

Key terms

density, hydrometer, volume

During your study of this topic you should have learned:

○ To recall and use $\rho = m/V$.

○ About an experiment to determine the density of a liquid and of a regularly shaped solid and how to make the necessary calculation.

○ **EXTENDED** How to describe the determination of the density of an irregularly shaped solid by finding its mass and then determining the volume of water that it will displace, and how to make the necessary calculation.

End of topic questions

Note: The marks given for these questions indicate the level of detail required in the answers. In the examination, the number of marks given for questions like these may be different.

1. The density of air is 1.3 kg/m³. Estimate the mass of air in the room you are in.

2. Ice floats on water.

    a) Which has the greater density, ice or water?

    b) Use particle ideas to suggest why this is surprising.

3. A king who has studied physics believes that his jeweller has given him a crown that is a mixture of gold and silver, not the 1.93 kg of pure gold that he paid for. He weighs the crown on a balance and finds that it has the correct mass of 1.93 kg. He then immerses it in a measuring jug where the water level was originally 800 cm³.

    a) If the crown is pure gold, what will the new water level be?   (2 marks)

    b) What will happen to the water level if the jeweller has cheated?   (2 marks)

Effects of forces

△ Fig. 1.23 These people are applying forces to drag the net from the sea.

INTRODUCTION

We live in a dynamic universe. There is constant motion around us all the time, from the vibrations of our atoms to the sweep of giant galaxies through space, and the motion is constantly changing. Objects themselves do not remain constant – some change size, others change shape. Atoms arrange and rearrange themselves into many different chemicals. Energy moves about through the motion of objects and through transfer by waves. All of this motion and change is driven by forces. This topic explores the forces behind the movement of objects.

WHAT ARE FORCES?

A **force** is a push or a pull. The way that an object behaves depends on all of the forces acting on it. A force may come from the pull of a chain or rope, the push of a jet engine, the push of a pillar holding up a ceiling, or the pull of the gravitational field around the Earth.

Effects of forces

It is unusual for a single force to be acting on an object. Usually there will be two or more. The sizes and directions of these forces determine whether the object will move and the direction it will move in.

Forces are measured in newtons (N). They take many forms and have many effects, including pushing, pulling, bending, stretching, squeezing and tearing. Forces can:

- change the speed of an object
- change the direction of movement of an object
- change the shape of an object.

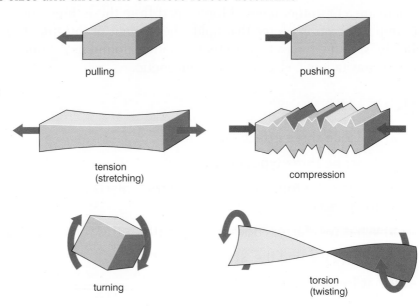

△ Fig. 1.24 Different types of force.

There are several different types of force. All objects in the Universe attract each other with the extremely feeble force of gravity. The strength of the attraction depends on the mass of the two objects and the distance between their centres. You may think that gravity is strong, but you are, after all, close to the Earth, which is a very massive object! The gravitational attraction between everyday objects is very small.

Electricity and magnetism both generate forces that are far stronger than gravity. You see magnetic forces and electric forces combining as an electromagnetic force used every day when an electric motor turns. A current in a magnetic field experiences a force. A motor has a coil of wire moving in a magnetic field. One side of the coil experiences an upwards force in the magnetic field, the other side experiences a downward force, and the coil turns in the magnetic field.

Electrostatic forces are the most important in our everyday lives. Electrostatic forces are those between charges, such as electrons. Like charges repel (so an electron will repel another electron) and unlike charges attract (so a negatively charged electron will attract a positively charged proton). The reason that you are not sinking into the floor at the moment is that the electrons on the outside of the atoms of your shoes are being repelled by the electrons on the outside of the atoms of the floor (Fig. 1.25).

The same force is used when your hand lifts something up, or when friction slows down a car. In fact, all of the forces in this section are either gravitational or – ultimately – electrostatic.

△ Fig. 1.25 Electrons on the surface of the floor repel electrons on the surface of the sole of the shoe.

And when you consider that it is an electrostatic force that allows a bulletproof coat to stop a speeding bullet, you'll probably agree that electrostatic forces are much stronger than gravity.

There are a few other types of force apart from these three, for example, the 'strong' force that holds the nucleus of the atom together. But most of the forces that we feel or notice around us are one of the three: gravitational, electrostatic or magnetic.

QUESTIONS

1. a) Describe three effects of a force.

   b) Describe three types of force.

2. a) What is the force that causes all the objects in the Universe to attract?

   b) What two factors does the strength of this force depend on?

3. Where do we find the 'strong' force?

4. What force is seen in a motor?

SCIENCE IN CONTEXT   **LINKING THE FUNDAMENTAL FORCES**

Scientists are working to find a theory that links all the fundamental forces: gravity, electromagnetism, strong nuclear force and weak nuclear force. Particle accelerators, in which high-energy collisions take place, are useful tools in this search. In 1963, Glashow, Salam and Weinberg predicted that the electromagnetic force and the weak nuclear force might combine (in what would be called the electroweak force) at energies of about 100 GeV or temperatures of about 10^{15} K, which would have occurred shortly after the Big Bang. This prediction was confirmed 20 years later in a particle accelerator.

Δ Fig. 1.26 Inside the Large Hadron Collider.

There are theories that predict that the electroweak and strong forces would combine at energies greater than 10^{15} GeV and that all the forces may combine at energies greater than 10^{19} GeV. At present, the largest particle accelerator is the Large Hadron Collider, at the European Organisation for Nuclear Research (CERN) in Switzerland. It is able to accelerate protons to 99.99% of the speed of light, and they can reach energies of 1.4×10^4 GeV, so it is still some way short of the energies needed to test the theory about combining the electroweak and strong forces, and all four forces.

However, science never stands still and it may be that these energies are reached in your lifetime!

ADDING FORCES

When two or more forces are pulling or pushing an object in the same direction, then the effect of the forces will add up; when they are pulling it in opposite directions, then the backwards forces can be subtracted.

Δ Fig. 1.27 These husky dogs are able to pull the sledge due to the low level of friction between the sledge and the snow.

Twelve husky dogs are pulling a sledge. The sledge is travelling to the right and each dog is pulling with a force of 50 N. There is a friction force of 250 N that is trying to slow the sledge, and therefore must be pointing to the left.

The total force to the right is (12×50) N = 600 N.

The total force to the left is 250 N.

The **resultant force** (the total added-up force) = 600 − 250 N to the right

$\qquad\qquad\qquad\qquad\qquad\qquad$ = 350 N to the right.

Note that you must give the direction of the resultant force.

Think very carefully about this. Having zero resultant force does *not* mean the object is stationary. What it *does* mean is that the object is not *accelerating* – if it is already moving it will continue to do so at constant speed in a straight line.

HOW ARE MATERIALS AFFECTED BY STRETCHING?

When weights are added to a length of wire, the wire will stretch. The graph in Fig. 1.28 shows how the amount that the wire stretches (the **extension**) varies with the load attached to it (the force). The wire will stretch in proportion to the load up to a certain point, which depends on the material from which the wire is made. Beyond this point, the extension is no longer proportional to the load, and so this point is called the **limit of proportionality**.

A string on a musical instrument, such as a guitar string, will behave as shown in the graph for wire, but will break shortly after the limit of proportionality is reached. This means that, when tuning a string on a musical instrument, we need to take care that we do not tighten the string too much or we risk it breaking.

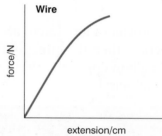

△ Fig. 1.28 Force–extension graph for a wire.

REMEMBER

A material shows elastic behaviour if it returns to its original length when any deforming forces have been removed. During elastic behaviour, the particles in the material are pulled apart a little, so they return to their original positions when the forces are removed. A material shows plastic behaviour if it remains deformed when a load is removed. During plastic behaviour, the particles slide past each other and the structure of the material is changed permanently.

SCIENCE IN CONTEXT

THE BEGINNINGS OF PLASTICS

In 1862, at the Great Exhibition in London, Alexander Parkes demonstrated an organic material that was derived from cellulose that, once it was heated, could be moulded and kept its moulded shape when it was cooled. This was the first plastic material, which was called Parkesine.

Six years later, John Wesley Hyatt invented celluloid, which again is derived from cellulose, as an alternative to ivory, which was then used to make billiard balls. Celluloid became more famous as the first flexible photographic film used for still photographs and moving pictures. By 1900 it had an expanding market in movie films.

The first fully synthetic resin to be commercially successful was Bakelite, which was invented in 1907 by Leo Hendrik Baekeland.

△ Fig. 1.29 Many early telephones were made from Bakelite.

Hooke's law

For a wire, there is a section of the force–extension graph that is linear. Like a wire, when a spring stretches, the extension of the spring is proportional to the force stretching it, provided the limit of proportionality (see Fig. 1.28) of the spring is not exceeded. This is **Hooke's law** and is shown by a straight line on a force–extension graph (Fig. 1.30).

The gradient of the line is a measure of the stiffness of the spring.

An experiment to show Hooke's law:

1. Assemble the apparatus (Fig. 1.31) and allow the spring to hang down. Measure the starting position of the bottom end of the spring on the ruler.

2. Take the first mass, which consists of the hook and base plate, typically of mass 100 g (a weight of 1 N), and hang it on the spring (take care to ensure that the mass can't fall onto anyone's feet). Measure the new position of the bottom end of the spring on the ruler. The difference in the readings is the extension of the spring.

3. Add masses one by one to the first one. Typically each mass is C-shaped, and adds an additional 100 g. Add the masses carefully so that the spring stretches slowly.

4. You should then reverse the experiment to see what happens as the masses are removed.

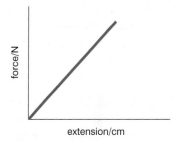
△ Fig. 1.30 Hooke's law in a spring is shown by a straight line.

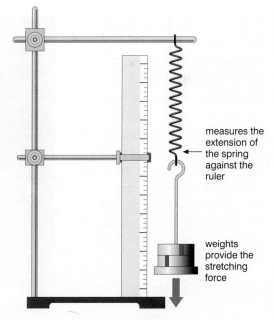

measures the extension of the spring against the ruler

weights provide the stretching force

△ Fig. 1.31 Apparatus to investigate Hooke's law.

5. Calculate the extension (Table 1.2) and plot a graph of extension against force (Fig. 1.32).

Mass/g	Force/N	Reading/ cm	Calculate the extension/ cm	Extension/ cm
0	0	15.2	–	–
100	1.0	16.8	16.8 – 15.2	1.6
200	2.0	18.5	18.5 – 15.2	3.3
300	3.0	19.9	19.9 – 15.2	4.7
400	4.0	21.6	21.6 – 15.2	6.4
etc.	etc.			

△ Table 1.2. Results of experiment.

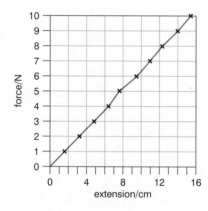

△ Fig. 1.32 Graph of results.

A spring that obeys Hooke's law shows '**proportional' behaviour**: the extension of the spring increases in proportion to the load on the spring. It also shows elastic behaviour – when the force is removed, the spring returns to its original length provided the elastic limit (the point at which the spring returns to its original length after the load is removed) has not been exceeded.

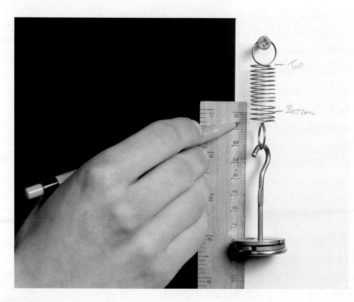

△ Fig. 1.33 Carrying out the experiment.

HOOKE'S LAW IN ACTION

Hooke's law applies to springs that are both extended and compressed by a load, so whenever a spring is used it is an application of Hooke's law. Examples are toys that use springs such as jack-in-the boxes, and trampolines, which rely on springs returning to their original length and then stretching again to give the 'bounce' required. Anyone who sleeps on a mattress that contains springs also experiences Hooke's law in action on a nightly basis – a mattress with springs that do not return to their original length after being compressed would be rather uncomfortable to sleep on!

Developing investigative skills

A student assembled the apparatus as in Fig. 1.31, allowing the spring to hang vertically. Wearing safety glasses, the student measured the initial length of the spring and then measured it again after hanging an additional 100 g onto it. The student continued adding 100 g masses, measuring the length of the spring after each one. His measurements are shown in the table.

Mass added/g	Force/N	Length of spring/cm	Extension of spring/cm
0		2.0	
100		6.0	
200		10.0	
300		14.0	
400		18.0	
500		22.0	
600		26.0	
700		30.0	
800		34.0	
900		38.0	
1000		42.0	
1100		46.0	
1200		52.0	
1300		59.0	
1400		77.0	

Using and organising techniques, apparatus and materials

❶ Why should the student wear eye protection during this experiment?

❷ Describe how any other safety risks can be minimised.

❸ The student carried out a preliminary experiment before deciding to use 100 g masses. Why is a preliminary experiment valuable?

LIMIT OF PROPORTIONALITY

When you stretch a spring too far, the line is no longer straight and Hooke's law is no longer true. This point at the end of the straight line is known as the 'limit of proportionality'.

The spring may (if you do not stretch it too far) be elastic and go back to its original length.

However, as you stretch the material beyond the limit of proportionality, different materials can behave in widely different ways.

As we have seen, the equation for Hooke's law is:

force = spring constant × extension of spring

$F = kx$

Where: F = force in newtons

k = spring constant in N/m

x = extension of the spring in m

Note that it is acceptable to use a spring constant in N/cm or N/mm, so long as the extension is measured in the same units.

This equation works for springs that are being stretched or compressed. The value of k will be the same for both, but note that some springs cannot be compressed (if, for example, the turns of the spring are already in contact).

You can use the triangle in Fig. 1.34 to help you to rearrange the equation. Cover the quantity you want to find and the form of the other two will show you how to write the equation. For example, to find x, cover it and you will see that the equation should be written as $x = F/k$.

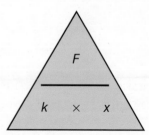

△ Fig. 1.34 The equation triangle for Hooke's law.

WORKED EXAMPLE

A motorbike has a single compression spring on the rear wheels. When the rider sits on the bike, she pushes on the rear wheel with 60% of her weight. Her mass is 50 kg, and the spring constant is 60 N/cm. How much does the spring compress when she sits on the bike?

The formula for the weight of the rider:	$W = mg$
Substitute the values for m and g:	$W = 50 \times 10$
Work out the answer and write down the units:	$W = 500\,N$
The force on the rear spring	$= 60\%$ of $500\,N$
	$= 0.6 \times 500\,N$
	$= 300\,N$

Write down the formula for the compression of the spring: $x = F/k$

Substitute the values for F and k: $\qquad x = 300/60$

Work out the answer and write down the units: $\qquad x = 5\,cm$

The spring compresses by 5 cm.

QUESTIONS

1. What force is required to stretch a spring with spring constant 0.2 N/m a distance of 5 cm?

2. A vertical spring stretches 5 cm under a load of 100 g. Determine the spring constant.

3. A force of 600 N compresses a spring with spring constant 30 N/cm. How far does the spring compress?

END OF EXTENDED

BALANCED FORCES

Usually there are at least two forces acting on an object. When these two forces are balanced, then the object will either be stationary or moving at a constant speed. If forces are balanced, there is no resultant force on the object. Two forces are balanced when their magnitude is the same but they act in opposite directions.

The book in Fig. 1.35 is stationary because the push upwards from the table is equal to the weight downwards. If the table stopped pushing upwards, the book would fall.

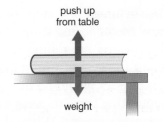

△ Fig. 1.35 The forces on this book are balanced.

The aircraft in Fig. 1.36 is flying 'straight and level' because the lift generated by the air flowing over the wings is equal and opposite to the weight of the aircraft. This diagram shows that the plane will neither climb nor dive, as it would if the forces were not equal.

This is **Newton's first law of motion**, which simply says you need a resultant force to change the way something is moving.

△ Fig. 1.36 The balanced forces on this aircraft mean that its direction of motion will not change.

UNBALANCED FORCES

For an object's speed or direction of movement to change, the forces acting on it must be unbalanced (there must be a resultant force). You can find the resultant of two unbalanced forces by adding them up, taking into account their direction. So, when the driving force on a car is 100 N to the left but the friction force is 50 N to the right, the resultant force is 50 N to the left.

WORKED EXAMPLES

1. Find the resultant force when a skydiver of mass 60 kg jumps from a plane and the air resistance is 10 N.

   Force downwards $= m \times g$
   $= 60 \times 10$
   $= 600$ N

   Force upwards $= -10$ N (if you take the downwards direction as positive)

   Resultant force $= 600 - 10$ N
   $= 590$ N downwards

2. Find the resultant force on a car when the driving force in 1500 N to the left and the friction force is 100 N to the right.

   Resultant force $= 1500 - 100$
   $= 1400$ N to the left

As a gymnast first steps on to a trampoline, his weight is much greater than the opposing supporting force of the trampoline, so he moves downwards, stretching the trampoline. (Note that we are not talking about the gymnast jumping onto the trampoline – if that were the case, the physics would be different!) As the trampoline stretches, its supporting force increases until the supporting force is equal to the gymnast's weight. When the two forces are balanced, the trampoline stops stretching. If an elephant stood on the trampoline, the trampoline would break because it could never produce a supporting force equal to the elephant's weight.

You see the same effect when you stand on snow or soft ground. When you stand on quicksand, the supporting force will not equal your weight, and you will continue to sink.

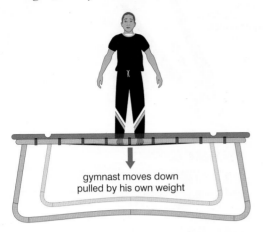
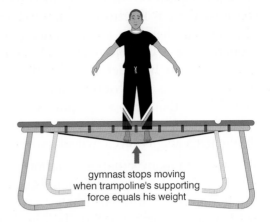

gymnast moves down pulled by his own weight

gymnast stops moving when trampoline's supporting force equals his weight

△ Fig. 1.37 A trampoline stretches until it supports the weight on it.

QUESTIONS

1. Describe the motion of an object when the forces on it are balanced.

2. Describe the motion of an object when the forces on it are unbalanced.

3. In Fig. 1.37, if the gymnast is standing at rest on the trampoline, what must the supporting force of the trampoline be equal to?

SCIENCE LINK

BIOLOGY – LIFE PROCESSES

- Forces changing the shapes of objects links to the idea of how muscles work and the effects they have on the body. For example, the heart is muscle tissue that changes its shape to maintain blood flow around the body.

- A second example is the action of the biceps and triceps muscles – an antagonistic pair – in the arm which also change shape to move the arm.

CHEMISTRY – METALS

- Metals are malleable – applying a force to a metal can cause its shape to change. Forces have this effect on metals due to the metallic bonding.

WHAT IS FRICTION?

A force that opposes motion may not be a bad thing. When you walk, your feet try to slide backwards on the ground. It is only because there is **friction** (working against this sliding) that you can move forwards. Where there is friction, heat is released. Just think how much harder it is to walk on a slippery (that is, low friction) surface such as ice. A force that opposes motion is also very useful in applications such as between the brake pads and a bicycle wheel.

However, in many situations friction can be a disadvantage. For example, there is some friction in the bearings of a bicycle wheel, which causes some energy to be transferred to unwanted thermal energy.

Air resistance is a form of friction. The air resistance experienced by a parachutist is a form of friction and is useful in this case because there is a large resistive force, which means that the terminal velocity of a parachutist is quite low and he or she can land relatively safely.

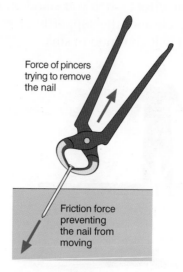

Force of pincers trying to remove the nail

Friction force preventing the nail from moving

△ Fig. 1.38 Friction can stop any movement occurring at all, and it is friction that stops a nail coming out of a piece of wood.

QUESTIONS

1. Give an example of where friction may be useful.

2. Give an example of where friction may be a disadvantage.

End of topic checklist

Key terms

electrostatic forces, extension, force, friction, Hooke's law, limit of proportionality, Newton's first law of motion, proportional behaviour, resultant force

During your study of this topic you should have learned:

◯ That a force may produce a change in size and shape of a body.

◯ **EXTENDED** How to plot load–extension graphs and describe the associated experimental procedure.

◯ **EXTENDED** That Hooke's law can be summarised as $F = kx$ where F is the load, k is the spring constant and x is the extension.

◯ **EXTENDED** The significance of the term 'limit of proportionality' for an extension–load graph.

◯ To describe the ways in which a force may change the motion of a body.

◯ How to find the resultant of two or more forces acting along the same line.

◯ That if there is no resultant force on a body it either remains at rest or continues at constant speed in a straight line.

◯ That friction is the force between two surfaces that impedes motion and results in heating.

◯ That air resistance is a form of friction.

End of topic questions

Note: The marks given for these questions indicate the level of detail required in the answers. In the examination, the number of marks given to questions like these may be different.

1. A student performed an experiment stretching a spring. She loaded masses onto the spring and measured its extension. Here are her results.

Extension/cm	0	4	8	12	16	20	24
Load/N	0	2.0	4.0	6.0	7.5	8.3	8.6

   a) On graph paper, plot a graph of load (vertical axis) against extension (horizontal axis). Draw a suitable line through your points. **(3 marks)**

   b) EXTENDED Mark on the graph the limit of proportionality, and indicate the region where proportional behaviour occurs and the region where the behaviour is probably plastic. **(3 marks)**

   c) EXTENDED How does she check whether the spring, after being loaded with 8.6 N, has shown plastic behaviour or purely elastic behaviour? **(1 mark)**

2. For each of the following situations, give two different examples:

   a) Forces causing objects to change speed

   b) Forces causing objects to change direction

   c) Forces causing objects to change shape.

3. For each of the examples you gave in your answer to question 2, describe what would happen after the force is removed.

4. EXTENDED Find the resultant of a force of 3 N acting vertically and a force of 4 N acting horizontally. **(4 marks)**

Pressure

INTRODUCTION

A snowmobile can travel over soft snow because its weight is spread over a large area of snow by the skis. If the rider got off and stood on the snow, he would probably sink into it up to his knees, even though he is much lighter than the snowmobile.

If a pair of shoes has narrow (stiletto) heels, the wearer can easily damage a wooden floor by making indentations in it. You can push a drawing pin into a notice board by the pressure your thumb exerts on the sharp end.

△ Fig. 1.39 The skis underneath stop the snowmobile from sinking into deep snow.

In each case, the question is not just what force is used, but also what area it is spread over.

KNOWLEDGE CHECK

✓ Know how to calculate areas of regular shapes, such as squares and rectangles.
✓ Know how to calculate the volume of regular objects, such as cubes and cylinders.
✓ Understand the concept of force.

LEARNING OBJECTIVES

✓ Be able to relate (without calculation) pressure to force and area, using appropriate examples.
✓ EXTENDED Recall and use the equation $p = F/A$.

MEASURING PRESSURE

Pressure is the ratio of force to area. Where we have a large force over a small area we have a high pressure, and a small force over a large area gives us a low pressure.

Pressure is measured in newtons per square metre (N/m^2), usually called pascals (Pa). So $1\ Pa = 1\ N/m^2$.

△ Fig. 1.40 When a drawing pin is placed pin-side down, the pressure on the surface is much greater than when it is placed head-side down.

To measure how 'spread out' a force is, use this formula:

pressure = force/area

$$p = F/A$$

Where: p = pressure in pascals, Pa (or newtons per square metre, N/m^2)

F = force in newtons, N

A = area in m^2

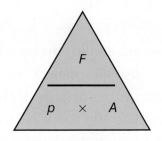

△ Fig. 1.41 the equation triangle for force, pressure and area.

WORKED EXAMPLE

What pressure on the snow does a snowmobile make when it has a weight of 800 N and the runners have an area of $0.2\,m^2$?

Write down the formula: $p = F/A$

Confirm that F is in N and A is in m^2.

Substitute the values for F and A: $p = 800/0.2$

Work out the answer and write down the units: $p = 4000\,Pa$ or $4\,kPa$

Note that $4\,kPa$ is a very low pressure. When you stand on the ground in basketball shoes, the pressure on the ground will be around $20\,kPa$. The wheel of a car generates a pressure on the ground of around $200\,kPa$. Pressures can be quite high, so the kPa is often used.

QUESTIONS

1. Why can you push a drawing pin into a surface using your thumb when you can't push your thumb into the same surface?

2. EXTENDED Calculate the pressure exerted by a 100 N force acting on an area of $0.2\,m^2$.

3. EXTENDED A pressure of 40 Pa is exerted over an area of $2\,m^2$. Calculate the force involved.

4. EXTENDED A force of 500 N produces a pressure of 640 Pa. Over what area is the force acting?

Developing investigative skills

A student has been reading about how scientists can gain information about the mass of dinosaurs from the depth of their fossilised footprints. He decides to investigate how far a wooden block sinks into sand when the pressure on it changes. The student loads 100 g masses onto the block one at a time (to a maximum of six masses) and measures how deeply the block is pushed into the sand.

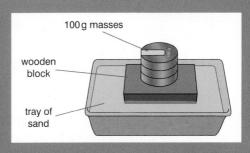

△ Fig. 1.42 Apparatus for the investigation.

The student finds very little evidence of a pattern in his measurements. He thinks this is because the block tends to tip over, rather than standing straight, which means that the sand is not equally pushed down. He also thinks that the sand does not push down very much anyway – it just gets pushed to the side.

To extend his experiment, the student has an idea about investigating if the 'wetness' of the sand makes a difference to the way the block behaves, but he has not yet devised a plan to test this.

Using and organising techniques, apparatus and materials

❶ Explain how using the block in different ways and using different numbers of 100 g masses allows the student to test a variety of different pressures.

❷ Devise a method to measure the depth to which the block sinks in the sand. You should name any equipment needed. Remember that the block may not sink equally in all directions.

❸ What is the independent variable in this investigation? What is the dependent variable?

Observing, measuring and recording

❹ Use ideas about particles to explain the student's observation that the sand 'just gets pushed to the side'.

Handling experimental observations and data

❺ Suggest how the student could measure the 'wetness' of the sand in a reliable way.

❻ Give an example of a situation where the idea of pressure can be used to explain why an object does not sink into a material such as sand.

Key term

pressure

During your study of this topic you should have learned:

○ **EXTENDED** That $p = F/A$.

○ That pressure is related to force and area and be able to give appropriate examples.

End of topic questions

Note: The marks given for these questions indicate the level of detail required in the answers. In the examination, the number of marks given to questions like these may be different.

1. **a)** EXTENDED Calculate the pressure on the floor caused by:

    i) an ordinary shoe heel (person of mass 40 kg, heel 5 cm × 5 cm) when all the person's weight is on one heel   **(2 marks)**

    ii) an elephant (mass 500 kg, area of one foot 300 cm^2) when all four feet are on the ground   **(2 marks)**

    iii) a high-heeled shoe (worn by a person of mass 40 kg, heel area 0.5 cm^2) when all the person's weight is on one heel.   **(2 marks)**

   b) Which of the situations described in part **a)** will damage a wooden floor that starts to yield at a pressure of 4000 kPa?   **(2 marks)**

   (Note: to convert from cm^2 to m^2 you need to divide by 10 000.)

2. A skater glides on one skate. The mass of the skater is 65 kg and the area of the skate is 9×10^{-4} m^2. What pressure is exerted on the ice by the skate?   **(2 marks)**

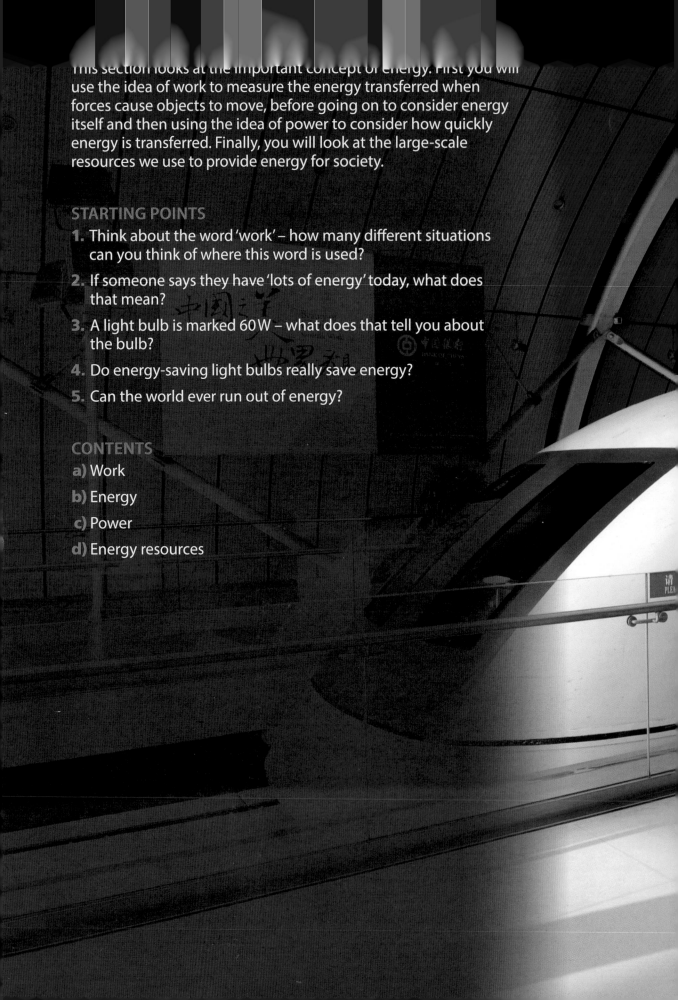

This section looks at the important concept of energy. First you will use the idea of work to measure the energy transferred when forces cause objects to move, before going on to consider energy itself and then using the idea of power to consider how quickly energy is transferred. Finally, you will look at the large-scale resources we use to provide energy for society.

STARTING POINTS

1. Think about the word 'work' – how many different situations can you think of where this word is used?

2. If someone says they have 'lots of energy' today, what does that mean?

3. A light bulb is marked 60 W – what does that tell you about the bulb?

4. Do energy-saving light bulbs really save energy?

5. Can the world ever run out of energy?

CONTENTS

a) Work

b) Energy

c) Power

d) Energy resources

2
Work, energy and power

△ The Maglev train in Shanghai, China is the
first commercially operated high-speed magnetic
levitation line in the world.

△ Fig. 2.1 The mobile crane is doing work as it lifts the girder into place.

Work

INTRODUCTION

How much work did you do today?

Be careful how you answer – this is an example of when the words we use in physics are the same words that we use in everyday life, but in physics we use the words to have a very specific meaning.

In physics, 'work' is related to the particular situation where a force causes an object to move. It allows us to calculate a number – this is important when comparing different situations.

KNOWLEDGE CHECK

✓ Be able to describe situations where forces cause objects to move.

✓ Know that the weight of objects is caused by the force of gravity.

✓ Know how to use rulers and forcemeters.

LEARNING OBJECTIVES

✓ Relate (without calculation) work done to the magnitude of a force and the distance moved in the direction of the force.

✓ **EXTENDED** Recall and use $W = F \times d = \Delta E$.

WORK

Work is done when the application of a force results in movement. The amount of work done depends on the magnitude of the force and the distance moved in the direction of the force. Work can only be done when the object or system has energy. When work is done, energy is transferred.

Look at Fig. 2.2. In this position, the gymnast is not doing any work against his body weight – he is not moving (he will be doing work pumping blood around his body though).

The gymnast in Fig. 2.3 is doing work. He is moving upwards against his weight. Energy is being transferred as he does the work.

△ Fig. 2.2 A gymnast.

△ Fig. 2.3 A gymnast doing work against his own body weight.

EXTENDED

Work done is equal to the amount of energy transferred. It can be calculated using the following formula:

work done = force × distance moved in the direction of the force
$\qquad$ = energy transferred

$$W = F \times d = E$$

where: W = work done in joules (J)

$\quad F$ = force in newtons (N)

$\quad d$ = distance moved in the direction of the force in metres (m)

$\quad E$ = energy transferred in joules (J)

1 joule of energy (or work) will move a weight of 1 newton a distance of 1 metre.

WORKED EXAMPLES

1. A cyclist pedals along a flat road. He exerts a force of 60 N on the road surface and travels 150 m. Calculate the work done by the cyclist.

Write down the formula: $\qquad W = F \times d$

Substitute the values for F and d: $\qquad W = 60 \times 150$

Work out the answer and write down the unit: $\quad W = 9000\,\text{J}$

2. A person does 3000 J of work in pushing a supermarket trolley 50 m across a level car park. What force was the person exerting on the trolley?

Write down the formula with F as the subject: $\quad F = W/d$

Substitute the values for W and d: $\qquad F = 3000/50$

Work out the answer and write down the unit: $\quad F = 60\,\text{N}$

When something slows down because of friction, work is done. The kinetic energy of the motion is transferred to heat as the frictional forces slow the object down. For example, if you are riding your bike and you brake, the energy from your motion is converted to heat in the brake blocks.

QUESTIONS

1. Calculate the work done when a 50 N force moves an object 5 m.

2. Calculate the force required to move an object 8 m by transferring 4000 J of energy.

3. Calculate the work done when a force of 40 N moves a block 2 m.

4. How far does an object move when the force on it is 6 N and the work done is 300 J?

5. What force is needed to move a piano a distance of 2 m when the work done is 800 J?

6. EXTENDED The Space Shuttle uses friction to do work on its motion upon re-entry into the Earth's atmosphere.

△ Fig. 2.4 The space shuttle in orbit.

The shuttle has 8.45×10^{12} J of energy to transfer over an 8000 km flight path. What force is applied by the atmosphere?

7. EXTENDED Use the internet or books to research the shuttle landing and answer the following questions.

a) What happens to the transferred energy?

b) What temperatures are generated by the work being done, and how does this relate to the material used for the underside of the shuttle surface, such as why is it not made from aluminium or iron?

End of topic checklist

Key term

work

During your study of this topic you should have learned:

○ That work done is related to the magnitude of a force and the distance moved.

○ EXTENDED How to describe energy changes in terms of work done.

○ EXTENDED $\Delta W = Fd = \Delta E$

End of topic questions

Note: The marks given for these questions indicate the level of detail required in the answers. In the examination, the number of marks given to questions like these may be different.

1. EXTENDED 50 000 J of work are done as a crane lifts a load of 400 kg. How far did the crane lift the load? (Gravitational field strength, g, is 10 N/kg.) **(3 marks)**

2. EXTENDED Use the relationship between the work done, force and distance to complete the table.

Work done/J	Force/N	Distance/m
	100	2
750		375
9000	120	
	450	200
3000		30
60 000	150	

(6 marks)

3. EXTENDED Jumana and Maria went up the hill. Jumana's weight is 500 N and Maria's weight is 450 N. Who did the most work? **(3 marks)**

4. EXTENDED Explain why, however long you have been sitting writing, you have hardly done any work at all. **(3 marks)**

△ Fig. 2.5 These trams take electrical energy from the overhead wires and convert some of it into kinetic energy.

Energy

INTRODUCTION

The study of energy, how it moves about, what it does when it is transferred, is at the heart of physics. Energy is that 'stuff' that allows things to happen. But what actually is 'energy'?

Energy is surprisingly hard to pin down. We have an intuitive 'feel' that when we have lots of energy, we can get lots of things done. When we are feeling 'drained' of energy, then it is much harder.

Being able to track where the energy is moving, in all its 'disguises', is a key skill that will help you explain many aspects of physics.

KNOWLEDGE CHECK

✓ Know some everyday uses of energy.
✓ Be able to describe devices that transform energy from one form to another.

LEARNING OBJECTIVES

✓ Demonstrate understanding that work done = energy transferred.
✓ Demonstrate understanding that an object may have energy due to its motion (kinetic energy) or its position (potential energy) and that energy may be transferred and stored.
✓ Identify changes in kinetic, gravitational potential, chemical, elastic (strain), nuclear and internal energy that have occurred as a result of an event or process.
✓ Recognise that energy is transferred during events and processes, including examples of transfer by forces (mechanical working), by electrical currents (electrical working), by heating and by waves.
✓ Apply the principle of conservation of energy to simple examples.

✓ **EXTENDED** Recall and use the expressions kinetic energy = $\frac{1}{2}mv^2$ and change in gravitational energy = $mg\Delta h$.

ENERGY

A car will not move without using fuel. At present this fuel could be petrol, alcohol, diesel fuel or liquefied petroleum gas (LPG). In the past the fuel could, just possibly, have been coal; and in the future it could be hydrogen, or electricity stored in a battery. However, whatever fuel you use, you are buying something with the ability to make that car move. This stored ability is known as **potential energy**.

Fig. 2.6 Fuel, whatever form it comes in, gives a car the ability to move.

A clock needs energy to make the hands move, and this energy can be stored in a spring that you wind up with a key, in an electrical battery, or in weights that are raised up.

Potential energy is stored or hidden energy. In this context 'potential' means 'containing power'. When a spring is stretched or compressed, the spring will have elastic potential energy as shown in Fig. 2.7.

When a load is raised above the ground, it will have **gravitational potential energy**, as shown in Fig. 2.8. Gravitational potential energy is energy due to an object's position.

If the spring is released or the load moves back to the ground, the stored potential energy is transferred to movement energy, which is called **kinetic energy**.

△ Fig. 2.7 The spring in this vehicle suspension system stores elastic potential energy.

◁ Fig. 2.8 This cuckoo clock stores gravitational potential energy in two weights: one to run the mechanism that turns the hands, and one to make the cuckoo sing on each hour.

Potential energy can be used to make an object move, and so give it kinetic energy. Kinetic energy can also be transferred into potential energy, and this can be seen most clearly in the action of a pendulum, where at each end of its swing (at A and C in Fig. 2.9) the pendulum has a maximum amount of gravitational potential energy, and at the middle of its swing (at B) some of the potential energy has been transferred into kinetic energy (the pendulum is moving fastest), as shown in Fig. 2.9.

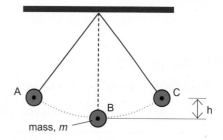

△ Fig. 2.9 Energy changes in the swing of a pendulum.

An object gains gravitational potential energy as it gains height. Work has to be done to increase the height of the object above the ground. Therefore: gain in gravitational potential energy of an object = work done on that object against gravity.

Delta notation

We use the Greek letter Δ (delta) to stand for 'the change in'. For example, Δh means 'the change in the height'. When you are using Δh in an equation, treat it as one symbol meaning 'the change in energy'; don't separate them.

EXTENDED

You can use the expression p.e. = $mg\Delta h$ to calculate the amount of potential energy an object has. In this expression, m is its mass, g is acceleration due to gravity (usually taken as $10 \, m/s^2$) and h is its height above the ground (zero level).

WORKED EXAMPLE

A skier has a mass of 70 kg and travels up in a ski lift that has a vertical height of 300 m. Calculate the change in the skier's gravitational potential energy.

Write down the formula: p.e. = $m \times g \times \Delta h$

Substitute values for m, g and h: p.e. = $70 \times 10 \times 300$

Work out the answer and write down the unit: p.e. = 210 000 J or 210 kJ

WORKED EXAMPLE

An ice skater has a mass of 50 kg and travels at a speed of 5 m/s. Calculate the ice skater's kinetic energy.

You can use the expression k.e. = $\frac{1}{2}mv^2$ to calculate the amount of potential energy an object has. In this expression, m is its mass and v is its velocity.

Write down the formula: k.e. = $\frac{1}{2}mv^2$

Substitute the values for m and v: k.e. = $\frac{1}{2} \times 50 \times 5 \times 5$

Work out the answer and write down the unit: k.e. = 625 J

QUESTIONS

1. Calculate the gravitational potential energy gained when a 5 kg mass is lifted 2 m.

2. Calculate the kinetic energy of a 2 kg ball rolling at 2 m/s.

END OF EXTENDED

DIFFERENT FORMS OF ENERGY

As shown by the pendulum, energy can either be stored or can be seen as a form of motion. The different types of stored energy are all forms of potential energy. Here are some important examples:

- **Gravitational potential energy:** This is energy stored by an object that has been raised up in a gravitational field, for example, a ball at the top of a hill.
- **Elastic (strain) energy:** The word 'strain' means stretched. 'Strain energy' can be stored in springs (in clocks, for example) and in bows when they are drawn back before the arrow is released.
- **Chemical energy:** The energy stored in fuels, such as petrol and diesel, is usually called 'chemical energy'. In any object, the atoms are held together by forces that are called bonds. These bonds behave like springs. In some materials, such as fuels and explosives, the bonds are forced to be shorter or longer than they wish. This stores energy in the bonds that can be transferred by breaking up the structure of the fuel or the explosive.

  A battery is ready to turn 'chemical energy' into 'electrical energy', and a rechargeable battery is so called because every time that it is discharged it can be recharged by forcing electricity through it backwards. The 'electrical energy' that is transferred is stored as 'chemical energy'.

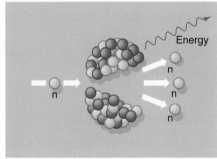

△ Fig. 2.10 Splitting a nucleus can release a lot of energy.

- Nuclear energy: The energy in a nucleus of an atom is stored in the extremely strong bonds between the particles of which the nucleus is made. Some of this energy can be released, in the case of uranium (and some other metals) by splitting the nucleus of the atom into two smaller nuclei. This can be done either slowly and for good purposes in a nuclear power station, or very rapidly in an atomic bomb.

Forms of kinetic energy

Here are some other important types of energy. They are actually all different sorts of kinetic energy, but this is far from obvious in some cases:

When people just use the words 'kinetic energy', they are referring to the energy of a visible moving object with k.e. = ½mv^2.

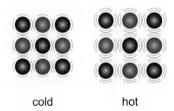

cold   hot

△ Fig. 2.11 The atoms in a hot object vibrate more because they have more energy.

- **Internal energy (heat):** This is contained within an object and makes the difference between the object being hot or cold. A hot object contains atoms that are moving fast or vibrating strongly.

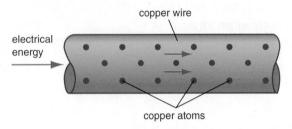

△ Fig. 2.12 Electrical energy in a wire.

- Electrical energy: Electrical currents carry electrical energy from one place to another. Electrical energy can easily be turned into kinetic energy in a motor or internal energy in a resistor, perhaps used as a heater.
- Light energy: A light wave carries 'light energy' as it travels, and this will be turned into internal energy in most objects when it strikes them. If the light hits a solar panel it can be made to generate electrical energy.

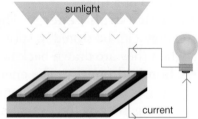

△ Fig. 2.13 A solar cell transfers light energy into electrical energy.

- Sound waves: These carry a very small amount of energy from the source of the noise. The source vibrates, setting air particles around it into vibration. These vibrations are passed through the air as a **longitudinal wave**. When the wave reaches the ear it sets the eardrum into vibration. (Do not confuse the 2000 W of electricity consumed by a band performing on stage with the 100 W of sound being emitted by the loudspeakers. The ear is extremely good at detecting sound.)

TRANSFER OF ENERGY

Any type of energy can be transferred into any other type of energy. In some cases, this transfer can be done efficiently, such as between kinetic energy and electrical energy. In other cases, the transfer is inefficient. One example of inefficient transfer of energy is the power station, Fig. 2.20.

In every case of transfer of energy, some of the energy is converted to internal energy. A light bulb transfers electrical energy to light energy but also gets hot; an electric motor transfers electrical energy to kinetic energy but also gets hot; a diesel engine transfers chemical energy to kinetic energy but also gets hot; a battery that is being charged gets hot. Even a pendulum eventually stops swinging because the movement of the pendulum through the air heats up the air due to resistance.

SCIENCE IN CONTEXT

ENERGY AND THE EARTH

Energy and matter are constantly interacting on our planet. Part of this interaction produces volcanoes, glaciers, mountain ranges, oceans and continents. The energy comes from two sources: energy from the Sun, which keeps the oceans and the atmospheric cycles (such as the water cycle) going; and the internal energy, which comes from radioactive decay in the Earth's core and is the driving force behind plate tectonics.

The amount of energy that moves through the system is huge: it is of the order of 1.74×10^{17} W. Most of this comes from the Sun. Figure 2.14 shows the energy transfers that take place in the Earth's system.

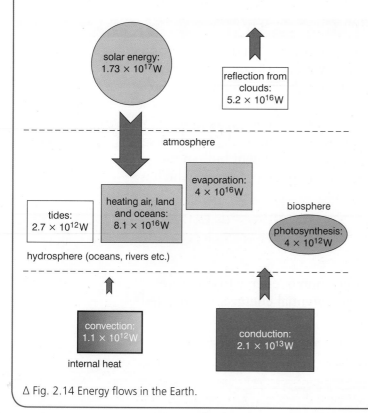

△ Fig. 2.14 Energy flows in the Earth.

QUESTIONS

1. Describe the energy changes that take place as a pendulum swings from one side to the other.

2. Where can elastic strain energy be found?

3. What is the source of chemical energy?

CONSERVATION OF ENERGY

The law of conservation of energy says that energy cannot be created or destroyed. Often the words 'conversion' and 'conservation' are misused.

Energy *conversion* is transferring one form of energy to another (such as electrical to light in a light bulb). In another example, a tram takes electrical energy and converts it mainly into kinetic energy, but also into internal energy and sound. Likewise, as a pendulum swings, some of its energy is transferred between kinetic and gravitational potential energy: but when you add up its total energy, you will find that the total stays almost the same. The movements of the pendulum slowly die away as energy is transferred to the air in the room and the air heats up slightly.

REMEMBER

You may need to describe how energy is transferred in different situations, but remember that total energy is always conserved: the energy at the start and at the end must have the same total value. So, you must account for all of the energy converted, and that includes the energy that will have been transferred as internal energy, as well perhaps as light or sound. For example, the amount of electrical energy that is put into a light bulb will all come out of the light bulb in the form of light (useful output) and some internal energy (heat), which is wasted output.

EXTENDED

You can use the principle of the conservation of energy to calculate what happens when kinetic energy and potential energy are converted from either one to the other.

So long as negligible energy is lost in the conversion, $mgh = \frac{1}{2}mv^2$.

QUESTIONS

1. State the law of conservation of energy.

2. Describe the energy transfers that take place in a light bulb.

3. Consider a tram.

    a) What form of energy is its input?

    b) What form of energy is its useful output?

    c) What form of energy is its 'waste' output?

WORKED EXAMPLE

A stone is thrown vertically upward and reaches a height of 6 m above the hand of the thrower. What speed was the stone travelling at when it left the person's hand?

The decrease in k.e. of the stone as it rises = the increase in the p.e. of the stone.

As the final k.e. of the stone = 0, the initial k.e. of the stone = the increase in the p.e. of the stone at the top of its flight.

Write down the formula: $\frac{1}{2}mv^2 = mgh$
$$\frac{1}{2}v^2 = gh$$

Note that the mass has cancelled out; the mass does not matter in this case.

Substitute values for g and h: $v^2 = gh \times 2$
$$= 10 \times 6 \times 2$$
$$= 120$$

Work out the answer and write down the unit:
$$v = \sqrt{120}$$
$$= 10.95 \, \text{m/s}$$

△ Fig. 2.15 The path of a stone when it is thrown.

The kinetic energy given to the stone when it is thrown is transferred to potential energy as it gains height and slows down. At the top of its flight a large part of the kinetic energy will have been converted into gravitational potential energy. A small amount of energy will have been lost due to friction between the stone and the air.

END OF EXTENDED

SCIENCE LINK

BIOLOGY – LIFE PROCESSES, ECOSYSTEMS

• Ideas about energy transfer allow us to describe the central purpose of life processes such as respiration and photosynthesis.

• The law of conservation of energy – that the total energy present at any stage is always the same – allows us to explain the relationships behind food chains and the number of organisms an ecosystem is able to support.

CHEMISTRY – ENERGY IN CHEMICAL REACTIONS

• Energy is conserved in all chemical reactions. Calculating the energies involved in bond-making and bond-breaking leads to an overall figure deciding whether any particular reaction will release energy (an exothermic reaction) or require energy to be input (an endothermic reaction).

End of topic checklist

Key terms

chemical energy, elastic strain energy, gravitational potential energy, internal energy, kinetic energy, longitudinal wave, potential energy

During your study of this topic you should have learned:

◯ That work done = energy transferred.
◯ About examples of energy in different forms, including kinetic, gravitational, chemical, strain, nuclear, internal, electrical, light and sound.
◯ That energy is transferred during events and processes, including examples of transfer by forces (mechanical working), by electrical currents (electrical working), by heating and by waves.
◯ How to apply the principle of conservation of energy to simple examples.
◯ **EXTENDED** That k.e. = $\frac{1}{2}mv^2$ and p.e. = $mg\Delta h$.

End of topic questions

Note: The marks given for these questions indicate the level of detail required in the answers. In the examination, the number of marks given to questions like these may be different.

1. **EXTENDED** A student is carrying out a personal fitness test. She steps on and off the 'step' 200 times. She transfers 30 J of energy each time she steps up. Calculate the energy transferred during the test. **(3 marks)**

2. **EXTENDED** A child of mass 35 kg climbed a 30 m high snow-covered hill.

    a) Calculate the change in the child's potential gravitational energy. **(3 marks)**

    b) The child then climbed onto a lightweight sledge and slid down the hill. Calculate the child's maximum speed at the bottom of the hill. (Ignore the mass of the sledge.) **(3 marks)**

    c) Explain why the actual speed at the bottom of the hill is likely to be less than the value calculated in part **b)**. **(3 marks)**

3. **EXTENDED** Calculate the potential energy of a piano of mass 300 kg lifted through a vertical height of 9 m. **(3 marks)**

4. **EXTENDED** Calculate the height climbed up a ladder when the person's mass is 70 kg and the gravitational potential energy gained is 2800 J. **(3 marks)**

5. **EXTENDED** A 1500 kg helicopter gains potential energy of 1.35 MJ in climbing from the ground. Calculate its height. **(3 marks)**

6. EXTENDED Use the relationship between kinetic energy, mass and velocity to complete the table.

Kinetic energy/J	Mass/kg	Speed/m/s
	84	9
196		1.4
50	1	
	950	13
62 500		250
6000	3000	

(6 marks)

7. EXTENDED What is the kinetic energy of a bird of mass 200 g flying at 6 m/s?

(3 marks)

8. EXTENDED What is the speed of a car of mass 1500 kg with a kinetic energy of 450 kJ?

(3 marks)

9. EXTENDED a) A skateboarder of mass 60 kg is 3.15 m above ground level travelling at 1 m/s. What is his kinetic energy?

(3 marks)

b) What is the gravitational potential energy of the skateboarder in part **a)**?

(3 marks)

c) What is the total energy (kinetic + gravitational) of the skateboarder in parts **a)** and **b)**?

(2 marks)

d) Assuming that no energy is lost in the descent, show that the skateboarder is travelling at about 8 m/s on reaching ground level after the descent down the 3.15 m slope.

(3 marks)

10. EXTENDED A man pushes a wheelbarrow up a 5 m long ramp onto a surface 1.6 m higher than his starting level. The weight of the barrow is 300 N.

a) How much work has been done in raising the barrow 1.6 m?

(3 marks)

b) The force he needed to push the barrow along the ramp is 100 N. How much work did he do?

(3 marks)

c) Why are the numbers in parts **a)** and **b)** different?

(2 marks)

d) Why are ramps useful?

(1 mark)

Power

INTRODUCTION

There are many situations where it is important to know how *quickly* energy is being transferred – a kettle is no use if it takes 5 hours to deliver the energy to heat some water. For this we need to introduce the concept of power.

△ Fig. 2.16 When designing a kettle it is important to think about how quickly the energy is transferred.

KNOWLEDGE CHECK

✓ Know that domestic appliances have power ratings.
✓ Be able to describe the energy transfers in a range of situations.
✓ Know that energy can be transferred at different rates.

LEARNING OBJECTIVES

✓ Relate (without calculation) power to work done and time taken, using appropriate examples.
✓ **EXTENDED** Recall and use the equation $P = \Delta E\, /\, t$ in simple systems.

POWER

A powerful engine in a car can take you up a road to the top of a mountain more quickly than a less-powerful engine. Both engines can do the same amount of work, given enough time, but the powerful engine can do the work more quickly. In the same way, a powerful electric motor on a cooling fan will move the air in the room more quickly; and a 'powerfully built' athlete will, by transferring more kinetic energy to it as it is launched, throw a javelin further.

Power is defined as the rate of doing work or the rate of transferring energy. The more powerful a machine is, the quicker it does a fixed amount of work or transfers a fixed amount of energy.

Since power is the rate of doing work or the rate of transferring energy, power can be calculated using the formula:

power = work done/time taken = energy transferred/time taken

$$P = W/t \text{ or } P = \Delta E/t$$

Where: P = power in joules per second or watts (W)

ΔE = energy transferred in joules (J)

W = work done in joules (J)

t = time taken in seconds (s)

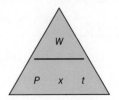

△ Fig. 2.17 The equation triangle for work done, power and time. You can use this triangle to help you rearrange the formula.

1 watt of power is 1 joule of work being done every second.

WORKED EXAMPLES

1. A crane lifts a 100 kg girder for a skyscraper by 20 m in 40 s. Hence it does 20 000 J of work in 40 s. Calculate its power over this time. Note: this calculation tells you the power of the electric motor that the crane needs.

Write down the formula: $P = W/t$

Substitute the values for W and t: $P = 20\,000/40$

Work out the answer and write down the unit: $P = 500\,W$

2. A student with a weight of 600 N runs up the flight of stairs, a distance of 5 m, shown in the diagram (right) in 6 s. Calculate the student's power.

Write down the formula for work done: $W = Fd$

Substitute the values for F and d: $W = 600 \times 5 = 3000\,J$

Write down the formula for power: $P = W/t$

Substitute the values for W and t: $P = 3000/6$

Work out the answer and write down the unit: $P = 500\,W$

△ Fig. 2.18 A student running up a flight of stairs.

REMEMBER

The student is lifting his body against the force of gravity, which acts in a vertical direction. The distance measured must be in the direction of the force (that is, the vertical height).

QUESTIONS

1. A man (70 kg) and a boy (35 kg) run up a set of stairs in the same time. Explain why the man is twice as powerful.

2. When a machine is called 'powerful', what does it mean?

3. **EXTENDED** What is the unit of power?

4. **EXTENDED** Calculate the power of a motor that transfers 1200 J of energy every 5 s.

5. **EXTENDED** **a)** A crane lifts a mass of 60 kg to a height of 5 m. How much work does it do?

   b) The crane takes 1 minute to do this. Calculate the power of the crane.

End of topic checklist

Key term

power

During your study of this topic you should have learned:

○ That power is related to work done and time taken and be able to give appropriate examples.

○ EXTENDED That $P = \Delta E/t$ in simple systems.

End of topic questions

Note: The marks given for these questions indicate the level of detail required in the answers. In the examination, the number of marks given to questions like these may be different.

1. EXTENDED Peter and Paul walk home from school together up a hill. Peter is heavier than Paul.

   a) Who does more work? (2 marks)

   b) Who produces more power? (2 marks)

2. EXTENDED A crane takes 10 s to lift a load of 5000 N a distance of 20 m. What is its power? (4 marks)

3. EXTENDED Calculate the work done by a 75 kW tractor in 20 s. (3 marks)

Energy resources

INTRODUCTION

Our everyday lives depend on energy being transferred to us so that we can then 'use it' to power our modern society. But where does the energy come from?

The original source of almost all our energy is nuclear fusion reactions in the Sun. This energy has either been locked away over millions of years in non-renewable resources, such as coal, or drives renewable energy resources such as the wind.

This topic looks at the energy resources we use in more detail.

△ Fig. 2.19 Wind farms are used around the world to generate electricity.

ENERGY RESOURCES

Fossil fuels

Most of the energy we use is obtained from **fossil fuels** – coal, oil and natural gas.

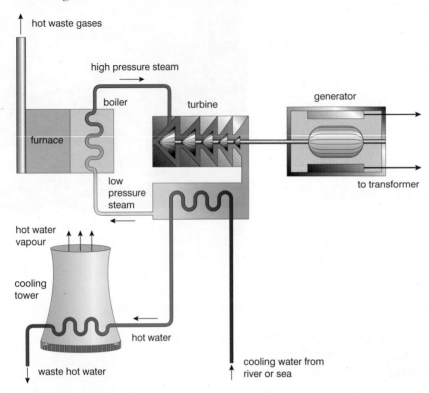

△ Fig. 2.20 How a power station works. The most common fuels used in power stations are coal, oil and gas.

Many power stations use fossil fuels (coal, oil, natural gas) to produce electricity that is supplied to homes and factories. Other power stations burn alternative fuels to produce this electricity, but the basic method of producing power is generally the same:

- Fuel is burned and steam is produced in a boiler.
- The steam turns a **turbine**.
- The turbine drives a generator.
- The generator produces electricity.
- The electricity is supplied to homes and industry.

Once supplies of fossil fuels have been used up they cannot be replaced – they are **non-renewable**. At current levels of use, oil and gas supplies will probably last for about another 40 years, and coal supplies for no more than a few hundred years from now. The development of **renewable** sources of energy is therefore becoming increasingly important.

Wind power

The wind is used to turn windmill-like turbines that generate electricity directly from the rotating motion of their blades. Modern wind turbines are efficient, but it takes about one thousand of them to produce the same amount of energy as a modern gas, coal or oil-burning power station, and that is only when the wind is blowing favourably.

△ Fig. 2.21 On a windy day, a very large wind turbine generates 2000 kW of electricity. That's enough to meet the needs of about 1200 families.

Developing investigative skills

You are going to plan an investigation to evaluate wind power as an energy source. You have the following equipment:

- model wind turbine
- multimeter to measure the voltage generated
- anemometer to measure wind speed
- hairdryer to generate wind power (note: set hairdryer setting to cold)
- metre rule to measure distance.

Using and organising techniques, apparatus and materials

❶ Plan your experiment, describing clearly the following:

a) the aim of your investigation

b) what you will measure

c) the number and range of readings that you will take

d) the independent variable

e) the dependent variable

f) the control variables

g) how you will make your experiment a fair test.

❷ Draw out a results table that you would use in your investigation.

Planning and evaluating investigations

❸ Write an evaluation identifying aspects of your experiment where modifications are possible.

Dams can be used to store water, which is allowed to fall in a controlled way that generates electricity. This is particularly useful in hilly regions for generating hydroelectric power. When demand for electricity is low, surplus electricity can be used to pump water back up into the high dam for use in times of high demand.

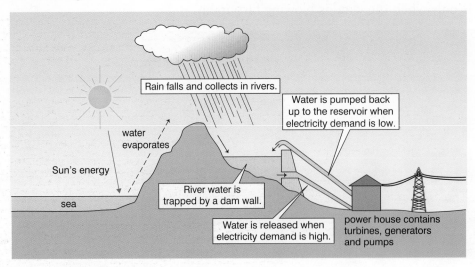

Rain falls and collects in rivers.

Water is pumped back up to the reservoir when electricity demand is low.

water evaporates

Sun's energy

sea

River water is trapped by a dam wall.

Water is released when electricity demand is high.

power house contains turbines, generators and pumps

Δ Fig. 2.22 A pumped storage hydroelectric power station.

Solar power

Solar cells can be used to convert light energy from the Sun directly into electricity. This electricity can be stored, often in batteries, to be used when convenient. Electricity generated in this way uses a renewable source. These panels are commonly referred to as solar PV (photovoltaic) panels to distinguish them from solar heating panels.

In solar panels, the energy from the Sun is used simply to heat water that is pumped through black pipes in a panel, often on the roof of a house. Heating the water in this way reduces the demand on other energy resources. Again, the energy can be stored in the water for later use.

Δ Fig. 2.23 Solar (photovoltaic) panels on a roof.

Make sure you don't confuse the two systems for using the Sun's energy. One heats water; the other generates electricity.

Solar power is energy from the Sun, which itself is powered by nuclear fusion reactions (where the small nuclei of hydrogen atoms join to make larger nuclei that are, in fact, helium) and an enormous amount of energy is released.

Although there have been several attempts to reproduce this continuous release of energy on Earth, so far they have been unsuccessful. The Sun is the source of energy for all our energy resources except geothermal, nuclear and tidal.

Geothermal power

Geothermal power is obtained using the heat of the Earth. In certain parts of the world, water forms hot springs that can be used directly for heating. Water can also be pumped deep into the ground to be heated.

Nuclear fission

A nuclear power station uses the heat generated by a controlled fission process to convert water to steam. This drives a turbine as in a conventional power station. However, a typical power plant produces 20 metric tonnes of waste per year. People disagree over whether this radioactive waste is more hazardous than the gases emitted by coal-fired power stations.

△ Fig. 2.24 A nuclear power station in the Czech Republic.

Water power

The motion of waves can be used to move large floats and generate electricity. A very large number of floats are needed to produce a significant amount of electricity.

Δ Fig. 2.25 An artist's impression of a wave generator.

Dams on tidal estuaries trap the water at high tide. When the water is allowed to flow back at low tide, tidal power can be generated. This obviously limits the use of the estuary for shipping and can cause environmental damage along the shoreline.

EXTENDED

The River Severn Barrage is a proposed project to build a huge dam on the estuary of the River Severn in the UK. The project is expected to cost around $20 billion and, if completed, will produce a clean, sustainable source of electricity for the next 120 years.

In the area behind the dam are huge areas of mud that are exposed at low tide. These mud flats contain many small animals and are a significant source of food for many species of birds. If the dam is built, these mud flats could be disrupted and it may not be easy for the birds to feed on the small animals in the mud.

Imagine that you are called as an expert witness as part of an environmental group to evaluate the benefits, disadvantages and environmental impact of constructing the barrage. Write a report in preparation for a press release. It should be approximately 200 words long.

END OF EXTENDED

QUESTIONS

1. What energy transfers take place in a solar cell?

2. What energy transfers take place in a wind turbine?

3. Describe the process used to generate electricity in fossil-fuelled power stations.

4. How is electricity produced from geothermal sources?

End of topic checklist

Key terms

non-renewable resource, renewable resource, turbine

During your study of this topic you should have learned:

○ That electricity or other useful forms of energy may be obtained from:
- chemical energy stored in fuel
- water, including the energy stored in waves, in tides, and in water behind hydroelectric dams
- geothermal resources
- nuclear fission
- heat and light from the Sun (solar cells and panels).

○ About the advantages and disadvantages of each method in terms of renewability, cost, reliability, scale and environmental impact.

○ **EXTENDED** That the Sun is the source of energy for all our energy resources except geothermal, nuclear and tidal.

○ **EXTENDED** That energy is released by nuclear fusion in the Sun.

End of topic questions

Note: The marks given for these questions indicate the level of detail required in the answers. In the examination, the number of marks given to questions like these may be different.

1. **a)** What is meant by a non-renewable energy resource? **(2 marks)**

   b) Name three non-renewable energy sources. **(2 marks)**

   c) Which non-renewable energy source is likely to last the longest? **(2 marks)**

2. Draw up a table to compare renewable energy resources and non-renewables. Add columns to your table to describe at least one advantage and one disadvantage for each energy resource when it is used to provide large-scale electricity production. **(6 marks)**

3. Compare the effects on the environment of coal-fired power stations and nuclear power stations.

   a) Which of these power stations releases greenhouse gases? **(1 mark)**

   b) Which of the fuels used in these power stations will run out first? **(1 mark)**

4. Power stations need to be located on suitable sites. Write down *three* factors that a company may consider before choosing a site for a coal-fired power station. **(3 marks)**

Most things in the world are either a solid, a liquid or a gas. These are the three main states of matter. Substances can change from a solid to a liquid in a process called melting, and from liquid to gas in a process called evaporation. Gases change to liquids by condensation and liquids change to solids in solidification.

You are probably most familiar with these changes for water. It is possible for these states to exist at the same time: for example, there is a temperature at which ice, water and steam are all present. This is called the triple point of water and is used to define the kelvin scale of temperature, which you will meet later in this section.

STARTING POINTS

1. Describe how particles are arranged in: a) a solid; b) a liquid; and c) a gas.

2. What happens to the particles when a solid melts?

3. Explain what happens when a solid dissolves in a liquid.

4. Are evaporation and boiling the same thing? Give a reason for your answer.

5. What happens to the speed of gas molecules as temperature increases?

CONTENTS

3

Thermal physics

△ Water exists as a solid, a liquid and a gas in the world around us.

Simple kinetic molecular model of matter

△ Fig. 3.1 Water in all three states of matter.

INTRODUCTION

Almost all matter can be classified as a solid, a liquid or a gas. These are called the three states of matter.

The fourth state of matter is called 'plasma'. It only exists at high temperatures seldom seen on Earth, so we won't consider it further here, even though most of the matter in the Universe and most stars are made of plasma.

STATES OF MATTER

The three states of matter each have different properties:

• A solid has a fixed volume and shape, is not easily compressed (squashed) and does not flow easily.

- A liquid assumes the shape of the part of the container that it occupies, usually the lowest level, is not easily compressed and flows easily.
- A gas assumes the shape and volume of its container, occupying the whole volume, can be compressed and flows easily.

MOLECULAR MODEL

In this topic, these properties of matter are explained in terms of the molecular structure of the three states.

◁ Fig. 3.2 The main body of this rocket is filled with liquid oxygen and liquid hydrogen, which have to be kept at extremely low temperatures to prevent them from heating up and turning back into gas. If the fuel were made colder it would turn into a solid.

We now know that all materials are made of tiny particles called **atoms** that can attract each other. The atoms in a solid are locked together by the forces between them. However, even in a solid, the particles are not completely still. They vibrate constantly about their fixed positions. When the material is heated, it is given more internal energy, and the particles vibrate faster and further.

When the temperature is increased more, the vibrations of the particles increase to the point at which the forces are no longer strong enough to hold the structure together in the rigid order of a solid. The forces can no longer prevent the atoms moving around, but they do prevent them from flying apart from each other. This is what makes a liquid. The volume of the liquid is the volume occupied by the particles from which it is made.

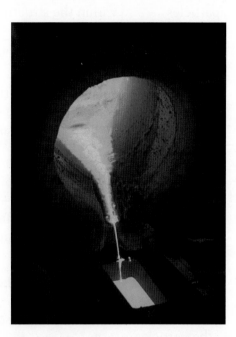

△ Fig. 3.3 The molten iron can be poured into a mould before it cools down and turns back into a solid.

When the temperature is increased even more, then the particles do fly apart. They now form a gas. The particles fly around at high speed – several hundred kilometres per hour. If they are in a container, they travel all over it, bouncing off the walls. The volume of a gas is not fixed; it just depends on the size of the container that the gas is put into. We use the **kinetic molecular model** to explain the behaviour of solids, liquids and gases. Table 3.1 summarises this model.

	Solid	**Liquid**	**Gas**
Arrangement of particles	Regular pattern, closely packed together, particles held in place.	Irregular, closely packed together, particles able to move past each other.	Irregular, widely spaced, particles able to move freely.
Diagram			
Motion of particles	Vibrate in place within the structure.	'Slide' over each other in a random motion.	Random motion, faster movement than the other states.

△ Table 3.1 The kinetic molecular model of matter.

The kinetic molecular model uses this idea that all materials are made up of atoms that behave rather like tiny balls. When the model is used to try to explain the behaviour of gases it is often called the **kinetic theory of gases**.

The arrangement of atoms can be used to explain the properties of solids, liquids and gases that you met earlier.

Solids:

- retain a fixed shape and volume because the particles are locked into place in the lattice
- are not easily compressed because there is little free space between the particles
- do not flow easily because the particles cannot move or slide past one another.

Liquids:

- assume the shape of the part of the container that they occupy because the particles can move or slide past one another
- are not easily compressed because there is little free space between the particles
- flow easily because the particles can move or slide past one another.

Gases:

- assume the shape and volume of their container because the particles move past one another continuously
- are compressible because there is lots of free space between the particles
- flow easily because the particles can move past one another.

END OF EXTENDED

QUESTIONS

1. What happens to the motion of atoms as the temperature increases?

2. Explain why it is easier to compress a gas than a liquid.

3. Describe the arrangement of particles in: **a)** a solid; **b)** a liquid; **c)** a gas.

4. What does the volume of a gas depend on?

What are molecules?

In Table 3.1, you can see that the particles in the liquid and the gas consist of single atoms. There are materials like this – elements such as helium and neon. In most materials, though, the particles that move around in the liquid or the gaseous state are groups of atoms called **molecules**. A water molecule is H_2O and a nitrogen molecule is N_2. This means that the particles moving around in liquid or gaseous water each consist of two hydrogen atoms and one oxygen atom. In liquid nitrogen or in nitrogen gas, the particles each consist of two nitrogen atoms.

Taking the idea of particles one step further, you can start to apply your knowledge of forces and motion to the molecules of a gas. This step is building up a theory – the kinetic theory of gases – to see if the predictions that come from our ideas match what happens when we experiment with gases.

It is quite a simple theory to begin with. We only look at gases where the particles are generally separated from each other and the maths is not too difficult. However, the ideas have proved to be remarkably successful in describing the behaviour of materials.

△ Fig. 3.4 This balloon rises because the gas inside it is less dense than the surrounding area.

The kinetic theory of gases builds up a set of ideas from the basic idea that a gas is made of many tiny particles called molecules. These ideas give a picture of what happens inside a gas (Table 3.2).

Observed feature of a gas	Related ideas from the kinetic theory
Gases have masses that can be measured.	The total mass of a gas is the sum of the masses of the individual molecules.
Gases have temperatures that can be measured.	The individual molecules are always moving. The faster they move (the more kinetic energy they have), the higher the temperature of the gas.
Gases have a pressure that can be measured.	When the molecules hit the walls of the container, they exert a force on it. It is this force, divided by the surface area of the container, that we observe when measuring pressure.
Gases have volumes that can be measured.	Although the volume of each molecule is only tiny, they are always moving about and spread out throughout the container.
Temperature has an absolute zero.	As the temperature falls, the speed of the molecules (and their kinetic energy) becomes less. At absolute zero the molecules would have stopped moving.

△ Table 3.2 What happens inside a gas.

These ideas help to explain Boyle's law.

Boyle's law states that when the temperature of the gas stays constant, the volume of the gas is inversely proportional to the pressure. The link to kinetic theory is as follows. The temperature stays constant, so the average speed of the molecules stays constant. If the volume of the gas is reduced by half, then the molecules make the same number of collisions with half the surface area of the wall, so the pressure (= force/area) must be doubled. This is **inverse proportionality**.

Gases only follow Boyle's law if the mass of the gas remains constant (that is, no particles move in or out of the system) and the gases are ideal, that is they do not liquefy or solidify.

QUESTIONS

1. How does the kinetic theory explain the measurable volume of a gas?

2. A fixed mass of gas is at a constant temperature. What happens to the volume when you increase the pressure?

3. What conditions must be met for gases to follow Boyle's law?

EVAPORATION

When particles break away from the surface of a liquid and form a vapour, this is known as **evaporation**. The more energetic molecules of the liquid escape from the surface, as shown in Fig. 3.5. This reduces the average energy of the molecules remaining in the liquid, and so the temperature of the liquid falls.

Therefore, evaporation causes cooling. The evaporation of sweat helps to keep your body cool in hot weather. The more energetic molecules of liquid sweat escape from the surface of your skin and so the average energy and therefore temperature of the remaining molecules falls and your skin cools down. This is what happens when any body is in contact with an evaporating liquid. The body cools down. The cooling in a refrigerator is also due to evaporation of a special liquid inside the freezing compartment at the top of the refrigerator. The vapour is collected and compressed back into liquid inside the condenser behind the refrigerator. The liquid is circulated by a pump and recycled.

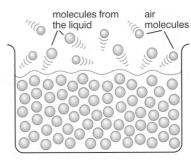

△ Fig. 3.5 Evaporation.

EXTENDED

The rate of evaporation is increased at higher temperatures. It is also increased by a strong flow of air across the surface of the liquid, as in this way the evaporating molecules are carried away quickly. A certain amount of water will also evaporate more quickly when you increase its surface area. Tea or coffee in a shallow, wide cup cools down much more quickly than in a tall, narrow mug because the large surface area of the cup allows more evaporation.

Imagine you are a particle that has experienced evaporation. Write a letter to your friend describing the experience. Your letter should answer the following questions.

1. What change of state did you go through?

2. How close to your neighbours were you in your original state?

3. What was given to you to make you change state?

4. How did you change state?

QUESTIONS

1. What factors increase the rate of evaporation?

2. Why does tea in a narrow mug cool down more slowly than tea in a wide mug?

END OF EXTENDED

BIOLOGY – TRANSPORT

- Evaporation of water drives the process of transpiration in plants. This vital process allows chemicals to be transported around the plant and accounts for most of the water a plant uses. The factors that control the rate of evaporation of the water therefore control the rate at which transpiration will happen and plants are adapted to allow for this.

CHEMISTRY – THE PARTICLE NATURE OF MATTER

- Evaporation involves a change of state from liquid to gas and so particle ideas are essential in helping to understand the process.

- Particle ideas explain why evaporation happens at a range of temperatures and why the rate of evaporation is affected by factors such as temperature and humidity.

MELTING AND BOILING

We experience melting and boiling many times in everyday life. This part explores what happens at the molecular level when materials melt and boil.

The melting point of a substance is the temperature at which it changes from a solid to a liquid. Solidification is when a substance changes from a liquid to a solid. The molecules in the substance come close together. The boiling point of a substance is the temperature at which it changes from a liquid to a gas. The molecules in the substance get further apart. Condensation is when a substance changes from a gas to a liquid.

End of topic checklist

Key terms

atom, evaporation, inverse proportionality, kinetic molecular model, kinetic theory of gases, molecules

During your study of this topic you should have learned:

○ About the distinguishing properties of solids, liquids and gases.

○ About the molecular structure of solids, liquids and gases.

○ EXTENDED How the motion of particles and the forces and distances between them relate to their properties.

○ How temperature affects the way gases behave.

○ How pressure affects the way gases behave.

○ About the effect of a change in temperature on the pressure of a gas at constant volume.

○ That the random motion of particles in a suspension gives evidence for the kinetic molecular model of matter.

○ How evaporation can be described in terms of escape of more-energetic molecules from the surface of a liquid.

○ How evaporation is related to the consequent cooling of a liquid.

○ EXTENDED That evaporation is affected by temperature, surface area and draught over a surface.

○ How a change in the volume of a gas is related to a change in pressure applied to the gas at constant temperature.

End of topic questions

Note: The marks given for these questions indicate the level of detail required in the answers. In the examination, the number of marks given to questions like these may be different.

1. Give an example of a material for each state of matter that demonstrates the properties of that state. **(3 marks)**

2. Use ideas about particles to explain why:

    a) solids keep their shape, but liquids and gases don't **(3 marks)**

    b) solids and liquids have a fixed volume, but gases fill their container. **(3 marks)**

3. How does kinetic theory explain the existence of absolute zero? **(3 marks)**

4. Use the kinetic molecular model to explain the following observations in detail:

    a) It is possible to keep a bottle of drink cold by standing it in a bowl and covering it with a wet cloth. **(3 marks)**

    b) EXTENDED The drink gets even colder when you place the bowl in a strong draught. **(3 marks)**

5. How is the speed of a gas molecule linked to the temperature of the gas? **(2 marks)**

Matter and thermal properties

INTRODUCTION

Some things expand when they are heated. In this topic you will find out why. You will also find out how thermometers work, and learn what happens at the molecular level when there is a change of state.

△ Fig. 3.6 This expansion joint on a bridge stops the bridge from buckling when it expands in hot weather.

KNOWLEDGE CHECK

✓ Know the kinetic model for the states of matter.
✓ Know the equation energy = $mc\Delta T$.
✓ Know thermal capacity = mc.

LEARNING OBJECTIVES

✓ Be able to describe qualitatively the thermal expansion of solids, liquids and gases.
✓ EXTENDED Explain, in terms of the motion and arrangement of molecules, the relative order of magnitude of the expansion of solids, liquids and gases.
✓ Be able to identify and explain some of the everyday applications and consequences of thermal expansion.
✓ Be able to use and describe the use of thermometers to measure temperature on the Celsius scale.

THERMAL EXPANSION OF SOLIDS, LIQUIDS AND GASES

With only two or three exceptions, all materials (solids, liquids and gases) expand as they become warmer. In the case of solids, the atoms vibrate more as the temperature goes up. So, even though they stay joined together, they move slightly further apart, and the solid expands a little in all directions.

The effect is small but not trivial. A metre rule that is heated from 0 °C to 100 °C (from the freezing point of water to its boiling point) will increase in length by 1–2 mm depending on what material it is made of. Some plastics do not make good metre rules as they get up to 10 mm longer when heated.

On a hot day, a 1000 km railway track can possibly become more than 300 m longer. In the case of a track that has joints, there is a gap of a few millimetres every 20 m, to allow the rails to expand. Modern long welded tracks have no expansion gaps of this type, although as can be seen from the photo in

△ Fig. 3.7 This track has to be held very firmly to stop it from bending sideways when it gets hot.

Fig. 3.7, the track has to be held extremely firmly to stop it moving. This track is on a curve, so it will try to bend sideways to the left when it gets hot.

Liquids expand for the same reason. The atoms vibrate as they move around, and get slightly further apart. This means that the volume of liquids increase as the temperature increases.

It is very difficult to restrict the thermal expansion of solids and liquids, as very large forces will be created in the material if it is not allowed to expand. So, for example, a large bridge is always built with expansion joints to allow it to get longer on a hot day (Fig. 3.8).

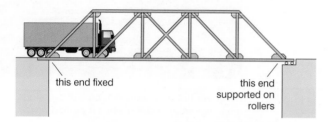

this end fixed

this end supported on rollers

Δ Fig. 3.8 Expansion joints on a bridge.

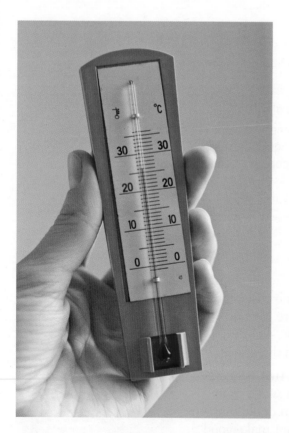

◁ Fig. 3.9 This thermometer has a bulb of coloured alcohol at the base, attached to a very narrow tube that is half full of alcohol (ethanol). As the alcohol expands and contracts with change in temperature, the length of the column of alcohol goes up and down. The top of the tube is sealed off to prevent the alcohol from evaporating.

However, the effect of expanding metal can also be useful. Metals expand at different rates as their temperatures rise. So, when strips of two metals are bound closely together, and are warmed, they bend as one metal expands more than the other. **Bimetallic strips** like this can be used to control the temperature in a heating system such as an electric iron.

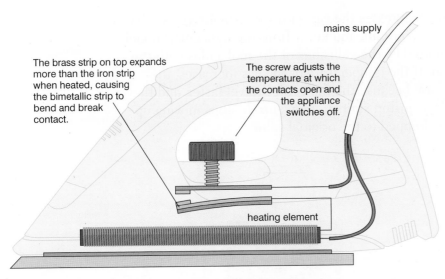

mains supply

The brass strip on top expands more than the iron strip when heated, causing the bimetallic strip to bend and break contact.

The screw adjusts the temperature at which the contacts open and the appliance switches off.

heating element

△ Fig. 3.10 The temperature control mechanism in an electric iron.

Like other liquids, water contracts as its temperature falls and its density increases. Unlike other liquids, when its temperature falls below 4 °C, water begins to expand again, and becomes less dense. This is called the anomalous expansion of water.

The density falls even further as it freezes, because the water molecules form an open crystal structure in the solid state. So, ice is less dense than water, while almost all other materials are more dense in the solid state than as a liquid.

REMEMBER

The properties of water are very strange. Not only does it require a great deal of heat to change its temperature, it is also unique in that it expands as it freezes. This makes ice less dense than liquid water, so ice floats on water. This has been vital to evolution – life can survive at the bottom of ponds, where in very cold weather the water stays liquid, even when the surface has frozen.

Gases behave completely differently. First, there is no need to allow the gas to expand if it gets hotter; if you put it in a sealed container, then you can just allow the pressure to increase instead.

EXTENDED

Second, if a gas is allowed to expand, then it will increase in volume much more than solids or liquids do as it gets hotter. Between 0 °C and 100 °C it will expand by a third, so 300 cm³ of gas will become 400 cm³.

END OF EXTENDED

In Fig. 3.11 the piston compresses the gas with a constant force so that the pressure of the gas is constant. You know from the molecular model that the piston is supported by the collisions of the molecules with the underside of the piston. If the temperature of the gas increases, the pressure starts to increase because the molecules travel faster, and they hit the piston harder. The piston starts to move up. It stops moving up when the pressure has dropped to the original value.

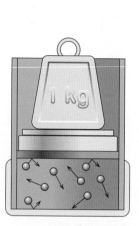 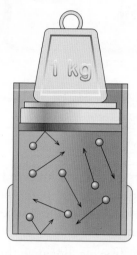

gas at low temperature

gas at high temperature

△ Fig. 3.11 The piston compresses the gas. As the gas is heated, the pressure increases and the piston moves up until the pressure returns to the original value.

The result is that the gas has been heated and its volume allowed to increase at constant pressure. Note that initially the pressure was caused by many molecules hitting the piston at moderate speeds. After the gas has heated up and expanded, the same pressure is now caused by the same number of molecules hitting the piston less frequently (because they are more spread out), but the molecules are moving faster. Each molecular collision produces a greater change of momentum and a bigger force.

EXTENDED

Gases expand more than liquids, which expand more than solids. This is because molecules in gases are spaced further apart than those in liquids, which are spaced further apart than those in solids.

END OF EXTENDED

QUESTIONS

1. Why do bimetallic strips bend?

2. Explain why a solid expands when it is heated.

3. What is the anomalous expansion of water?

4. How does the behaviour of gases differ from that of solids and liquids?

MEASUREMENT OF TEMPERATURE

Temperature can be measured using any suitable physical property that changes with temperature.

Common examples in use include:

• volume of a liquid – mercury-in-glass or alcohol-in-glass thermometer
• length of a solid – bimetallic strip in a thermostat.

All thermometers need calibrating before they are first used. In the case of a liquid-in-glass thermometer, this means that a scale must be fixed to it in the right place. To do this, two fixed points are needed. This type of thermometer has a bulb of the liquid (such as mercury or coloured alcohol) at the base attached to a very narrow tube that is half full of the liquid. As the liquid expands and contracts with change in temperature, the length of the column of liquid goes up and down. The top of the tube is sealed off (Fig. 3.12).

The Celsius scale is used in science, although other scales are used elsewhere. The two fixed points used by the Celsius scale, as originally defined, are the melting point of ice, defined as $0\,°C$, and the boiling point of water at standard atmospheric pressure, defined as $100\,°C$. (At lower pressures water boils at a lower temperature.)

To calibrate the mercury-in-glass thermometer at these two fixed points, the thermometer is immersed first in a funnel containing melting ice, and the $0\,°C$ point is marked. It is then immersed in the steam from a boiling kettle and the $100\,°C$ point is marked. Finally, the distance between the marks (distance h in Fig. 3.12) is divided into 100 equal distances, each corresponding to 1 degree Celsius. The scale can be extended beyond $100\,°C$ to measure higher temperatures, and below $0\,°C$ to measure negative temperatures.

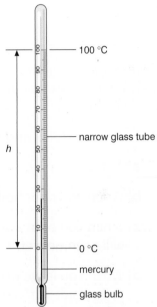

△ Fig. 3.12 A mercury-in-glass thermometer.

Key term

bimetallic strip

During your study of this topic you should have learned:

○ How to describe the thermal expansion of solids, liquids and gases.

○ **EXTENDED** About the relative order of magnitude of the expansion of solids, liquids and gases.

○ About some of the everyday applications and consequences of thermal expansion.

End of topic questions

Note: The marks given for these questions indicate the level of detail required in the answers. In the examination, the number of marks given to questions like these may be different.

1. Explain the following observations:

a) A steel ruler is often marked 'Use at 20 °C'. **(2 marks)**

b) When you pour boiling water into a drinking glass, the glass may crack. **(2 marks)**

c) When you pour a very cold drink into a drinking glass, the outside of the glass will become wet. **(2 marks)**

d) When you leave frozen food in a freezer for several weeks without covering it, the outside surface of the food will suffer from what is called 'freezer burn' and will look dry and unpleasant. **(2 marks)**

2. **EXTENDED** A bimetallic strip consists of a thin strip of aluminium, 100 mm × 10 mm, attached to a thin strip of stainless steel of the same size.

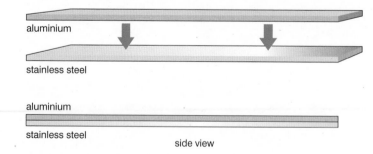

The two strips are joined together face to face, to give a thicker strip that is still 100 mm × 10 mm. They are joined together at 20 °C using epoxy glue.

The strip is fixed to a block of metal at one end. What happens to the other end at each change of temperature if the temperature goes to 100 °C, then to −10 °C, and finally back to 20 °C? (Note that aluminium expands more than stainless steel when equal lengths are exposed to the same change in temperature.) **(3 marks)**

b) Explain how this strip could be used as a thermometer. **(2 marks)**

c) Design an electrical circuit that will use the bimetallic strip to switch on a warning light if the temperature of the bimetallic strip drops below room temperature and approaches the freezing point of water. (Such a device is known as a 'frost stat' and is used to prevent damage from freezing.) **(3 marks)**

d) Explain what would happen if the warning light heated the bimetallic strip. **(2 marks)**

Thermal processes

△ Fig. 3.13 These paragliders stay in the air longer by using of convection currents in the atmosphere to take them higher.

INTRODUCTION

Energy will always try to flow from areas at high temperatures to areas at low temperatures. This is called thermal transfer.

Thermal energy can be transferred by conduction, convection and radiation.

In this topic we shall consider all three methods of heat transfer.

KNOWLEDGE CHECK

✓ Know and be able to explain the molecular structure of solids, liquids and gases.
✓ Know what is meant by energy.
✓ Know that internal energy (heat) can be transferred from one place to another.

LEARNING OBJECTIVES

✓ Recognise and name typical conductors.
✓ Be able to describe experiments to demonstrate the properties of good and bad conductors of heat.
✓ **EXTENDED** Be able to give a simple molecular account of heat transfer in solids.
✓ Be able to recognise convection as an important method of thermal transfer in fluids.
✓ Be able to interpret and describe experiments designed to illustrate convection in liquids and gases (fluids).
✓ **EXTENDED** Be able to relate convection in fluids to density changes.
✓ Be able to identify infrared radiation as part of the electromagnetic spectrum.
✓ Be able to recognise that radiation does not require a medium.
✓ Be able to describe the effect of surface colour (black or white) and texture (dull or shiny) on the emission, absorption and reflection of radiation.
✓ **EXTENDED** Be able to describe experiments to show the properties of good and bad emitters and good and bad absorbers of infrared radiation.
✓ **EXTENDED** Be able to show understanding that the amount of radiation emitted also depends on the surface temperature and surface area of a body.
✓ Be able to identify and explain some of the everyday applications and consequences of conduction, convection and radiation.

CONDUCTION

Materials that allow thermal energy to transfer through them quickly are called **thermal conductors**. Those that do not are called **thermal insulators**.

EXTENDED

If one end of a conductor is heated, the atoms that make up its structure start to vibrate more vigorously. As the atoms in a solid are linked together by chemical bonds, the increased vibrations can be passed on to other atoms. The energy of movement (kinetic energy) passes through the whole material.

Metals are particularly good thermal conductors because they contain freely moving electrons that transfer energy very rapidly.

As the electrons travel through the piece of metal, they take the thermal energy with them. This is in addition to the thermal energy that is transferred by vibrations of the atoms making up the structure of the metal. Fig. 3.14 shows **conduction** in a solid. Particles in the hot part of a solid (top) vibrate further and faster than particles in the cold part (bottom). The vibrations are passed on through the bonds from particle to particle. It is the movement of the free electrons that makes metals act as good thermal conductors.

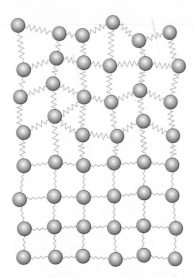

△ Fig. 3.14 Conduction in a solid.

Conduction cannot occur when there are no particles present, so a vacuum is a perfect insulator. Gases and liquids are poor heat conductors because their particles are so far apart.

END OF EXTENDED

Conduction can be demonstrated using the equipment shown in Fig. 3.15.

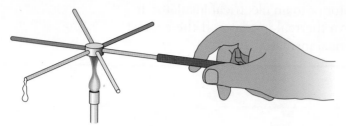

Another way to demonstrate the properties of good and bad conductors of thermal energy is to wrap a piece of paper around a bar made of wood on one side and copper on the other side. The bar is then held just above a Bunsen flame. The paper only chars on the wood side of the bar, as copper is a good thermal conductor so heat travels away from the paper on the copper side. Wood is a thermal insulator so doesn't conduct the heat away.

Another example is the fact that metal handlebars of a bike always feel colder than the plastic grips – heat is conducted away from your hands by the metal but not by the plastic.

QUESTIONS

1. Why are metals particularly good conductors?

2. Why is outer space a perfect insulator?

3. How is heat energy transferred in a thermal conductor?

4. Describe an experiment to demonstrate conduction.

5. EXTENDED Devise an experiment to find out whether or not the rate of energy transfer varies along a strip of copper. Explain how you could tell if any change is linear.

CONVECTION

Convection is the main method of **thermal transfer** in fluids.

EXTENDED

Convection occurs in liquids and gases because these materials flow (they are fluids). The particles in a fluid move all the time. When a fluid is heated, energy is transferred to the particles, causing them to move faster and further apart. This makes the heated fluid less dense than the unheated fluid. The less dense, warmer fluid rises above the more dense, colder fluid, causing the fluid to circulate as shown in Fig. 3.16. This **convection** current is how the thermal energy is transferred.

◁ Fig. 3.16 Potassium manganate(VII) crystals in water demonstrate convection currents. The warmer water expands, becomes less dense and rises, making a trail as some of the dissolved potassium permanganate is carried along as well. Colder water sinks and replaces the warmer water that has risen.

END OF EXTENDED

If a fluid's movement is restricted, energy cannot be transferred. That is why many insulators, such as ceiling tiles, contain trapped air pockets. Wall cavities in some houses are filled with fibre to prevent air from circulating and transferring thermal energy by convection.

QUESTIONS

1. Why does convection only occur in liquids and gases?

2. EXTENDED Explain why warm air rises.

3. How could you demonstrate convection in a laboratory?

4. Describe how cavity-wall insulation in houses reduces heat loss by convection.

RADIATION

Radiation, unlike conduction and convection, *does not need particles* at all. Radiation can travel through a vacuum. This is clearly shown by the radiation that arrives at the Earth from the Sun. Radiated heat energy is carried mainly by **infrared radiation**, which is part of the **electromagnetic spectrum**. Infrared radiation is similar to light, but has a longer wavelength.

△ Fig. 3.17 Thermogram of a house.

All objects take in and give out infrared radiation all the time. Hot objects radiate more infrared than cold objects. The amount of radiation given out or absorbed by an object depends on its temperature and on its surface. Figure 3.17 shows a thermogram. Thermograms give a visual representation of the amount of infrared radiation that is given out by an object at any particular moment.

Type of surface	As an emitter of radiation	As an absorber of radiation	Examples
Dull black	Good	Good	Emitter: cooling fins on the back of a refrigerator are dull black to radiate away more energy. Absorber: the surface of a black bitumen road gets far hotter on a sunny day than the surface of a white concrete one.
Bright shiny	Poor	Poor	Emitter: marathon runners, at the end of a race, wrap themselves in shiny blankets to prevent them from cooling down too quickly by radiation (and convection). Absorber: fuel storage tanks are sprayed with shiny silver or white paint to reflect radiation from the Sun.

△ Table 3.3 Comparison of different surfaces as emitters or absorbers of infrared radiation.

EXTENDED

Using a Leslie's cube

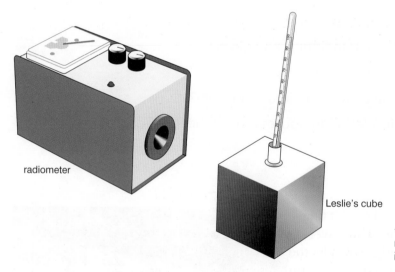

radiometer

Leslie's cube

◁ Fig. 3.18 A Leslie's cube. The meter measures the amount of radiation that is emitted by each surface.

To show the properties of good and bad emitters of infrared radiation, you can use a Leslie's cube (Fig. 3.18), which is filled with boiling water. Its sides have different surfaces – shiny, dull, dark, light – to show how they emit thermal radiation at different rates. Because all sides of the cube are heated by the same water inside the cube, any differences in the way they radiate energy can only be due to the differences in their surfaces.

To show how different surfaces absorb radiation, you can use boiling tubes covered with foil or with a matt black surface heated by radiation from an infrared (IR) bulb and measure the temperature rise in each case.

The amount of radiation emitted by a body varies according to its temperature. The diagram shows the amount of radiation emitted for a body at different temperatures.

The amount of radiation emitted also depends on the surface area – a larger surface area will emit more radiation.

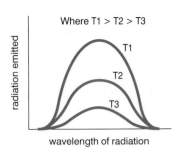

△ Fig. 3.19 Amount of radiation emitted at different temperatures.

QUESTIONS

1. How does radiation differ from conduction and convection?

2. Which is the better emitter of infrared radiation: a hot object or a cold object?

3. State two factors that affect the amount of thermal radiation emitted by an object.

4. EXTENDED Which side of a Leslie's cube will emit thermal radiation at the greatest rate?

5. EXTENDED Two bodies, A and B, are compared. A is at 200 K and B is at 400 K. Which will emit more radiation?

CONSEQUENCES OF ENERGY TRANSFER

This part of the topic considers some everyday consequences of energy transfer.

Radiators

Radiators are used to heat homes in countries that have cool winters. A radiator does radiate some heat, and if you stand near a hot radiator your hands can feel the infrared radiation being emitted. However, this is only around one quarter of the heat being released by the radiator. *Three quarters* of the heat is taken away by the hot air that rises from the radiator. Colder air from the room flows in to replace this hot air, and a convection current is formed as shown in Fig. 3.20. So, a 'radiator' is mainly a convection heater.

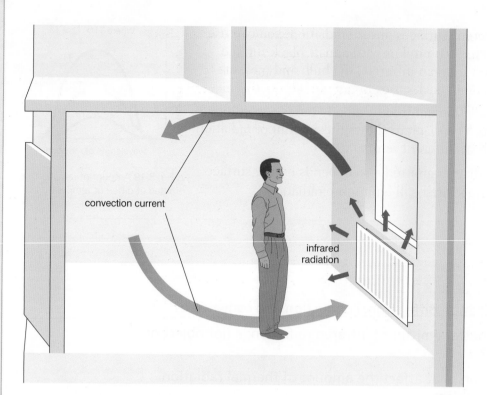

△ Fig. 3.20 A side view of a room with a hot-water radiator underneath the window. You will see from this that the convection current is far more efficient at heating the top of the room than it is at heating the person standing in front of the radiator.

Vacuum flask

Another example of thermal transfers in everyday life is a vacuum flask (Fig. 3.21). A vacuum flask will keep a hot drink hot or a cold drink cold for hours by almost completely eliminating the flow of heat out or in.

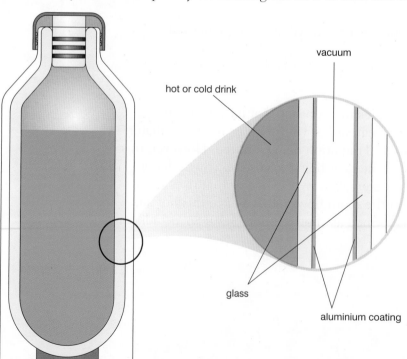

△ Fig. 3.21 A vacuum flask.

Conduction is almost entirely eliminated by making sure that any heat flowing out must travel along the glass of the neck of the flask. The path is a long one, the glass is thin, and glass is a very poor conductor of heat. Energy cannot be lost by conduction across the vacuum space between the two walls of the glass flask. The bung in the top of the flask must also be a very poor conductor of heat – cork or expanded polystyrene are good materials to use.

Convection is eliminated because the space between the inner wall and the outer wall of the flask is evacuated so that there is no air to form convection currents.

When the contents are hot, radiation is greatly reduced because the inner walls of the flask are coated with pure aluminium. Because the aluminium is in a vacuum, it stays extremely shiny forever, so the wall in contact with the hot liquid emits very little infrared radiation.

When the contents are cold, the shiny outer surface of the aluminium-coated glass reflects almost all the infrared radiation falling on it, so hardly any is absorbed.

QUESTIONS

1. What is the main method of heat transfer in a radiator?

2. Which part of a room is heated most efficiently by a radiator?

3. For a vacuum flask, describe which features reduce the energy transfer by:

   a) conduction

   b) convection

   c) radiation.

4. Explain why vacuum flasks are good at keeping hot drinks hot *and* cold drinks cold.

Developing investigative skills

A student heats water in a beaker until it is just boiling. As the water cools, the student measures its temperature and obtains the results shown in the graph.

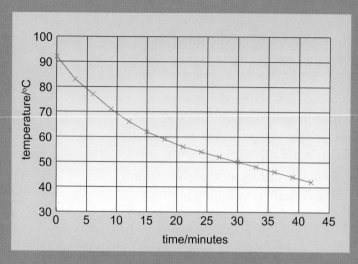

△ Fig. 3.22 Results of measuring the water temperature as it cools.

Observing, measuring and recording

❶ Why should the student record the room temperature as well?

❷ How often did the student measure the temperature of the water?

❸ How would the graph have changed if the student had measured the temperature:

  a) more frequently

  b) less frequently?

Planning and evaluating investigations

❹ How could the student have reduced the rate of cooling of the water?

❺ How could the student investigate the hypothesis 'the bigger the temperature difference between an object and its surroundings, the greater the rate of energy transfer between them'?

Using insulation to reduce energy transfers

There are a number of ways of reducing wasteful energy transfers in a house. Figure. 3.23 shows some of them.

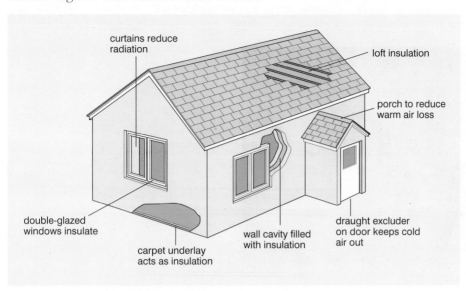

Δ Fig. 3.23 There are many ways to insulate a house.

Source of energy wastage	% of energy wasted	Insulation technique
Walls	35	*Cavity wall insulation.* Modern houses have cavity walls, that is, two single walls separated by an air cavity. The air reduces thermal energy transfer by conduction but not by convection as the air is free to move within the cavity. Fibre or polystyrene bead insulation is inserted into the cavity to prevent the air from moving, so reducing convection.
Roof	25	*Loft insulation.* Fibre insulation is placed on top of the ceiling and between the wooden joists in the loft. Air is trapped between the fibres, reducing thermal energy transfer by conduction and convection.
Floors	15	*Carpets.* Carpets and underlay prevent thermal energy loss by conduction and convection. In some modern houses foam blocks are placed under the floors.
Draughts	15	*Draught excluders.* Cold air can get into the home through gaps between windows and doors and their frames. Draught excluder tape can be used to block these gaps.
Windows	10	*Double glazing.* Energy is transferred through glass by conduction and radiation. Double glazing has two panes of glass with a layer of air between the panes. It reduces energy transfer by conduction but not by radiation. Radiation can be reduced by drawing the curtains.

Δ Table 3.4 How different types of insulation in a house prevent thermal energy transfers.

Mountaineers and other people who need clothing to protect them from extreme cold know that they need to wear several layers of clothing, with each layer full of trapped air. The whole aim is to have a thick layer of air around the body, because air is a poor conductor of heat.

△ Fig. 3.24 Mountaineers need well-insulated clothing.

The fibres of clothes, especially newer extremely fine spun-polyester fibres, do a very good job at stopping the air from moving, thus preventing convection. Mountaineers do not use metallised layers to prevent radiation, because such layers would trap perspiration and could interfere with movement. However, metallised plastic layers are used in the emergency survival bags that mountaineers carry.

QUESTIONS

1. Explain why the walls and the roof are the most important features of a house to insulate.

2. Air is a bad thermal conductor (a good insulator). Describe how this is used to reduce energy loss from a house.

3. Describe how trapping a layer of air near the body helps a person keep warm.

4. Explain why loose clothing can be an advantage in hotter climates.

End of topic checklist

Key terms

conduction, convection, electromagnetic spectrum, infrared, radiation, thermal conductor, thermal insulator, thermal transfer

During your study of this topic you should have learned:

○ About experiments to demonstrate the properties of good and bad conductors of heat.

○ **EXTENDED** How to describe heat transfer in solids in terms of molecules.

○ How to recognise convection as an important method of thermal transfer in fluids.

○ **EXTENDED** About the relationship between convection in fluids and density changes and how to describe experiments to illustrate convection.

○ That infrared radiation is part of the electromagnetic spectrum.

○ That radiation does not require a medium.

○ How to describe the effect of surface colour (black or white) and texture (dull or shiny) on the emission, absorption and reflection of radiation.

○ **EXTENDED** About experiments to show the properties of good and bad emitters and good and bad absorbers of infrared radiation.

○ **EXTENDED** That the amount of radiation emitted also depends on the surface temperature and surface area of a body.

○ About some everyday applications and consequences of thermal energy transfer.

End of topic questions

Note: The marks given for these questions indicate the level of detail required in the answers. In the examination, the number of marks given to questions like these may be different.

1. Why are several thin layers of clothing more likely to reduce thermal transfer than one thick layer of clothing? **(3 marks)**

2. The diagram shows a cross-section of a steel radiator positioned in a room next to a wall.

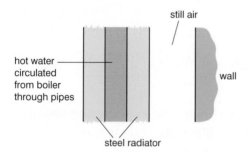

Describe how energy from the hot water reaches the wall behind the radiator.

(6 marks)

3. Suggest a colour for a firefighter's uniform. Explain your choice. **(3 marks)**

4. **EXTENDED** Discuss how you might design a solar cooker to heat water using infrared radiation from the Sun. **(3 marks)**

5. Explain why seawater absorbs infrared radiation faster than snow and ice.

(4 marks)

6. Imagine that you are a local councillor. You are deciding whether or not to give grants for installing home insulation. Discuss all the factors you would consider and what other information you would need before making a decision. **(6 marks)**

7. Explain why solids transfer energy mainly by conduction. (3 marks)

8. Describe how a convection heater warms a room. (4 marks)

9. Using conduction in your answer, suggest why serving dishes are usually made from glass or china. (3 marks)

10. Which factors affect the rate at which an object transfers energy? (3 marks)

What is the connection between the waves you see on water and light? Light is a wave that behaves in a similar way to water waves. Sound is another type of wave, as you will learn later in this section. Studying the behaviour of waves will help you to understand many of your everyday experiences, ranging from how you see objects to how you hear sounds.

You should already know that energy can be transferred as sound and light. White light is made up of a range of different colours and that light can be reflected and refracted. You should also know how the frequency and amplitude of a sound wave are related to the pitch and loudness of the sound.

STARTING POINTS

1. Explain why a red object looks red.

2. Describe the pitch and loudness of the sound you hear when the sound wave has a large amplitude and the frequency is low.

3. Explain the meaning of the words translucent, transparent and opaque.

4. How could you demonstrate the difference between a light wave and a sound wave using: a) a rope; b) a spring; c) water? (If you cannot use one or more of these to demonstrate the difference, explain why.)

5. How does light travel through space?

6. Draw a diagram to show how light is reflected by a plane mirror.

CONTENTS

a) General wave properties

b) Light

c) Electromagnetic spectrum

d) Sound

4
Properties of waves, including light and sound

△ Light is a wave and has many properties in common with a wave in the sea.

△ Fig. 4.1 This photo shows many examples of waves in action.

General wave properties

INTRODUCTION

The behaviour of waves affects us every second of our lives. Waves are reaching us constantly: sound waves, light waves, infrared (heat), television and mobile phone microwave signals, radio waves, and so on. The study of waves is one of the central subjects of physics.

The woman in Fig. 4.1 is surrounded by waves: she can feel the heat waves from the Sun coming in through the windows; she can hear the sound waves of her friend on the phone; the phone uses microwaves; and she can see around her with light waves.

KNOWLEDGE CHECK

✓ Know some simple examples of wave motion.
✓ Be able to measure lengths and times.
✓ Be familiar with everyday examples of reflection.

LEARNING OBJECTIVES

✓ Be able to demonstrate an understanding that waves transfer energy without transferring matter.
✓ Be able to describe what is meant by wave motion as illustrated by vibration in ropes and springs, and by experiments using water waves.
✓ Be able to give the meaning of *speed, frequency, wavelength* and *amplitude*.
✓ **EXTENDED** Be able to distinguish between transverse and longitudinal waves and give suitable examples.
✓ Be able to describe the use of water waves to show reflection and refraction.
✓ **EXTENDED** Understand that refraction is caused by a change in speed when a wave moves from one material to another.
✓ **EXTENDED** Be able to recall and use the equation $v = f\lambda$.

LONGITUDINAL AND TRANSVERSE WAVES

All waves transfer energy without transferring matter. Wave motion can be illustrated by vibrations in ropes and springs. Fig. 4.2 shows two types of wave motion illustrated by a spring.

EXTENDED

In a **longitudinal wave**, the vibrations are in the direction of travel of the wave. This type of wave can be shown by pushing and pulling a

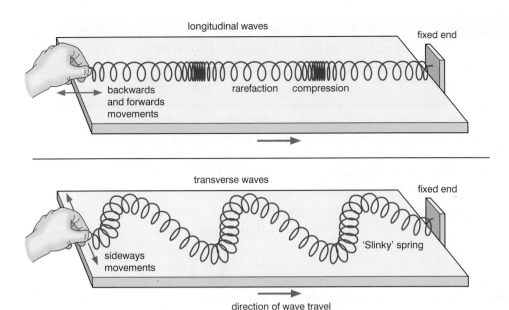

△ Fig. 4.2 Longitudinal and transverse waves are made by vibrations. Both types of wave have a repeating shape or pattern.

spring. The spring stretches in places and squashes in others. The stretching produces regions of **rarefaction**, where the coils spread out, while the squashing produces regions of **compression**. Sound is an example of a longitudinal wave.

In a **transverse wave**, the vibrations are at right angles to the direction of travel of the wave. Light, radio and other **electromagnetic waves** are transverse waves.

In the examples in Fig. 4.2 the waves are very narrow, and are confined to the spring or the string that they are travelling down. Most waves are not confined in this way. Clearly a single wave on the sea, for example, can be hundreds of metres wide as it moves along.

Longitudinal and transverse waves are made by vibrations. Both types of wave have a repeating shape or pattern.

END OF EXTENDED

Amplitude, frequency, wavelength and period

Waves have a wavelength, frequency, amplitude and time period.

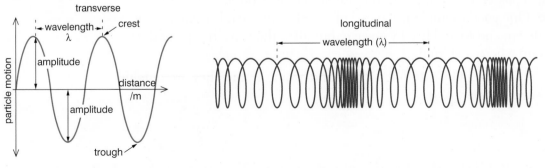

△ Fig. 4.3 The wavelength and amplitude of a transverse wave and the wavelength of a longitudinal wave.

Take care with graphs or diagrams like the one in Fig. 4.3. Make sure you notice if it is 'distance' or 'time' along the *x*-axis. Some labels only apply to one type of graph.

- The **wavelength** of a transverse wave is the distance between two adjacent peaks or, if you prefer, the distance between two adjacent troughs of the wave. In the case of longitudinal waves, it is the distance between two consecutive points of maximum compression, or the distance between two consecutive points of minimum compression.
- The **frequency** is the number of complete waves that go past each second (measured in Hz).
- The time period is the time taken for each complete cycle of the wave motion.
- The **amplitude** is the maximum particle displacement of the medium's vibration from the undisturbed position. In transverse waves, this is half the crest-to-trough height.
- The **speed** of the wave is the distance the wave travels in 1 s. The speed depends on the substance or **medium** the wave is passing through.

The largest ocean wave measured accurately had a wavelength of 340 m, a frequency of 0.067 Hz (that is to say, one peak every 15 s), and a speed of 23 m/s. The amplitude of the wave was 17 m, so the ship that was measuring the wave was going 17 m above the level of a smooth sea and then 17 m below. (The wave went down 34 m from crest to trough.)

Waves transfer energy and information

A wave carries energy and can also carry information. You can feel the energy in infrared waves from the Sun as they strike your hands; you can see the energy contained in the ocean waves from a typhoon as they reach the coast after travelling hundreds of miles. And you can see the information contained in the light reaching your eyes from this page, or from a movie screen.

Note that in none of these cases has any object or matter travelled by vibrations from the source of the waves to the destination. Instead the wave is passed on from point to point along the route taken by the wave. One good example is a piece of wood in the sea. It is shaken up and down, and to and fro, by a wave, but after the wave has passed it ends up where it started.

Surfers can travel by catching a wave and 'riding' it, but they are outside the wave, not part of it.

QUESTIONS

1. Describe how the vibrations travel in:

   a) a longitudinal wave

   b) a transverse wave.

2. What is: **a)** the wavelength, **b)** the frequency, **c)** the amplitude of a wave?

3. How far does a wave with speed 5 m/s travel in 3 s?

4. What do all waves transfer?

EXTENDED

Relationship between speed, frequency and wavelength

The speed of a wave in a given medium is constant. When you change the wavelength, the frequency *must* change as well. If you imagine that some waves are going past you on a spring or on a rope, then they will be going at a constant speed. When the waves get closer together, then more waves must go past you each second, and that means that the frequency has increased. The speed, frequency and wavelength of a wave are related by the equation:

wave speed = frequency × wavelength

$$v = f \times \lambda$$

where: v = wave speed, usually measured in metres/second (m/s)

   f = frequency, measured in cycles per second or hertz (Hz)

   λ = wavelength, usually measured in metres (m)

END OF EXTENDED

WORKED EXAMPLES

1. A loudspeaker makes sound waves with a frequency of 300 Hz. The waves have a wavelength of 1.13 m. Calculate the speed of the waves.

Write down the formula: $\qquad\qquad\qquad\qquad v = f \times \lambda$

Substitute the values for f and λ: $\qquad\qquad v = 300 \times 1.13$

Work out the answer and write down the unit: $v = 339$ m/s

△ Fig. 4.4 The equation triangle for wave speed, frequency and wavelength.

2. A radio station broadcasts on a wavelength of 250 m. The speed of the radio waves is 3×10^8 m/s. Calculate the frequency.

Write down the formula with f as the subject: $\quad f = \dfrac{v}{\lambda}$

Substitute the values for v and λ: $\qquad\qquad f = \dfrac{3 \times 10^8}{250}$

Work out the answer and write down the unit: $\quad f = 1\,200\,000$ Hz or 1200 kHz

Using water waves to show reflection, refraction and diffraction

A ripple tank can be used to show reflection, refraction and diffraction (Fig. 4.5).

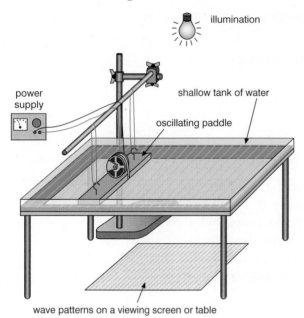

illumination

power supply

shallow tank of water

oscillating paddle

wave patterns on a viewing screen or table

◁ Fig. 4.5 A ripple tank.

To show **reflection**, you put a plane surface in the tank some distance from the paddle. To show **refraction**, you put a thin glass sheet in the water to change the depth of the water in a given region. To show **diffraction**, you put a plane surface with a gap approximately the same width as the wavelength of the water waves.

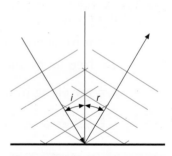

Fig. 4.6 shows how water waves can be used to explain reflection (see Reflection of light in the next topic) at a plane surface. Waves hit a barrier at an angle of incidence, i. The waves bounce off with the angle of incidence, i, equal to the angle of reflection, r. The reflected wave is the same shape as the incident wave.

△ Fig. 4.6 Reflection at a plane surface.

When a wave moves from one medium into another, it will either speed up or slow down. For example, a wave going along a rope will speed up if the rope becomes thinner. (This is why you can 'crack' a whip.) And sound waves going from cold air to hotter air will speed up. When a wave slows down, the *wavelength gets smaller*.

Fig. 4.7 shows this happening as a wave moves from deep water to shallow water. The wave slows down as it travels across the boundary to a more dense medium. The frequency of the wave

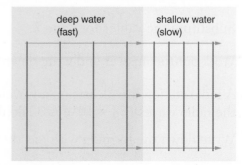

deep water (fast)   shallow water (slow)

△ Fig. 4.7 When waves slow down, their wavelength gets shorter.

does not change as it crosses the boundary, so if the speed decreases and the frequency stays the same, the wavelength must also decrease. When a wave speeds up, the *wavelength gets larger*. If the speed increases but the frequency stays the same, then the wavelength must also increase. Note that in both cases the same number of waves will pass you per second; the wavelength may have changed, but the frequency has not.

When waves slow down, their wavelength gets shorter. When a wave enters a new medium at an angle then the direction of the wave changes. This is known as refraction (see Refraction of light in the following topic). The amount that the wave is bent by depends on the change in speed. Water waves are slower in shallower water than in deep water, so water waves will refract when the depth changes as shown in Fig. 4.8.

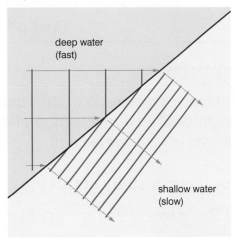

△ Fig. 4.8 If waves cross into a new medium at an angle, their wavelength and direction change.

QUESTIONS

1. A wave changes direction as it moves from one medium to another. What affects the amount of change and direction?

END OF EXTENDED

End of topic checklist

Key terms

amplitude, compression, diffraction, electromagnetic wave, frequency, longitudinal wave, medium, rarefaction, reflection, refraction, speed, transverse wave, wavelength

During your study of this topic you should have learned:

○ That waves transfer energy without transferring matter.

○ What is meant by wave motion, as illustrated by vibration in ropes and springs and by experiments using water waves.

○ The meaning of speed, frequency, wavelength and amplitude.

○ The difference between transverse and longitudinal waves, with suitable examples.

○ About the use of water waves to show: reflection at a plane surface; refraction due to a change of speed.

○ EXTENDED To use the equation $v = f\lambda$.

End of topic questions

Note: The marks given for these questions indicate the level of detail required in the answers. In the examination, the number of marks given to questions like these may be different.

1. A sound wave is displayed using a cathode ray oscilloscope. It has a simple, smooth repeating pattern.

   a) Draw a trace of what is seen on the screen and label it to indicate: the crest of the wave, the wavelength of the wave and the amplitude of the wave. **(6 marks)**

   b) The frequency of the wave is 512 Hz. How many waves are produced each second? **(2 marks)**

2. **EXTENDED** Radio waves of frequency 900 MHz are used to send information to and from a mobile phone. The speed of the waves is 3×10^8 m/s. Calculate the wavelength of the waves.
(1 MHz = 1 000 000 Hz, 3×10^8 = 300 000 000) **(2 marks)**

3. **EXTENDED** Calculate the wavelength of waves of speed 3×108 m/s with frequency of 400 MHz. **(2 marks)**

4. **EXTENDED** What is the frequency of waves of speed 3×10^8 m/s and wavelength 0.1 m? **(2 marks)**

5. **EXTENDED** What is the speed of a wave of 20 Hz with a wavelength of 4 m? **(2 marks)**

6. **EXTENDED** Copy the table and calculate the missing quantities.

Speed/m/s	Frequency/Hz	Wavelength/m	Period/s
	10	2	
20	5		
		6	0.2
340		34	
160	8		
		12	0.004

(12 marks)

Light

INTRODUCTION

Visible light is just part of the electromagnetic spectrum, but is an integral part of much human interaction. Just imagine how much communication takes place using visible light, from encounters with your friends in the street through to television and movies. An understanding of how visible light behaves will be helpful in physics and beyond.

△ Fig. 4.9 A music concert uses both light and sound.

✓ Know how to describe waves using key words such as wavelength, amplitude and frequency.
✓ Know the difference between longitudinal and transverse waves.
✓ **EXTENDED** Be able to describe the processes of reflection and refraction in terms of wavefronts.

LEARNING OBJECTIVES

✓ Be able to describe the formation of an optical image by a plane mirror, and give its characteristics.
✓ Be able to use the law angle of incidence = angle of reflection.

✓ **EXTENDED** Recall that the image in a plane mirror is virtual.
✓ Be able to perform simple constructions, measurements and calculations for reflection by plane mirrors.
✓ Be able to describe an experimental demonstration of the refraction of light.
✓ Be able to describe the action of a thin converging lens on a beam of light.
✓ Be able to use the terms 'principal focus' and 'focal length'.
✓ Be able to draw ray diagrams to illustrate the formation of a real image by a single lens.
✓ **EXTENDED** Be able to use and describe the use of a single lens as a magnifying glass.

REFLECTION OF LIGHT

Perhaps the most familiar type of wave in everyday life (along with water waves) is light. Light does not need a medium to travel through as it is an electromagnetic wave (see the Electromagnetic spectrum in the following topic).

Light waves have all of the properties of waves. You have already learned about their speed and wavelength. In addition, they are transverse waves.

Like all other waves, light can be reflected, refracted and diffracted. Reflection and refraction are easy to demonstrate with a mirror and a glass of water. The effects caused by diffraction of light waves are very hard to see. The effects are small because the wavelength of light is so short.

Reflection of light and ray diagrams

When you look in a plane mirror you see an image of yourself (Fig. 4.10). The image is said to be laterally inverted because when you raise your right hand your image raises what you would call its left hand. The image is formed as far behind the mirror as you are in front of it and is the same size as you. The image cannot be projected onto a screen. It is known as a **virtual image**.

△ Fig. 4.10 As you look at the face of the girl and her image in the mirror, you can see that every part of her face is directly opposite its image in the mirror, and that each part is the same distance away from the mirror as its image.

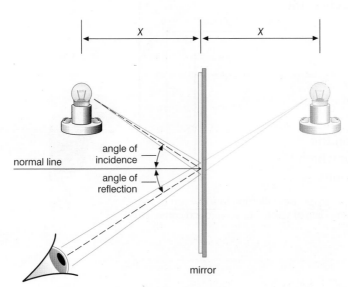

◁ Fig. 4.11 Rays of light travel outwards from the lamp in all directions. Here just two rays are drawn to show how light goes from the lamp to the observer's eye. After the rays have reflected from the mirror, they travel along lines that *look* as if they started from the image. The eye is tricked into thinking that the light really did start from the image.

A ray of light is a line drawn to show the path that the light waves take. We need to study what happens when an incident light ray (a light ray that is going to fall on a surface) hits a mirror and is reflected. Light rays are reflected from mirrors in such a way that:

angle of incidence (i) = angle of reflection (r)

The angles are measured to an imaginary line at 90° to the surface of the mirror. This line is called the **normal**.

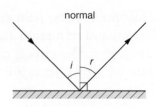

△ Fig. 4.12 The angles of incidence and reflection are the same when a mirror reflects light.

The image in a plane mirror is virtual (it does not really exist). When you are given the position of the object, you can construct a diagram to show the position of the image. The diagrams in Fig. 4.13 show you how. Notice that in each case the angle of incidence (angle between normal and blue ray) is the same as the angle of reflection (angle between normal and red ray).

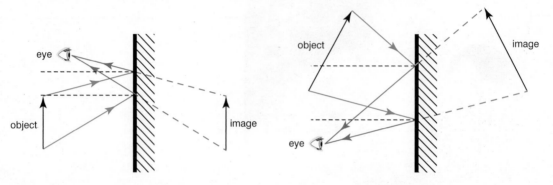

△ Fig. 4.13 Calculating the position of an image. The blue rays are the incident rays and the red rays are the reflected rays.

In a plane mirror the image is always the same size as the object. Examples of plane mirrors include household mirrors, security mirrors for checking under vehicles and periscopes.

An image formed by a plane mirror, such as the one in Fig. 4.13, is:

- virtual
- laterally inverted
- the same size as the object
- the same distance behind the mirror as the object is in front of the mirror.

Note that the image is never formed on the surface of the mirror. This is a mistake which is often made by candidates in examinations.

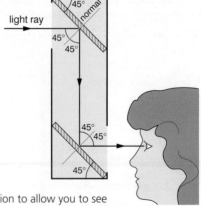

▷ Fig. 4.14 A periscope uses reflection to allow you to see above your normal line of vision – or even round corners.

Developing investigative skills

A student wants to find the position of an image in a plane mirror. To do this, she sets up a plane mirror with an object pin placed vertically a few centimetres in front of it (Fig. 4.15). To find the image, the student looks into the mirror at an angle a little further along the mirror. She can see the image of the object pin in the mirror and she places two 'sighting pins' in line with the image in the mirror. The student repeats this process from a slightly different angle, again putting in two sighting pins.

Using and organising techniques, apparatus and materials

❶ How can the student use the positions of these sighting pins to find the position of the image?

❷ How can the student use the positions of the sighting pins to show that the angle of incidence is equal to the angle of reflection for light reflecting off the mirror?

Observing, measuring and recording

❸ Draw a plan view diagram to show this experiment. Include the object pin, sighting pins and position of the image.

❹ On your diagram, check that the image is as far behind the mirror as the object is in front. Check also that the line joining the object and image cuts through the mirror at 90°.

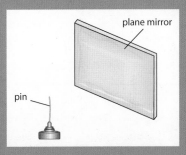

△ Fig. 4.15 The apparatus needed for the investigation.

Handling experimental observations and data

❺ Apart from checking the measurements as in your diagram, how could the student check that she has marked the position of the image correctly?

❻ Would this method for finding the image work with a curved mirror? Explain your answer.

REFRACTION OF LIGHT

The view through some windows is deliberately obscure – the image that you see is distorted. This is because the glass has a different thickness in different places. Rays of light passing through the window are bent to a different extent. Plain windows in old houses can give slightly distorted images because it used to be much harder to make large, flat panes of glass.

The bending of light is the phenomenon known as refraction, which we shall study in this part of the topic.

Light waves *slow down* when they travel from air into glass. When they are at an angle to the glass, light rays bend *towards* the normal as they enter the glass. When the light rays travel out of the glass into the air, their speed increases and they bend *away* from the normal. When the block of glass has parallel sides, the light resumes its original direction after passing through. This is why a sheet of window glass has so little effect on the view beyond. However, the view is shifted slightly sideways when you look through the glass at an angle.

Figure 4.16 shows how you can demonstrate the refraction of light through a glass block.

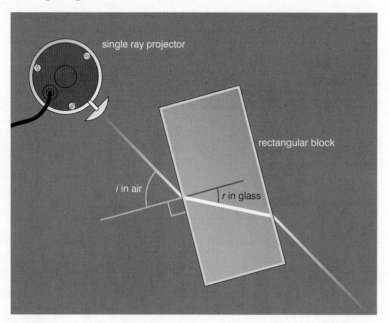

single ray projector

rectangular block

i in air

r in glass

△ Fig.4.16 Refraction of light.

The angle of incidence, *i*, is the angle between the incident light ray and the normal to the surface. The angle of refraction, *r*, is the angle between the refracted light ray and the normal to the surface inside the material.

(Note that people tend to use the letter *r* both for the angle of reflection and the angle of refraction. It should be clear from the context whether they are talking about reflection or refraction.)

QUESTIONS

1. Refraction of light at the surface of a pond can make the pond look shallower than it really is. Explain why.

2. Sound waves can be refracted when they travel through balloons filled with different gases. How would the motion of a sound wave be changed if it travelled through a balloon filled with carbon dioxide?

THIN CONVERGING LENS

Convex (converging, positive) lenses cause parallel rays of light to converge (Fig. 4.17). The light rays are bent by refraction, as described on pages 531–532, except that the amount of refraction increases from the middle of the lens (where there is no refraction) to a maximum at the outer edge of the lens. The point where the parallel light rays arriving along the axis of the lens all cross over is known as the principal focus, F, and the focal length, f, is the most important feature of the lens. Note that there are two principal foci. (The word 'foci' is the plural of 'focus'.) The foci are each side of the lens, at the same distance away from it.

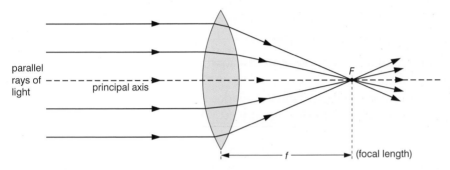

△ Fig. 4.17 Rays through a convex lens converge to a focus.

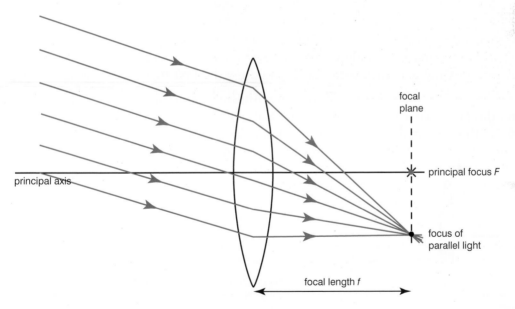

△ Fig. 4.18 Incoming rays converge to a point in the focal plane.

The principal axis is the line through the middle of the lens and at right angles to it. The principal foci lie on this line.

Converging lenses are used to form images in magnifying glasses, cameras, telescopes, binoculars, microscopes, film projectors and spectacles for long-sighted people.

You can check your spectacles to see if they act as a weak magnifying glass, or if they make things seem smaller. If they magnify, then you are long-sighted and you see distant things more clearly. If they do the opposite, then you are short-sighted and see close-up things more clearly. Long sight is corrected with a converging lens. Short sight is corrected with a different lens that diverges light and is not covered in this course.

When the object is further from the lens than the focal length, the converging lens forms a real image. For example, a cinema projector makes a real image on the screen of the film inside the projector, and a camera makes a real image of the object being photographed on the film or the digital sensor inside. In the first case the image is bigger than the object; in the second case it is smaller.

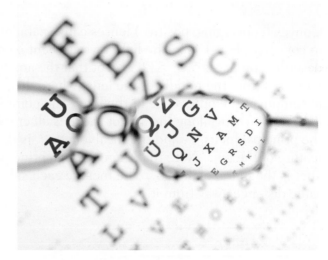

△ Fig. 4.19 What type of lenses do these spectacles have?

To find the position of an image

To find the position of the image of an object formed by a converging lens, you can draw a ray diagram. There are three standard rays that you can use to do this. A standard ray is one whose complete path you know. In ray diagrams, you need any two of the standard rays to find the position and size of the image. In addition, it is a wise precaution to draw the third ray to check the accuracy with which you drew the first two.

For convenience, ray diagrams are usually drawn with the bottom of the object on the principal axis. This means that the bottom of the image is also on the principal axis, and you need only locate the top of the image. Figures 4.20, 4.21 and 4.22 show the standard rays.

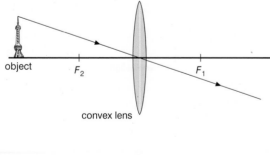

⊲ Fig. 4.20 A ray from the top of the object, straight through the centre of the lens.

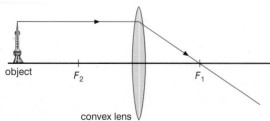

⊲ Fig. 4.21 A ray from the top of the object, parallel to the principal axis until it reaches the lens, and then down through the principal focus F_1 on the far side of the lens.

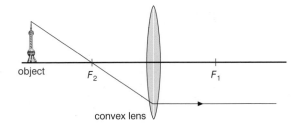

◁ Fig. 4.22 A ray from the top of the object through the principal focus F_2 on the near side of the lens, down to the lens and then parallel to the axis.

You can use a suitable combination of these three rays to locate images. Three examples are given here.

In Fig. 4.23 the image is real, inverted, smaller than the object and closer to the lens than the distance of the object from the lens. A real image is one through which the rays actually pass, and which could be picked up on a suitable screen. Examples of the formation of this type of image are in the eye and in a camera.

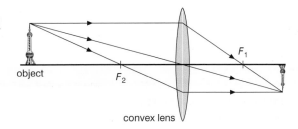

△ Fig. 4.23 The object is a long way from the lens (more than twice the focal length).

In Fig. 4.24 the image is real, inverted, larger than the object and further from the lens than the distance of the object from the lens. Examples of the formation of this type of image are in a film projector and in a photographic enlarger.

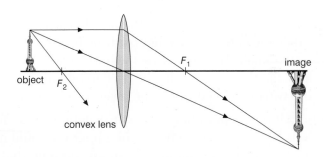

△ Fig. 4.24 The object is closer to the lens (between ×1 and ×2 the focal length)

EXTENDED

In Fig. 4.25 the image is virtual (does not really exist), upright, larger than the object and further away from the lens than the distance of the object from the lens. A virtual image is one through which the rays do not actually pass, but from which they appear to come. Such an image cannot be picked up on a screen. An example where this type of image is formed is in the magnifying glass.

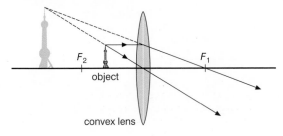

△ Fig. 4.25 The object is closer to the lens than the focal length.

Figure 4.26 shows how the eye sees the image through a magnifying glass.

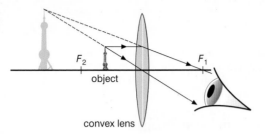

△ Fig. 4.26 Looking at an image through a magnifying glass. Note that to see the image, you have to look through the lens.

END OF EXTENDED

REMEMBER

The image in all these cases is never on the lens itself. This is a mistake that students often make in examinations.

Let's find the position of the image in Fig. 4.27 a. As you can see in Fig. 4.27 b, we use the two rays to find where the image would be. It is clear that one of the rays cannot really be followed by light because the lens is too small. The lines are some of the real light rays that go through the lens, but these rays cannot be used to find the position of the image. We use two rays to find where the image would be, as shown in Fig. 4.27 b.

Note that these special rays are 'construction lines'. You may find that if the lens has a smaller diameter, the upper ray could miss the lens completely. This does not matter, simply pretend that the lens has a large enough diameter and work out where the image is. That is where all of the real rays will go. The image will not move just because the real lens has a smaller diameter.

a

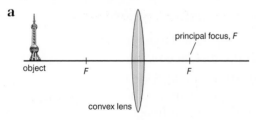

b

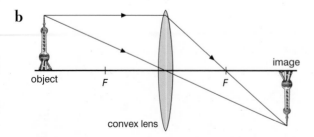

△ Fig. 4.27 The camera forms a small inverted image of the object in front of it on the digital sensor or the film at the back.

End of topic checklist

Key terms

angle of incidence, angle of reflection, normal, virtual image

During your study of this topic you should have learned:

○ How to describe the formation of an optical image by a plane mirror and give its characteristics.

○ About the law, angle of incidence = angle of reflection.

○ How to perform simple constructions, measurements and calculations for reflection by plane mirrors.

○ About an experimental demonstration of the refraction of light.

○ How to describe the action of a thin converging lens on a beam of light.

○ The terms 'principal focus' and 'focal length'.

○ How to draw ray diagrams to illustrate the formation of a real image by a single lens.

○ How to describe the nature of an image using the terms enlarged/same size/diminished and upright/inverted.

○ **EXTENDED** How to use and describe the use of a single lens as a magnifying glass.

End of topic questions

Note: The marks given for these questions indicate the level of detail required in the answers. In the examination, the number of marks given to questions like these may be different.

1. Write down all four characteristics that describe the image formed by a plane mirror. Use the correct scientific language. **(4 marks)**

2. Where must you place the object so that a convex lens gives an image with a magnification of more than 1? **(2 marks)**

3. Draw a diagram to show how a plane mirror forms an image of a simple object such as a candle. **(4 marks)**

4. Draw a diagram to show how a thin lens can be used as a magnifying glass. **(4 marks)**

Electromagnetic Spectrum

△ Fig. 4.28 Bats sleeping in a cave.

INTRODUCTION

A number of animals, such as dogs and bats, can hear sounds that are much too high for humans to hear. The vibrations of the sound waves repeat so quickly that our ear drums cannot detect these waves. The sounds humans can hear are only a small fraction of the range of sounds.

In just the same way, there are light waves that our eyes cannot detect. But those light waves do exist and scientists have been able to study them. We collect all the different light waves into a family we call the electromagnetic spectrum.

KNOWLEDGE CHECK

✓ Know that waves can transfer energy and information from place to place.
✓ Know that waves can be described using the key words wavelength, frequency, amplitude and wave speed.
✓ Be able to describe simple properties of light, such as reflection and refraction.

LEARNING OBJECTIVES

✓ Be able to describe the main features of the electromagnetic spectrum and state that all e.m. waves travel with the same high speed in a vacuum.
✓ Be able to describe the role of electromagnetic waves in: radio and television communications; satellite television and telephones; electrical appliances, remote controllers for TVs and intruder alarms; medicine and security.
✓ Be able to demonstrate an awareness of safety issues regarding the use of microwaves and X-rays.
✓ Be able to state the dangers of ultraviolet radiation, from the Sun or from tanning lamps.
✓ **EXTENDED** Be able to state the appropriate value of the speed of electromagnetic waves.

The electromagnetic spectrum is a 'family' of waves. Electromagnetic waves all travel at the same high speed in a vacuum. This high speed explains why you can have a phone call between China and New Zealand with a delay of only 0.1 s before you hear the reply from the person at the other end. It takes the infrared signal this long to travel there and back through an optical fibre.

EXTENDED

The speed of electromagnetic waves is 300 000 000 m/s in a vacuum, and approximately the same in air. This can be written more conveniently as 3×10^8 m/s.

END OF EXTENDED

However, for astronomical distances the delays quickly become longer. Even when Mars is at its nearest to Earth, it takes 10 minutes to send a message to a robot on the surface and receive a reply. Getting a reply from the nearest star (apart from the Sun) would take 8½ years.

Note that all electromagnetic waves can travel through a vacuum, which is why we can see the light and feel the heat coming from the Sun. Other waves, such as sound waves, cannot travel through a vacuum.

Order of the electromagnetic spectrum

The **visible spectrum** is only a small part of the full electromagnetic spectrum. The electromagnetic spectrum has waves of wavelength of the order of 10^4 m right down to wavelengths of the order of 10^{-12} m . Visible light has wavelengths ranging between 10^{-6} and 10^{-7} m. All electromagnetic waves travel at the same speed in a vacuum, which is 3×10^8 m/s. This means that the different wavelengths must have different frequencies, and that the longest wavelengths have the lowest frequencies. Fig. 4.31 shows the full electromagnetic spectrum.

△ Fig. 4.29 These satellite dishes are receiving high-energy radio waves, which are part of the electromagnetic spectrum.

△ Fig. 4.30 A prism splits white light into the colourful spectrum of visible light.

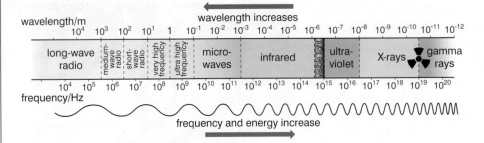

wavelength/m

△ Fig. 4.31 The electromagnetic spectrum.

The energy associated with an electromagnetic wave depends on its frequency. The waves with the higher frequencies are potentially the more hazardous ones, because they carry most energy.

The visible light region of the spectrum contains the colours ranged from red, through orange, yellow, green and blue to violet. Infrared lies next to red, and ultraviolet is next to violet.

QUESTIONS

1. EXTENDED What is the speed of electromagnetic waves in free space?

2. Which has greater frequency, microwaves or X-rays?

3. Put these electromagnetic radiations into order of increasing wavelength:

   microwaves   gamma rays   infrared
   visible light   ultraviolet

SCIENCE LINK   BIOLOGY – LIFE PROCESSES, TROPISM

- Most animals are able to sense electromagnetic radiation in their environment, although there are variations in the particular part of the spectrum involved. Knowledge of the range of wavelengths involved provides a linking theme across a range of adaptations, for example, the ability of bees to 'see' ultraviolet radiation or the ability of some snakes to find their prey using infrared radiation

- Plants also respond to electromagnetic radiation. Phototropism is the process where the direction in which a plant grows is affected by the direction from which the light is coming.

Uses and hazards of electromagnetic radiation

From Fig. 4.31 you will have seen that there are different types of electromagnetic radiation. Some of these types of radiation can have harmful effects on the human body when exposure to them is excessive.

Gamma rays, which, as you see in Fig. 4.31 are the electromagnetic waves with the shortest wavelength, and therefore the greatest frequency and energy, are produced by radioactive nuclei. They carry more energy than X-rays and can cause cancer or mutation in body cells. Gamma rays are frequently used in radiotherapy to kill cancer cells.

Radioactive substances that emit gamma rays are used as tracers. A tracer is something that can be used to track the flow of a substance, and is often used to track substances in biological systems. For example, if scientists want to know where in a plant the phosphorus goes, they can feed the plant a radioactive isotope of phosphorus and then measure the radioactivity that is given off by different parts of the plant.

X-rays, which have the next highest energy radiation after gamma on the electromagnetic spectrum, are produced when high-energy electrons are fired at a metal target. Bones absorb more X-rays than other body tissues. When a person is placed between an X-ray source and a photographic plate, the bones appear to be white on the developed photographic plate compared with the rest of the body. X-rays have high energy, as shown in Fig. 4.32, and can damage or destroy body cells. They may also cause cancer. An X-ray could save your life, but it is not free from danger. A doctor will only arrange for an X-ray when it is clear that the benefits to you are far greater than the tiny risk that it will make you ill.

However, X-rays are also used to treat cancer. X-rays are targeted at the tumour with the aim of destroying the tumour cells while leaving healthy cells surrounding the tumour untouched.

X-rays are also used to screen baggage at airports and other places. It works in the same way as a medical X-ray. Different materials absorb different amounts of X-rays.

X-ray scanners can also be used to scan bodies and much larger objects. They can even be used to scan trucks and shipping containers to see if there are any things such as weapons or people hidden in the trucks or containers.

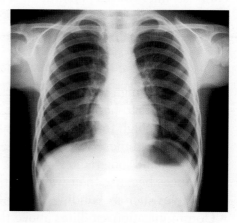

△ Fig. 4.32 An X-ray of the lungs.

Ultraviolet radiation (UV) is the component of the Sun's rays that gives you a suntan. It is also created in fluorescent light tubes after exciting the atoms in a mercury vapour. The UV radiation is then absorbed by the coating on the inside of the fluorescent tube and re-emitted as visible light. Fluorescent tubes are more efficient than light bulbs because they do not emit heat energy, so more energy is available to produce light. Ultraviolet can also damage the surface cells of the body, which can lead to skin cancer. It can also damage the eyes, leading to blindness.

All objects give out **infrared radiation** (IR). The hotter the object is, the more radiation it gives out. Thermograms are photographs taken to show the infrared radiation given out from objects. Infrared radiation grills and cooks our food in an ordinary oven and is used in remote controls to operate televisions and videos. Excessive exposure to infrared, such as when you get close to a very hot object, can burn skin and other body tissue.

Microwaves are high-frequency **radio waves**. They are used in radar to find the position of aircraft and ships. Metal objects reflect the microwaves back to the transmitter, enabling the distance between the object and the transmitter to be calculated. Microwaves are also used for cooking. Water particles in food absorb the energy carried by microwaves. They vibrate more, making the food much hotter. Microwaves penetrate several centimetres into the food and so speed up the cooking process. Because of their ability to penetrate several centimetres, which is useful when cooking food, microwaves can heat body tissue internally, so care must be taken to ensure they cannot escape from a microwave oven.

REMEMBER

Infrared radiation is absorbed by the surface of the food, then the energy is spread through the rest of the food by conduction. In contrast, microwaves penetrate a few centimetres into the food and then the energy is transferred throughout the food by conduction.

Radio waves have the longest wavelengths and lowest frequencies.

- UHF (ultra-high frequency) waves are used to transmit television programmes to homes.
- VHF (very high frequency) waves are used to transmit local radio programmes.
- Medium and long radio waves are used to transmit over longer distances because their wavelengths allow them to diffract around obstacles such as buildings and hills.
- Communication satellites above the Earth receive signals carried by high-frequency (short-wave) radio waves. These signals are amplified and re-transmitted to other parts of the world.

PROTECTION FROM ULTRAVIOLET RAYS

In 2007, scientists at the University of Virginia published a study about how cells protect themselves (or fail to protect themselves) from damage to the DNA caused by ultraviolet rays. The study showed that there is a simple switch mechanism inside cells, which is triggered by exposure to ultraviolet and helps cells to survive and even thrive after exposure to ultraviolet rays.

After DNA damage caused by exposure to ultraviolet rays, cells normally stop moving and responding to stimuli until they are repaired. If the repair work is not carried out properly, the result can be cancer, as the damaged cell keeps dividing.

QUESTIONS

1. Gamma rays and X-rays are both used in medical contexts. Describe how each is produced.

2. How is UV created in fluorescent tubes?

3. What is a thermogram?

4. How do microwaves cook food?

End of topic checklist

Key terms

gamma ray, infrared, microwave, radio wave, ultraviolet light, visible spectrum, X-ray

During your study of this topic you should have learned:

○ About the main features of the electromagnetic spectrum and that all e.m. waves travel with the same high speed in a vacuum.

○ **EXTENDED** About the approximate value of the speed of electromagnetic waves.

○ About the role of electromagnetic waves in:

- radio and television communications (radio waves)
- satellite television and telephones (microwaves)
- electrical appliances, remote controllers for televisions and intruder alarms (infrared)
- medicine and security (X-rays).

○ About safety issues regarding the use of microwaves, UV rays and X-rays.

End of topic questions

Note: The marks given for these questions indicate the level of detail required in the answers. In the examination, the number of marks given to questions like these may be different.

1. Here is a list of types of wave:

    gamma, infrared, microwaves, radio, ultraviolet, visible, X-rays

    Choose from the list the type of wave that best fits each of these descriptions:

    a) Stimulates the sensitive cells at the back of the eye. **(1 mark)**

    b) Necessary for a suntan. **(1 mark)**

    c) Used for rapid cooking in an oven. **(1 mark)**

    d) Used to take a photograph of the bones in a broken arm. **(1 mark)**

    e) Emitted by a TV remote control unit. **(1 mark)**

2. Gamma rays are part of the electromagnetic spectrum. Gamma rays are useful to us but can also be very dangerous.

    a) Explain how the properties of gamma rays make them useful to us. **(3 marks)**

    b) Explain why gamma rays can cause damage to people. **(3 marks)**

    c) Give one difference between microwaves and gamma rays. **(1 mark)**

    d) EXTENDED Microwaves travel at 300 000 000 m/s. At what speed do gamma rays travel? **(2 marks)**

3. **a)** Write down the parts of the electromagnetic spectrum in order of increasing wavelength. **(4 marks)**

    b) How would your list in part **a)** be different if you wrote it in order of increasing frequency? **(1 mark)**

End of topic questions continued

4. Copy and complete the table describing waves of the electromagnetic spectrum.

Wave type	Source of wave	Use	Property
Radio	Radio transmitter and aerial		
X-rays	X-ray tubes	Security at airports	
	Lamps, Sun, flames		Can be dispersed into seven colours
Ultraviolet	Mercury vapour lamps	Security markings	
		Thermal imaging	Can cause burns
Gamma	Radioactive substances		Very penetrating
Sound		Hearing	
	Magnetron	Heating food quickly	

(13 marks)

5. Why do microwave ovens take less time to cook food than normal ovens?

(3 marks)

6. a) Copy the electromagnetic spectrum shown and complete it by filling in the gaps.

Radio waves		Infrared	Light		X-rays	

(3 marks)

b) Which waves have the longest wavelength? **(1 mark)**

c) Which type of wave carries the greatest energy? **(1 mark)**

d) What do all of the waves in the electromagnetic spectrum have in common? **(2 marks)**

e) What part of the spectrum is detected by: **i)** the eyes; **ii)** the skin? **(2 marks)**

Sound

INTRODUCTION

Think about the importance of sound. Humans and other animals use sounds for communicating in a variety of situations – for greetings, to get information across, to attract a mate, to warn of danger – there are many examples.

Sound is a form of wave, like light. However, light is a transverse wave and sound is a longitudinal wave. Some of the properties of sound are the same as light, but some of them are different. In this section you are going to find out how these sound waves operate.

△ Fig. 4.33 Sound is caused by vibrations which can be seen by the motion recorded in this photo of active speakers.

KNOWLEDGE CHECK

✓ Know how to describe waves using key words such as wavelength, amplitude and frequency.
✓ Know the difference between longitudinal and transverse waves.
✓ Be able to describe the processes of reflection, refraction and diffraction.

LEARNING OBJECTIVES

✓ Be able to describe the production of sound by vibrating sources.
✓ **EXTENDED** Be able to describe the longitudinal nature of sound waves.
✓ Be able to state the approximate range of audible frequencies.
✓ Be able to show an understanding of the term ultrasound.
✓ Be able to show an understanding that a medium is needed to transmit sound waves.
✓ Be able to describe an experiment to determine the speed of sound in air.
✓ Be able to relate the loudness and pitch of sound waves to amplitude and frequency.
✓ Be able to describe how the reflection of sound may produce an echo.
✓ **EXTENDED** Be able to describe compression and rarefaction.
✓ **EXTENDED** Be able to state the order of magnitude of the speed of sound in air, liquids and solids.

WHAT IS SOUND?

Sound is caused by vibrations, for example, of the front of a violin or a cello, or of the column of air inside a trumpet. In the case of a loudspeaker it is particularly clear that the cone of the loudspeaker moves in and out and changes the pressure in the air in front of it. The sound travels as longitudinal waves.

The compressions (where vibrations are closer together on the wave) and rarefactions (where vibrations are further apart on the wave) of sound waves result in small differences in air pressure (see Fig. 4.2 in General wave properties)

Like other longitudinal waves, sound waves can be reflected, refracted or diffracted. Sound waves travel faster through liquids than through air. The speed of sound in air is 340 m/s. The speed of sound in water is 1484 m/s. Sound travels fastest through solids (in iron it travels at 5120 m/s). This is because particles are linked most strongly in solids. Note, however, that sound must have a medium through which to travel. Unlike electromagnetic waves, sound will not travel through a vacuum.

△ Fig. 4.34 This orchestra is creating a single longitudinal wave of very complicated shape. In ways that we barely understand, our brains can pick out the sounds of all the individual instruments that are playing together.

Sounds humans can hear

The human ear can detect sounds with pitches in the range 20 Hz to 20 000 Hz. Sound with frequencies above this range is known as **ultrasound**. Ultrasound is used by bats for navigation. It can also be used to build up images of organs within the body, since different tissues reflect ultrasound waves in different ways. By combining the various reflections, an image is generated. A well-known example of the use of ultrasound scanning is for checking the development of a fetus during pregnancy.

Measuring the speed of sound in air

The simplest method to measure the speed of sound in air uses two microphones and a fast recording device, such as a digital storage oscilloscope.

1. A sound source and the two microphones are arranged in a straight line, with the sound source beyond the first microphone.

2. The distance between the microphones (x), called the microphone basis, is measured.

3. The time of arrival between the signals (delay) reaching the different microphones (t) is measured.

4. Then speed of sound $= x/t$

EXTENDED

Speed of sound in air, liquids and solids

The speed of sound in solids is greater than that in liquids, which is greater than that in gases. Typical values are:

- steel: 3200 m/s
- fresh water: 1497 m/s
- dry air: 343 m/s.

END OF EXTENDED

SCIENCE IN CONTEXT **DAMAGING EARS**

The ear is far more easily damaged than most people realise. You should always take care, both with the volume of sound and the length of time that your ear is exposed to it. The damage is cumulative, so is not noticed at first. Many older rock musicians have serious hearing problems, and many of the younger ones now wear earplugs to prevent their own performances from damaging their hearing.

Earplugs worn by rock musicians are designed to reduce the range of audio frequencies equally, so that the wearer hears the upper and lower frequencies at the same relative levels as they would without the earplugs. (If this were not the case, then the bass guitarist, for example, may feel that he or she was not playing loudly enough to balance the vocalist, simply because the lower frequencies were reduced in level too much in relation to the higher vocal frequencies.)

This type of earplug usually has a tiny diaphragm to reduce low frequencies (100 to 50 Hz), and absorbent or damping material to reduce high frequencies (5 to 2 kHz). They are quite expensive and are intended to be used again and again. They reduce noise levels by about 20 **decibels** (dB) but are not intended to protect the wearer from noise levels above 105 dB (Fig. 4.35).

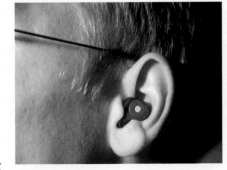

△ Fig. 4.35 A musician's earplug.

1. What type of wave is a sound wave?

2. **EXTENDED** What happens to the air as a result of compressions and rarefactions in a sound wave?

3. What is the frequency range for human hearing?

4. What is ultrasound?

5. Describe a simple method for measuring the speed of sound in air.

Using an oscilloscope and microphone to display sound waves

Sound waves can be represented on an **oscilloscope** by using a microphone and a loudspeaker as shown in Fig. 4.36. This produces a voltage–time graph for the sound wave on the screen of the oscilloscope. From the voltage–time graph, you can find the frequency of the sound wave, since you will know the time taken for one cycle of the wave (which is the frequency).

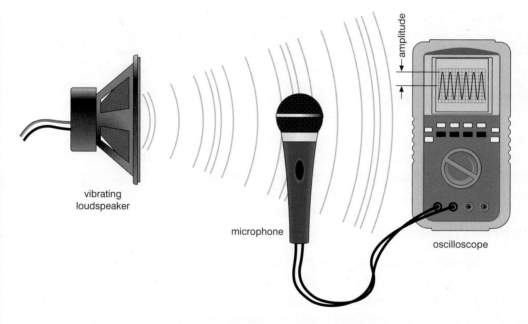

Δ Fig. 4.36 Displaying sound waves on an oscilloscope screen.

Pitch, frequency, amplitude and loudness

Sounds with a high **pitch** have a high frequency. Examples of high-pitch sounds include birdsong and all the sounds that you hear from someone else's personal stereo when they have set the volume too high. Low-pitch sounds have a low frequency. Examples of low-pitch sounds include the horn of a large ship and a bass guitar.

Loud sounds have large amplitude, whereas quiet sounds have small amplitude. The loudness of sounds can be compared using decibels. Typical sound wave patterns are shown in Fig. 4.37.

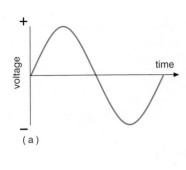
(a)

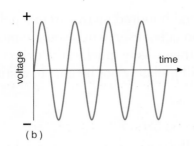
(b)

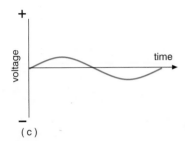
(c)

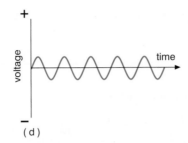
(d)

△ Fig. 4.37 Wave sound patterns: (a) A loud sound of low frequency; (b) A loud sound of high frequency; (c) A quiet sound of low frequency; (d) A quiet sound of high frequency.

WORKED EXAMPLE

Using this displacement–time graph for a sound wave, calculate the amplitude of the waveform.

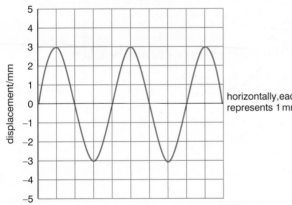

horizontally, each square represents 1 mm

◁ Fig. 4.38 Displacement–time graph for a sound wave.

The amplitude is the maximum displacement from the mean position, so can be read straight from the graph.

Amplitude = 3 mm.

Echoes

Hard surfaces reflect sound waves. An **echo** is a sound that has been reflected before you hear it. For an echo to be clearly heard, the obstacle needs to be large compared with the wavelength of the sound. For example, you will hear an echo when you make a loud noise when you are several hundred metres from a brick wall or a cliff. You will not hear an echo when you are several hundred metres from a pole stuck in the ground. There will still be an echo, even when you are much closer to the wall, but because sound travels very quickly, the echo will return in such a short time that you will probably not be able to distinguish it from the sound that caused it.

Measuring the speed of sound by an echo method

The following worked example illustrates how echoes may be used to measure the speed of sound.

WORKED EXAMPLE

Two students stand side by side at a distance of 480 m from the school wall. Student A has two flat pieces of wood, which make a loud sound when clapped together. Student B has a stopwatch.

As student A claps the boards together, student B starts the stopwatch. When student B hears the echo, he stops the stopwatch. The time recorded on the stopwatch is 2.9 s.

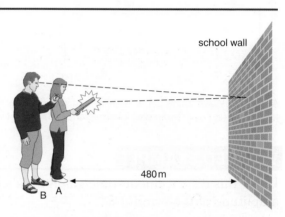

Δ Fig. 4.39 Two students carrying out an investigation into the speed of sound.

Calculate the speed of sound.

Write down the formula:	speed of sound	= distance travelled/time taken
Work out the distance:	distance to wall and back	= 2 × 480
		= 960 m

Record the time the sound took to travel there and back:

	time	= 2.9 s
Substitute the formula:	speed of sound	= 960/2.9
		= 331 m/s

See if you can think of some things that might be done to improve the accuracy of this experiment.

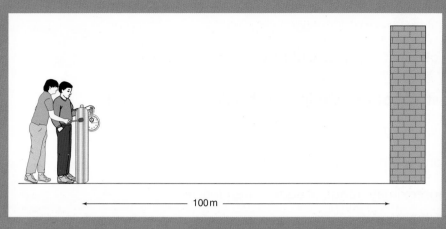

△ Fig. 4.40 Two students measuring the speed of sound.

Two students are finding the speed of sound. They stand 100 m from a large wall. The first student strikes a piece of metal with a hammer. The sound this makes reflects back from the wall and the student hears an echo. The student hits the metal again in time with the echo and continues to do so, tapping out a steady rhythm.

The second student is in charge of timing. He knows that in between each sound that the first student makes the sound travels 200 m (to the wall and back again). By timing the interval for a number of strikes, the second student can record the data he needs.

Observing, measuring and recording

❶ Explain whether the students should repeat their measurements for this experiment.

❷ The students stand 100 m away from the wall that provides the echo. Explain why this is a suitable distance to use.

❸ The students measure a time of 2.3 s from striking the metal to striking the metal at the fourth following echo. Calculate the speed of sound given by this measurement.

❹ A better way to find the speed of sound would be through the use of a graph. Describe how the students could collect suitable data and how a graph could be used to find the speed of sound.

❺ Explain why using a graph to find the speed of sound would improve the accuracy of the experiment.

Handling experimental observations and data

❻ What effect will reaction time have on the measurements made?

❼ Suggest how the timing in this experiment could be improved.

End of topic checklist

Key terms

decibel, echo, oscilloscope, pitch, ultrasound

During your study of this topic you should have learned:

○ How sound waves are produced by vibrating sources.

○ About the longitudinal nature of sound waves.

○ About the approximate range of audible frequencies.

○ About the term ultrasound.

○ That a medium is needed to transmit sound waves.

○ About an experiment to determine the speed of sound in air.

○ That loudness and pitch of sound waves are related to amplitude and frequency.

○ How to describe how the reflection of sound may produce an echo.

○ EXTENDED About compression and rarefaction.

○ EXTENDED About the order of magnitude of the speed of sound in air, liquids and solids.

End of topic questions

Note: The marks given for these questions indicate the level of detail required in the answers. In the examination, the number of marks given to questions like these may be different.

1. a) i) What causes a sound? **(2 marks)**

    ii) Explain how sound travels through the air. **(2 marks)**

  b) Astronauts in space cannot talk directly to each other – they have to speak to each other by radio. Explain why this is so. **(3 marks)**

2. Ayesha and Salma are carrying out an experiment to measure the speed of sound. They stand 150 m apart.

Ayesha starts the stopwatch when she sees Salma make a sound and she stops it when she hears the sound herself. She measures the time as 0.44 s. Calculate the speed of sound in air from this data. **(3 marks)**

3. EXTENDED The speed of sound is approximately 340 m/s.

  a) Calculate the wavelength of the musical note middle C, which has a frequency of 256 Hz. **(3 marks)**

  b) A student hears two echoes when she claps her hands. One echo is 0.5 s after the clap, and one echo is 1.0 s after the clap. She decides that the two echoes are from two buildings in front of her. How far apart are the buildings? **(3 marks)**

4. Draw an oscilloscope trace representing each of the following sounds.

  a) low-frequency quiet sound **(1 mark)**

  b) high-frequency loud sound **(1 mark)**

  c) high-frequency quiet sound **(1 mark)**

  d) low-frequency loud sound. **(1 mark)**

5. A polystyrene ball is suspended so that it is touching the prongs of a vibrating tuning fork. The ball kicks away from the tuning fork and then moves back to it.

  a) Explain the behaviour of the ball by:

    i) describing how the prongs of the tuning fork move **(2 marks)**

    ii) describing how a sound wave is created by the tuning fork **(2 marks)**

    iii) making a drawing of the sound wave, labelling the key features. **(2 marks)**

  b) Explain what you would see if the tips of the tuning fork were dipped into a beaker of water. **(3 marks)**

Have you ever rubbed a balloon and used static electricity to 'stick' the balloon to a wall? Or have you ever 'got a shock' when you touched a metal handrail? Has your phone ever run out of charge completely, leaving you unable to communicate?

Just think how your life would be if we had no understanding of electrical effects and how to use them.

In this section you will consider the origins of electric forces before moving on to study electric circuits. You will look at the key ideas involved and the key measurements we can make.

STARTING POINTS

1. Which parts of an atom are charged?

2. Why are electric cables made of metals?

3. What is a lightning conductor?

4. List five examples of energy changes caused by electrical circuits.

5. Why do bathrooms not have electric sockets but kitchens do?

CONTENTS

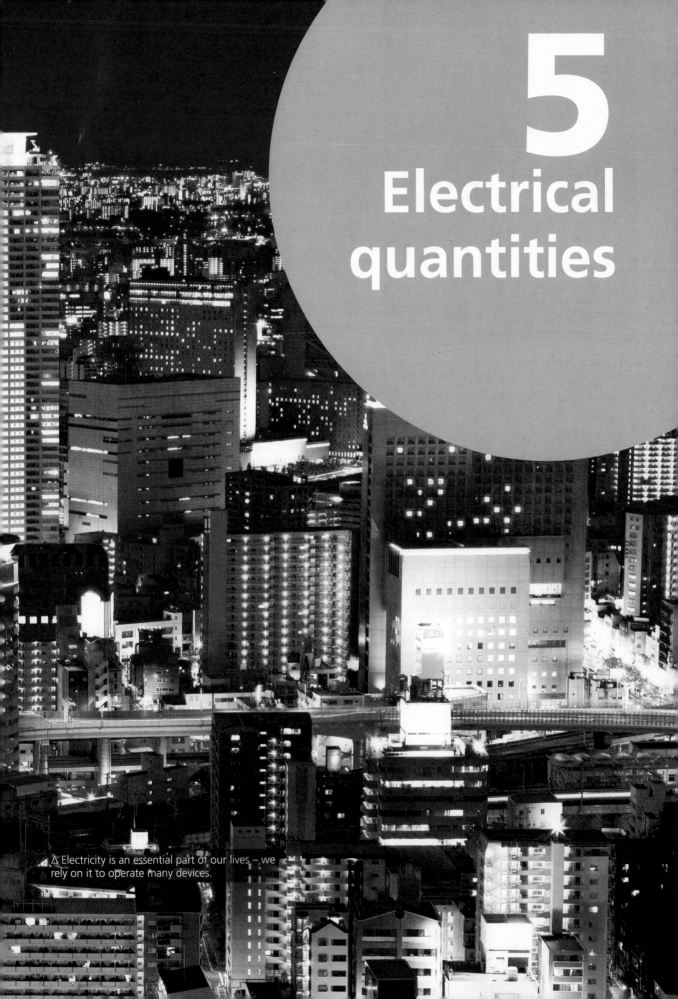

5

Electrical quantities

△ Electricity is an essential part of our lives – we rely on it to operate many devices.

Electric charge

INTRODUCTION

Have you ever brushed your hair and seen individual hairs standing on end and wondered why this happens? Would you like to know how to stick a balloon to a surface without glue? This section explores how and why these things happen. They are both the result of movement of electric charge.

△ Fig. 5.1 A plasma globe. The blue light is caused by a flow of electric charge.

KNOWLEDGE CHECK

✔ Know that common electric effects are caused by imbalances in the number of electrons present.
✔ Know that forces can be attractive or repulsive.

LEARNING OBJECTIVES

✔ Describe simple experiments to show the production and detection of electrostatic charges.
✔ State that there are positive and negative charges.
✔ State that unlike charges attract and that like charges repel.
✔ State that charging a body involves the addition or removal of electrons.
✔ Distinguish between electrical insulators and conductors and give typical examples.

ELECTRIC CHARGE

Individual hairs stand on end after brushing and balloons stick to certain surfaces without glue as a result of movement of electrostatic **charge**.

Conductors and insulators

Substances that easily allow electric energy to pass through them are called **conductors**; those that do not are called **insulators**.

Metals are conductors. In a metal structure, the metal atoms exist as ions surrounded by a 'sea' of electrons. (Fig. 5.2).

When the electrons are moving through the metal structure, they bump into the metal ions and experience **resistance** to the electron flow or current. In different conductors, the ease of flow of the electrons is different and so the conductors have different resistances. For example, copper is a better conductor than iron.

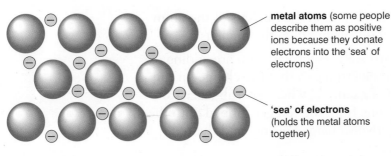

metal atoms (some people describe them as positive ions because they donate electrons into the 'sea' of electrons)

'sea' of electrons (holds the metal atoms together)

△ Fig. 5.2 In a metal structure the metal ions are surrounded by a 'sea' of electrons.

Table 5.1 lists materials ranging from the best conductor to the best insulator. The range in the resistance of different materials is truly amazing. Silver is about 10^{27} times better at conducting charges than the plastic Teflon. So to replace a 1 mm diameter wire of silver or copper in an electrical circuit, you would need a bar of Teflon far larger in diameter than the Moon's orbit around the Earth.

Name	Metal/non-metal	Conductor/ insulator
Silver	Metal	Conductor (best)
Copper	Metal	Conductor
Aluminium	Metal	Conductor
Iron	Metal	Conductor
Graphite	Non-metal	Conductor
Silicon	Non-metal	Semiconductor
Most Plastics	Non-metal	Insulator
Oil	Non-metal	Insulator
Glass	Non-metal	Insulator
Teflon	Non-metal	Insulator (best)

Table 5.1 Comparison of conductors and insulators.

Electrostatic charges: charging insulators by friction

Materials such as glass, acetate and polythene can only become charged when they are rubbed. This is because they are insulators. Electrons do not move easily through insulating materials, so when extra electrons are added, they stay on the surface instead of flowing away, and the surface stays negatively charged. Similarly, when electrons are removed, electrons from other parts of the material do not flow in to replace them, so the surface stays positively charged. Conductors, such as metals, cannot be charged by rubbing.

Charge as the loss or gain of electrons

All atoms are made up of three main kinds of particles, called electrons, protons and neutrons. Electrons are the tiniest of these and have a negative charge. Protons and neutrons have about the same mass, but protons are positively charged while neutrons have no charge. Protons and neutrons are found in the nucleus of the atom and electrons are found as a cloud surrounding the nucleus.

When you charge an object with static electricity, you are adding or taking away negatively charged electrons, so that the charge on the object overall is unbalanced. For example, when you rub a glass or acetate rod with a cloth, electrons from the rod get rubbed onto the cloth. The cloth becomes negatively charged overall and the rod is left with an overall positive charge.

When you rub a polythene rod with a cloth, electrons from the cloth get transferred to the rod, so the polythene carries a negative charge overall and the cloth carries a positive charge (Fig. 5.3).

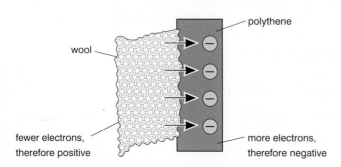

△ Fig. 5.3 Transferring charge from a cloth to a rod.

QUESTIONS

1. Have you ever 'got a shock' from a metal handrail? Explain why it was really the other way round – you gave a shock to the handrail.

2. Storm clouds can become charged, leading to lightning. Suggest how a cloud can become charged.

Attraction and repulsion

Every proton and electron produces an **electric field**. So, around any object in which the charges are not balanced, there is an electric field.

When a charged particle moves into the field, it feels a force toward or away from the other particle (see Fig. 5.4). The strength of the force depends on:

- how close the particles are: the closer they are, the larger the force
- how much electrical charge they carry: the more charge, the larger the force.

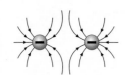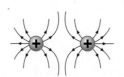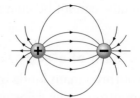

Like charges repel each other.   Unlike charges attract each other.

△ Fig. 5.4 Field lines show the shape of an electric field.

When an unbalanced charge collects on the surface of an object, the charge is called **static charge**. ('Static' means 'not moving'.) When electrons move, or flow, from one place to another, they produce an electric current.

Simple experiments using static electricity

When you suspend charged polythene and acetate rods so they can move freely, and bring the two close together, they will attract each other, since unlike charges attract.

Similarly, when a balloon is rubbed against clothing it will 'stick' to a wall or ceiling. This is because of the attraction between the negative charges on the balloon and the induced positive charges on the ceiling (Fig. 5.6).

△ Fig. 5.5 Because the static charge on each hair is similar, the hairs repel each other and stick up in all directions.

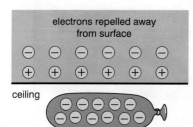

◁ Fig. 5.6 The balloon induces a charge on the ceiling's surface.

All the phenomena of electrostatics can be explained in terms of moving negative or moving positive charges. Early scientists did not know what was moving, and it was only towards the end of the nineteenth century that they became sure that it was the negative charges that moved.

QUESTIONS

1. The key discovery at the end of the nineteenth century was the electron. From your knowledge of atoms, explain why it is negative charges that move to create electrostatic effects.

2. Describe how electrostatic effects stop you from falling through the floor.

3. a) Complete the diagrams to show what happens when a cloth is rubbed on a material that gains electrons, and when a cloth is rubbed on a material that loses electrons.

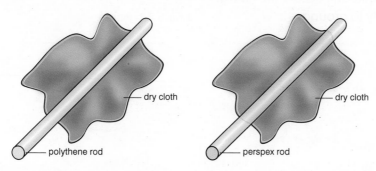

   b) What can you say about the sizes of the positive or negative charges on the cloth and on the rod in each of the diagrams?

HAZARDS OF ELECTROSTATICS

The sudden discharge of electricity caused by friction between two insulators can cause shocks in everyday situations, for example:

- combing your hair
- pulling clothes over your head
- ironing synthetic fabrics
- getting out of a car.

You may have noticed that you can get a nasty spark from your finger if you touch a metal object after rubbing your feet on a nylon carpet. This is similar to the effect you can sometimes feel if you touch a metal door handle. It is for this reason that workers who make sensitive electronic devices connect themselves to ground using devices such as antistatic wrist straps (Fig. 5.7), which link to a grounding point (a point that is connected to 0 V) so that any static charge can discharge safely via the wrist strap and not damage the equipment before starting work. Also, to protect against sparks of this type, aircraft are connected to the ground by a special wire before refuelling starts.

△ Fig. 5.7 An antistatic wrist strap in use.

Lightning is a spectacular example of electrostatics in action. Scientists believe that the electrical charge is generated by induction when ice particles in clouds collide. One bolt of lightning carries about 5 C of electrical charge. Lightning conductors on buildings usually prevent lightning strikes by discharging the cloud above, but if a strike still occurs the charge should be carried safely to ground.

△ Fig. 5.8 Lightning.

End of topic checklist

Key terms

conductor, electric charge, electron, insulator, ion, resistance, static electricity

During your study of this topic you should have learned:

○ About simple experiments which can produce static electric charges

○ That there are positive and negative charges and that these will attract if the charges are different but repel if the charges are the same

○ That charging an object involves adding or removing electrons

○ To carry out and describe simple experiments to show the production and detection of electrostatic charges.

○ How to distinguish between electrical insulators and conductors and give typical examples of each.

End of topic questions

Note: The marks given for these questions indicate the level of detail required in the answers. In the examination, the number of marks given to questions like these may be different.

1. A plastic rod is rubbed with a cloth.

   a) How does the plastic become positively charged? **(2 marks)**

   b) The charged plastic rod attracts small pieces of paper. Explain why this attraction occurs. **(2 marks)**

2. **a)** A car stops and one of the passengers gets out. When she touches a metal post she feels an electric shock. Explain why she feels this shock. **(2 marks)**

   b) Write down two other situations where people might get this type of shock. **(2 marks)**

△ Fig. 5.9 How many of these do you use in your life?

Current, potential difference and electromotive force (e.m.f.)

INTRODUCTION

You have seen the effects of static electric charges in the previous topic. However, when the charges move in an organised way a number of different effects are seen – these are the effects of electric currents. We can build upon our atomic model of electric charges to describe, measure and explain how electric circuits have these effects. This will allow us to design electric circuits to perform a wide variety of useful functions, from lighting our homes to controlling a heart pacemaker.

KNOWLEDGE CHECK

✓ Know that electric charges can be positive or negative.
✓ Know the electric charges produce forces of attraction and repulsion.

LEARNING OBJECTIVES

✓ State that current is related to flow of charge.
✓ Use and describe the use of an ammeter and a voltmeter, both analogue and digital.
✓ State that current in metals is due to a flow of electrons.
✓ State that the potential difference (p.d.) across a circuit component is measured in volts.
✓ State that the electromotive force (e.m.f.) of an electrical source of energy is measured in volts.

✓ **EXTENDED** show an understanding that current is a rate of flow of charge and recall and use the equation $I = Q/t$.

CURRENT

When there is no current in a conductor, the free electrons move randomly between atoms, with no overall movement. When you connect it in an electrical circuit with a power source such as a battery, there is a current in the conductor. Now the electrons drift in one direction, while still moving in a random way as well. The drift speed is very slow, often only a few millimetres each second. There can only be a current in a conductor when it is connected in a complete circuit. When the circuit is broken, the current stops.

BIOLOGY – LIFE PROCESSES

- An electric current is a rate of flow of electric charge. The idea of a 'rate of flow', how much a particular quantity changes in a measured amount of time, is a key concept that is used in many situations as we describe the world around us. For example, we might be interested in the rate of growth of a tree or an animal, or the rate at which bacteria multiply.

CHEMISTRY – CHEMICAL REACTIONS

- Rates of change are also important in chemistry. We may need to know how quickly a chemical change will happen, or if the rate of energy release will be enough to provide the power we need in a car engine.

The size of an electric current depends on the number of electrons that are moving and how fast they are moving. However, instead of measuring the actual number of electrons we use the total charge carried by the electrons round the circuit each second. So, current is a flow of charge.

Electric current is measured in **amperes**, or **amps** (A).

EXTENDED

When there is a current of 1A in a wire, then one **coulomb** of charge is passing any point on the circuit each second. (1A = 1 C/s.)

END OF EXTENDED

You use an **ammeter** to measure current in an electrical circuit. When the current is very small, you might use a milliammeter, which measures current in milliamps (1 mA = 0.001A). Even smaller currents are measured with a microammeter.

When you want to measure the current in a particular component, such as a lamp or motor, the ammeter must be connected in series with the component. In a series circuit, the current is the same at all points so it does not matter where the ammeter is put. This is not the case with a parallel circuit (see Section 6, Electric circuits for more about series and parallel circuits).

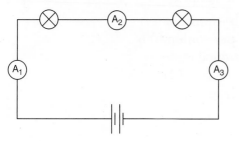

◁ Fig. 5.10 In this series circuit, the current will be the same throughout the circuit so $A_1 = A_2 = A_3$.

The electric current is the amount of charge flowing every second (the number of coulombs per second):

$$I = Q / t$$

Where: I = current in amperes (A)

Q = charge in coulombs (C)

t = time in seconds (s)

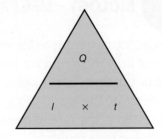

△ Fig. 5.11 Equation triangle for charge, current and time.

QUESTIONS

1. Calculate the flow of charge in the following:

a) current 3 A for 5 s

b) current 2 A for 10 s

c) current 4 A for 23 s

d) current 1.5 A for 0.5 minutes.

2. A charge of 120 C flows for 4 minutes. What is the current?

3. A charge of 60 C produces a current of 0.5 A. How long does this take?

ELECTROMOTIVE FORCE

The battery in an electrical circuit can be thought of as pushing electrical charge around the circuit to make a current. It also transfers energy to the electrical charge. The **electromotive force** (e.m.f.) of the battery, measured in volts, measures how much 'push' it can provide and how much energy it can transfer to the charge.

POTENTIAL DIFFERENCE

The electrons moving around a circuit have some kinetic energy, which can be referred to as electrical energy. As electrons pass through the battery, or other power supply, they are given potential energy, and as they move around a circuit, they transfer energy to the various components in the circuit. For example, when the electrons move through a lamp they transfer some of their energy to the lamp.

The amount of energy that a unit of charge (a coulomb) transfers between one point and another (the number of joules per coulomb) is called the **potential difference** (p.d.). Potential difference is measured in **volts**, so it is often referred to as **voltage** (Fig. 5.12).

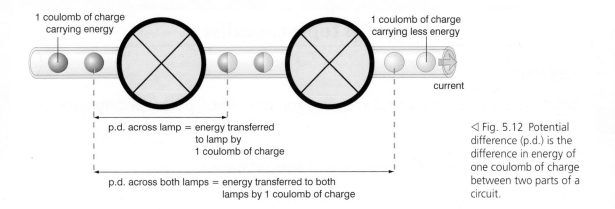

1 coulomb of charge carrying energy

1 coulomb of charge carrying less energy

current

p.d. across lamp = energy transferred to lamp by 1 coulomb of charge

p.d. across both lamps = energy transferred to both lamps by 1 coulomb of charge

◁ Fig. 5.12 Potential difference (p.d.) is the difference in energy of one coulomb of charge between two parts of a circuit.

Measuring electricity

Potential difference is measured using a voltmeter. When you want to measure the p.d. across a component then the voltmeter must be connected in parallel across that component. Testing with a voltmeter does not interfere with the circuit provided the voltmeter has a high resistance.

A voltmeter can be used to show how the potential difference varies in different parts of a circuit. In a series circuit you find different values of the voltage depending on where you attach the voltmeter. You can assume that energy is only transferred when the current passes through electrical components such as lamps and motors – the energy transfer to thermal energy as the current passes through copper connecting wire is very small indeed. Therefore, it is only possible to measure a p.d. or voltage across a component.

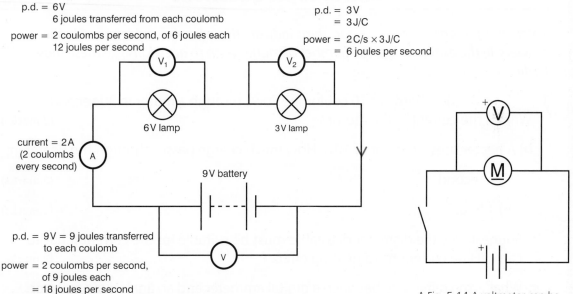

p.d. = 6 V
6 joules transferred from each coulomb

power = 2 coulombs per second, of 6 joules each
12 joules per second

V_1

6 V lamp

current = 2 A
(2 coulombs every second)

A

9 V battery

p.d. = 3 V
= 3 J/C

power = 2 C/s × 3 J/C
= 6 joules per second

V_2

3 V lamp

p.d. = 9 V = 9 joules transferred to each coulomb

power = 2 coulombs per second, of 9 joules each
= 18 joules per second

V

△ Fig. 5.13 The potential difference across the battery equals the sum of the potential differences across each lamp. That is $V = V_1 + V_2$.

V

M

△ Fig. 5.14 A voltmeter can be added after the rest of the circuit has been connected.

End of topic checklist

Key terms

ammeter, ampere (amp), coulomb, electromotive force, potential difference, voltage, volt

During your study of this topic you should have learned:

○ About current, potential difference and e.m.f.

○ That current is related to flow of charge.

○ How to use and describe the use of an ammeter and a voltmeter, both analogue and digital.

○ That current in metals is due to a flow of electrons.

○ That the e.m.f. of a source of energy and potential difference are both measured in volts.

○ EXTENDED That current is a rate of flow of charge and how to use the equation $I = Q/t$.

End of topic questions

Note: The marks given for these questions indicate the level of detail required in the answers. In the examination, the number of marks given to questions like these may be different.

1. EXTENDED **a)** A charge of 10 coulombs flows through a motor in 30 seconds. What is the current in the motor? **(2 marks)**

   b) A heater uses a current of 10 A. How much charge passes through the heater in:

   i) 1 second **(2 marks)**

   ii) 1 hour? **(2 marks)**

2. Suggest why the diameter of a cable must be suitable for the current it has to carry. (Hint: current has a heating effect.) **(2 marks)**

3. What is the difference between a digital ammeter and an analogue ammeter? **(2 marks)**

4. Electromotive force (e.m.f.) and potential difference (p.d.) are both measured in volts. Using ideas about energy, describe one similarity between e.m.f. and p.d. and one difference. **(2 marks)**

Resistance

INTRODUCTION

Having built our model on the ideas that current measures how much electric charge is moving per second and that p.d. and e.m.f. are related to the energy transfers involved, the third key feature of circuits that we need to consider is the material the charges are moving through. This brings in the idea of electrical resistance.

△ Fig. 5.15 The cables that carry electric current to household appliances have a very low resistance.

KNOWLEDGE CHECK

✓ Electric current is due to the movement of charges.
✓ How to measure currents with an ammeter.
✓ How to measure potential differences with a voltmeter.

LEARNING OBJECTIVES

✓ State that resistance = p.d. / current and understand qualitatively how changes in p.d. or resistance affect current.
✓ Recall and use the equation $R = V/I$.
✓ **EXTENDED** Recall and use quantitatively the proportionality between the resistance and length, and the inverse proportionality between resistance and cross-sectional area of a wire.

RESISTANCE

All components in an electrical circuit have a resistance to the current in them. The relationship between voltage, current and resistance in electrical circuits is given by this equation:

$V = IR$

where: V = potential difference in volts (V)

I = current in amps (A)

R = resistance in ohms (Ω).

It is important to be able to rearrange this equation when performing calculations. Use the triangle in Fig. 5.16 to help you.

From this equation, you find that $R = V / I$. This is **Ohm's law**.

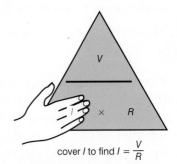

cover I to find $I = \dfrac{V}{R}$

△ Fig. 5.16 Equation triangle for voltage, current and resistance.

WORKED EXAMPLES

1. A heater element is connected to a 230 V supply. The current in the heater is 10A. Calculate the resistance of the heater.

Write down the formula in terms of R:	$R = V/I$
Substitute the values for V and I:	$R = 230/10$
Work out the answer and write down the unit:	$R = 23\,\Omega$

2. A 6 V supply is applied to 1000 Ω resistor. What will be the current?

Write down the formula in terms of I:	$I = V/R$
Substitute the values for V and R:	$I = 6/1000$
Work out the answer and write down the unit:	$I = 0.006\text{A}$

QUESTIONS

1. Use the Ohm's law equation to calculate the potential difference across a 5 Ω resistor, which has a current of 2 A in it.

2. A lamp has a potential difference of 3.0 V across it and a current of 0.5 A in it. What is its resistance?

Measuring resistance

The resistance of a component can be found using the circuit in Fig. 5.13 in the previous topic. The component (lamp, resistor or whatever) is placed in a circuit with an ammeter to measure the current in the component and a voltmeter to measure the potential difference across it. To take readings, the circuit is switched on and readings are made of the p.d. and the current.

The resistance is calculated from the following equation:

$R = V/I$

Note that the readings may change a little over the first few seconds. If so, this is probably because the component is heating up and its resistance is changing. If this happens, you would have to decide whether to take the readings before the component has heated up, and so measure the resistance at room temperature, or to wait until the readings have stopped changing. This would give you the 'steady-state' resistance with the component at its usual running temperature. 1 V is equivalent to 1 J/C.

You may wish to change the e.m.f. of the battery by changing the number of cells (or you may adjust the output of the power supply). When the component is a perfect resistor, then you will get the same answer for the resistance; but you will often find that the resistance of the component varies. (Modern multimeters measure resistance automatically and give a reading in ohms.)

For components such as **resistors** and thick wires, the current through the component doubles when you double the voltage and triples when you triple the voltage. The resistance of the component to the passage of electricity does not change, and the extra current is caused solely by the increased pressure of the extra voltage. The current is directly proportional to the voltage and the graph is a straight line.

Take care. If you try using thick wire as the component, the current will be extremely high for a very low voltage. The wire can get very hot very quickly and there is a risk of injury.

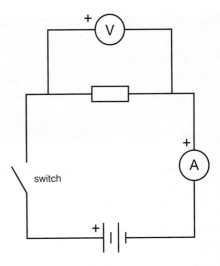

△ Fig. 5.17 Measuring the resistance of a component. To power the circuit you could use a battery as shown (with a maximum of 3 V), or you could use a power supply with a suitable output. For safety, switch on for a maximum of 10 seconds only.

HOW CHANGING RESISTANCE AFFECTS CURRENT

Figure 5.18 (left) shows the voltage–current graph for a component where the resistance remains constant, as shown by the constant gradient of the voltage–current graph. This would be the graph for an 'ohmic' resistor, such as carbon. For such a resistor, Ohm's law (see page 569) applies and the voltage is directly proportional to the current – a straight line is obtained. However, the resistance of most conductors becomes higher if the temperature of the conductor increases. As the temperature rises, the particles in the conductor vibrate more and provide greater resistance to the flow of electrons. For example, the resistance of a filament lamp becomes greater as the voltage is increased and the lamp gets hotter. Ohm's law is not obeyed because the heating of the lamp changes its resistance. Figure 5.18 (right) shows the voltage–current graph for such a component. You can see that the gradient of the graph increases with increasing voltage.

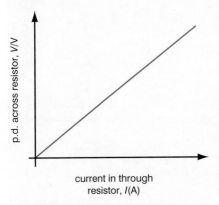

△ Fig. 5.18 Left: Voltage–current graph for a component where resistance remains constant. Right: Voltage current graph for a component where resistance increases as voltage increases.

Developing investigative skills

A student decides to find the resistance of a piece of fuse wire. He sets up the circuit in Fig. 5.19 and makes a note of the readings on the ammeter and the voltmeter for five different settings of the **variable resistor**. His measurements are shown in the table.

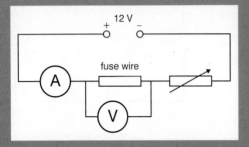

△ Fig. 5.19 A circuit diagram for the investigation.

Potential difference/V	Current /mA
0.0	0
0.5	44
1.0	88.5
1.5	137.5
2.0	181.5
2.5	225.5

Using and organising techniques, apparatus and materials

❶ The student did not measure the voltage of the supply or the particular settings of the variable resistor. Explain why these measurements were not required.

❷ The student should check the ammeter and voltmeter for zero errors. What are these?

Observing, measuring and recording

❸ Draw a graph of the student's results.

❹ Use your graph, with p.d. on the y-axis and current on the x-axis, to find the resistance of the wire.

Handling experimental observations and data

❺ Another student suggests that drawing a graph is not necessary. They say that you could use the equation $R = V / I$ for each pair of measurements and then find a mean of these values. Explain why calculating the gradient of the graph is a better method.

❻ To get an accurate value for the resistance of the wire, the student needed to avoid any heating effects in the wire. Describe how the student could reduce heating effects when carrying out the experiment.

Effects of length and cross-sectional area

For a particular conductor, the resistance is *proportional to length*. The longer the conductor, the further the electrons have to travel, the more likely they are to collide with the metal ions and so the greater the resistance. So, a wire that is twice as long will have twice as much resistance.

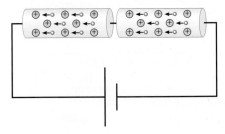

△ Fig. 5.20 Two wires in series are like one long wire, because the electrons have to travel twice as far.

Resistance is *inversely proportional to the cross-sectional area* of the wire. The greater the cross-sectional area of the conductor, the more electrons there are available to carry the charge along the conductor's length, so the lower the resistance. Therefore, a wire with twice the cross-sectional area will have half the resistance.

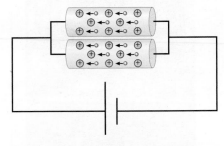

△ Fig. 5.21 Two wires in parallel are like one thick wire, so the electrons have more routes to travel along the same distance.

REMEMBER

If the wire is of twice the diameter, then its cross-sectional area will be four times greater, so the resistance of the wire will be one-quarter as much.

SUPERCONDUCTORS

In 1908, the Dutch physicist Heike Kamerlingh Onnes became the first person to produce liquid helium, which meant reaching temperatures less than −269 °C, the boiling point of helium. Having such a cold liquid meant that other low-temperature experiments became possible, as he could now cool down the apparatus sufficiently.

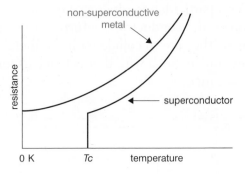

△ Fig 5.22 How resistance varies with temperature in a superconductor and a non-superconductor.

In particular, Kamerlingh Onnes investigated passing electric currents through extremely cold metals and in 1911 measured the resistance of a sample of mercury. He found that below a particular temperature, called the critical temperature, the mercury behaved as if it had no electrical resistance at all – he had discovered superconductivity.

Following this discovery, many more metallic elements were found to have superconducting properties, but it wasn't until the 1950s that a theory to explain their behaviour was developed. It requires energy for the electrons to scatter as they move through the metal lattice (these scatterings are the 'collisions' that lead to heating in a resistance) and at such low temperatures this energy is not available, so the electrons move smoothly – with zero resistance. The two key features of this are that no energy is wasted through heating the conductor, which leads to the ability to produce very large magnetic fields.

The search for superconductors has continued, with breakthroughs coming in the study of alloys rather than elements.

△ Fig. 5.23 A cross-section of a superconductor at CERN (Central European Organisation for Nuclear Research) in Geneva, Switzerland.

A particular milestone came in the discovery of materials that demonstrated superconductivity at temperatures up to −183 °C, as this meant that liquid nitrogen could be used as the coolant – and liquid nitrogen is readily available commercially. The search for materials that superconduct at higher temperatures continues.

Superconductors are used in a variety of applications. They produce the strong magnetic fields required for MRI scanning in medicine and to confine beams of particles in accelerators such as the Large Hadron Collider. They even provide magnetic fields to support Maglev trains that 'float' above the track. On a small scale, superconductors are used in SQUID (superconducting quantum interference device) magnetometers, which can measure the fine magnetic fields associated with activity in the brain.

QUESTIONS

1. Explain why the resistance of a conductor is proportional to length.

2. Explain why the resistance of a conductor is inversely proportional to cross-sectional area.

3. a) EXTENDED The length of a resistor A is x. Resistor B is made of the same material and is of the same thickness, but its length is $3x$. The resistance of A is R. What is the resistance of B?

   b) EXTENDED The area of cross-section of a resistor A is x. Resistor B is made of the same material, but its area of cross-section is $3x$. The resistance of A is R. What is the resistance of B?

End of topic checklist

Key terms

Ohm's law, resistance, resistor, variable resistor

During your study of this topic you should have learned:

○ That resistance = p.d./current and understand qualitatively how changes in p.d. or resistance affect current

○ How to use the equation $R = V/I$.

○ **EXTENDED** How to use quantitatively the proportionality between the resistance and length, and the inverse proportionality between resistance and cross-sectional area of a wire

End of topic questions

Note: The marks given for these questions indicate the level of detail required in the answers. In the examination, the number of marks given to questions like these may be different.

1. Calculate the following:

a) the potential difference required to produce a current of 2 A in a 12 Ω resistor (2 marks)

b) the potential difference required to produce a current of 0.1 A in a 200 Ω resistor (2 marks)

c) the current produced when a potential difference of 12 V is applied to a 100 Ω resistor (2 marks)

d) the current produced when a potential difference of 230 V is applied to a 10 Ω resistor (2 marks)

e) the resistance of a wire that under a potential difference of 6 V carries a current of 0.1 A (2 marks)

f) the resistance of a heater, which under a potential difference of 230 V carries a current of 10 A. (2 marks)

Studying the behaviour and properties of electric charges, either when they are stationary or when they are moving, is very interesting and it is a mental challenge to imagine a model to explain the observation we make. However, it is when we use our understanding to design the wide variety of circuits that we meet in everyday life that a knowledge of electricity becomes particularly useful.

In this section you will make sure you understand the international system for drawing electric circuits before studying the properties of two particularly key types of circuits – series and parallel. You will then go on to consider how to calculate the energy transferred by circuits, finishing by considering some of the factors we need to think about if we are to use circuits safely.

STARTING POINTS

1. Why is it important to have an agreed system for drawing electric circuits?

2. Think about where you live – do all the lights have a separate switch or can some be turned on or off by multiple switches?

3. If you have an electric cooker at home, it is likely to have its own, separate, electric circuit. Why is this?

4. Why might the thickness of the wires used in a circuit be important?

5. Does it matter which rating of fuse is used in a plug?

6. Why should care be taken with electrical circuits near water?

CONTENTS

6

Electric circuits

An abstract representation of an electric circuit.

Circuit diagrams

INTRODUCTION

Whenever you use an electrical appliance, electrical circuits operate. Some are visible to the eye but some have been etched onto microchips and are microscopic. The basic operation of all circuits relies on connecting **components**, the nature of the **components** and the energy supplied to the **circuit**. In this topic you will learn about different ways of connecting **components** in circuits, how to draw **circuit** diagrams that can be followed by anyone, anywhere in the world, and how to carry out calculations to choose the right values for the components in a circuit.

△ Fig. 6.1 You can investigate electrical circuits in the classroom.

KNOWLEDGE CHECK

✓ Be able to make simple circuits

LEARNING OBJECTIVES

✓ Draw and interpret circuit diagrams containing sources, switches, resistors, lamps, ammeters, voltmeters and fuses.

CIRCUIT DIAGRAMS

When people started using electricity, they quickly found that it was not convenient to draw accurate pictures of the circuits that they made. It was much easier to understand how the **circuit** worked, and to correct any faults, when they used standard symbols for the parts. It was also much easier when the wires were drawn in straight lines, rather than trying to copy the exact route taken.

Study the circuits used in this topic, and learn the symbols and what they represent.

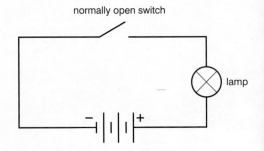

△ Fig. 6.2 A circuit diagram for a torch.

Figure 6.2 is a simple **circuit** diagram that shows how a torch is powered by a battery consisting of three 1.5 V cells, giving a total of 4.5 V.

$1.5\,V + 1.5\,V + 1.5\,V = 4.5\,V$

When powering a torch, the cells are put in separately, but in a 9 V battery, for example, the six cells are pre-assembled by the manufacturer. The word 'battery' means an assembly of several cells, but people often use the word to refer to a single cell.

The '+' terminal of the cell is indicated by the long line, and the '−' terminal by the short line. To help you remember, imagine yourself cutting the long line into two shorter pieces and turning them into a + sign.

The other symbols in the **circuit** are the normally open switch, and the lamp.

Figure 6.3 shows the **circuit symbols** that you need to know.

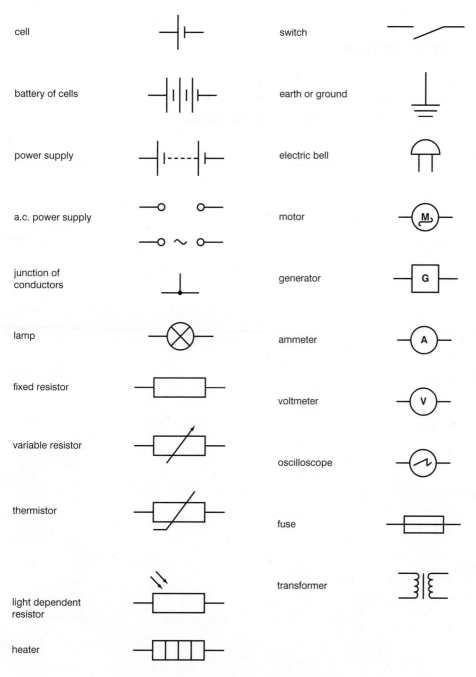

cell		switch
battery of cells		earth or ground
power supply		electric bell
a.c. power supply		motor
junction of conductors		generator
lamp		ammeter
fixed resistor		voltmeter
variable resistor		oscilloscope
thermistor		fuse
light dependent resistor		transformer
heater		

△ Fig. 6.3 Important circuit symbols for you to learn.

CHEMISTRY – METALS

- Electric circuits are usually constructed from wires made from metals. This links to two key properties of metals – they are good electrical conductors and they are ductile (they can be drawn out into thin wires). To understand why metals have these properties we need to follow the link to the structure of metals and how the atoms are joined together through metallic bonding.

BIOLOGY – NERVOUS SYSTEM

- The behaviour of electric circuits links to the properties of the nervous system. In both cases energy is transferred through the motion of charged particles, electrons in the case of circuits and electrons and ions in the case of nerves.

End of topic checklist

Key terms

circuit, circuit symbol, component

During your study of this topic you should have learned:

○ How to draw and interpret circuit diagrams containing a range of components, including sources, switches, resistors (fixed and variable), lamps, ammeters, voltmeters and fuses

End of topic questions

Note: The marks given for these questions indicate the level of detail required in the answers. In the examination, the number of marks given to questions like these may be different.

1. Draw the circuit symbols for a cell, a battery, a fixed resistor, a variable resistor, an ammeter and a voltmeter.

2. In a simple torch, two cells are connected to a switch and a lamp. Draw a circuit diagram to show this.

Series and parallel circuits

INTRODUCTION

There are two different ways of connecting two lamps (or other components) to the same battery (or other power source). Two very different kinds of circuit can be made. These circuits are called **series** and **parallel circuits**.

△ Fig. 6.4 A simple parallel circuit.

KNOWLEDGE CHECK

✓ In electric circuits, current, potential difference and resistance are related.
✓ How to measure current and potential difference.
✓ How to calculate resistance.

LEARNING OBJECTIVES

✓ Understand that the current at every point in a series circuit is the same.
✓ Calculate the combined resistance of two or more resistors in series.
✓ State that, for a parallel circuit, the current from the source is larger than the current in each branch.
✓ State that the combined resistance of two resistors in parallel is less than that of either resistance by itself.
✓ **EXTENDED** Calculate the combined resistance of two resistors in parallel.
✓ State the advantages of connecting lamps in parallel in a circuit.
✓ **EXTENDED** Recall and use the fact that the sum of the potential differences across the components in a series circuit is equal to the total p.d. across the supply.
✓ **EXTENDED** Recall and use the fact that the current from the source is the sum of the currents in the separate branches of a parallel circuit.

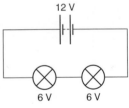

△ Fig. 6.5 A simple series circuit diagram.

CIRCUITS – IN SERIES AND PARALLEL

When components are in series, then there is exactly the same electric current in each of the components in the circuit. The voltage is shared between the units in the circuit. Thus, it is possible to join in series two identical lamps designed for 6 V and then to connect them to a 12 V battery (Fig. 6.5).

In a parallel circuit (Fig. 6.6), the current splits, with part of it going through each component. All of the appliances in a house are connected in parallel to the mains supply, and each one receives

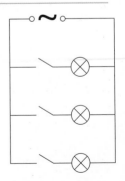

△ Fig. 6.6 A parallel circuit diagram.

the full 110 V or 230 V of the mains supply when it is switched on. The two great advantages of the parallel arrangement are that each appliance can be designed to work with the mains voltage supply, and that the appliances can be switched on and off individually.

△ Fig. 6.7 All of these lights are in parallel. If they were in series then they would all go off if any one of them failed or was switched off.

	Series	**Parallel**
Circuit diagram		
Appearance of the lamps	Both lamps have the same brightness, both lamps are dim.	Both lamps have the same brightness, both lamps are bright.
Battery	The battery is having a hard time pushing the same charge first through one bulb, then another. This means less charge flows each second, so there is a low current and energy is transferred slowly from the battery.	The battery pushes the charge along two alternative paths. This means that more charge can flow around the circuit each second, so energy is transferred quickly from the battery.
Switches	The lamps cannot be switched on and off independently.	The lamps can be switched on and off independently by putting switches in the parallel branches.
Advantages/ disadvantages	A very simple circuit to make. The battery will last longer. If one lamp 'blows' then the circuit is broken so the other one goes out too.	If one lamp 'blows', the other one will keep working. The battery will not last as long.
Examples	Strings of lights for decoration are often connected in series.	Electric lights in the home are connected in parallel.

△ Table 6.1 A comparison of series and parallel circuits for two identical 3 V lamps supplied from a 3 V battery.

QUESTIONS

1. If one bulb in a string of lights does not work, why does the rest of the string not work either?

2. A battery running two bulbs in parallel runs out of energy before the same battery running the same two bulbs in series. Explain why.

3. Why are electric lights in the home connected in parallel?

4. What is the difference in brightness in two bulbs connected:
   a) in series; b) in parallel with a given battery? Give a reason for your answer.

Current in a series circuit

The current in a circuit can be measured using an ammeter. When you want to measure the current in a particular component, such as a lamp or motor, the ammeter must be connected in series with the component. Figure 6.8 shows an ammeter in series with a motor. In a series circuit, the current is the same no matter where the ammeter is placed in the circuit. This is not the case with a parallel circuit.

The voltage across a component can be measured using a voltmeter, as shown in Fig. 6.9.

In a series circuit, the current is the same throughout the circuit (Fig. 6.10). In a parallel circuit, the current splits between the two branches of the parallel circuit. This means that the current from the source is larger than the current in each branch.

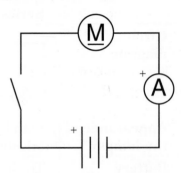

△ Fig. 6.8 A circuit has to be broken to connect an ammeter.

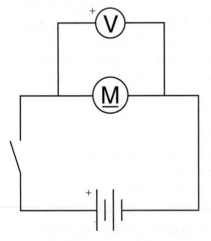

△ Fig. 6.9 A voltmeter can be connected across the motor after the rest of the circuit has been completed.

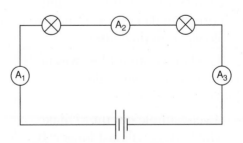

△ Fig. 6.10 In this series circuit, the current will be the same throughout the circuit so $A_1 = A_2 = A_3$.

In a parallel circuit, the current from the source is the sum of the currents in the separate branches of the circuit, so in Fig. 6.11 $A_1 = A_2 + A_3$.

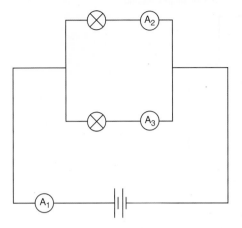

◁ Fig. 6.11 The current splits between the two branches of the parallel circuit so $A_1 = A_2 + A_3$.

Combining resistors

Two resistors can be replaced by a single resistor that has the same effect in the circuit. Calculating the value of the resistor needed depends on whether the original resistors are connected in series or parallel.

When the resistors R_1 and R_2 are in series, then the combined resistance, R_C, is:

$$R_C = R_1 + R_2$$

For resistors in parallel, the combined resistance is less than the value of either of the two resistors. This is because there are more paths for the charge to pass along, so the current is higher.

When the resistors R_1 and R_2 are in parallel, then R_C is:

$$\frac{1}{R_C} = \frac{1}{R_1} + \frac{1}{R_2} \text{ or } R_C = \frac{R_1 \times R_2}{R_1 + R_2}$$

In a series circuit, the potential difference across the battery equals the sum of the potential differences across each lamp.

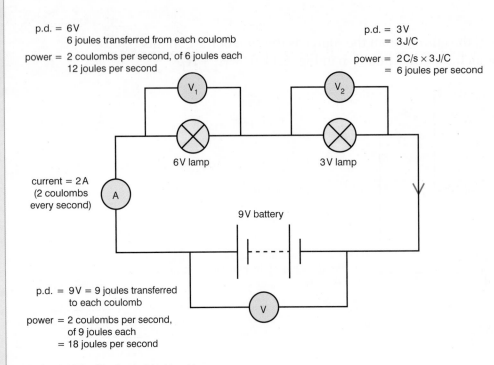

p.d. = 6V
6 joules transferred from each coulomb

power = 2 coulombs per second, of 6 joules each
12 joules per second

p.d. = 3V
= 3J/C

power = 2C/s × 3J/C
= 6 joules per second

current = 2A
(2 coulombs every second)

6V lamp

3V lamp

9V battery

p.d. = 9V = 9 joules transferred to each coulomb

power = 2 coulombs per second, of 9 joules each
= 18 joules per second

△ Fig. 6.12 In this circuit, $V = V_1 + V_2$.

WORKED EXAMPLES

1. Two resistors, $1000\,\Omega$ and $3000\,\Omega$, are connected in series. What is their combined resistance?

Write down the formula: $R_C = R_1 + R_2$

Substitute the values: $R_C = 1000 + 3000$

Work out the answer and add the unit: $R_C = 4000\,\Omega$

Note that the combined resistance is greater than the value of either of the two resistors.

2. EXTENDED Two resistors, $20\,\Omega$ and $30\,\Omega$, are connected in parallel. What is their combined resistance?

Write down the formula: $R_C = \dfrac{R_1 \times R_2}{R_1 + R_2}$

Substitute the values $R_C = \dfrac{20 \times 30}{20 + 30}$

Work out the answer and add the unit: $R_C = \dfrac{600}{50}\,\Omega$

$= 12\,\Omega$

QUESTIONS

1. In a series circuit, what can you say about the current at different points in the circuit?

2. Give two advantages of a parallel circuit in a house.

3. Two resistors, 100 Ω and 22 Ω, are connected in series. What is their combined resistance?

4. EXTENDED Two resistors, 147 kΩ and 220 kΩ, are connected in parallel. What is their combined resistance?

End of topic checklist

Key terms

parallel circuit, series circuit

During your study of this topic you should have learned:

◯ That the current at every point in a series circuit is the same.

◯ How to calculate the combined resistance of two or more resistors in series.

◯ That, for a parallel circuit, the current from the source is larger than the current in each branch.

◯ That the combined resistance of two resistors in parallel is less than that of either resistance by itself.

◯ EXTENDED How to calculate the combined resistance of two resistors in parallel.

◯ The advantages of connecting lamps in parallel in a circuit.

◯ EXTENDED That the sum of the potential differences across the components in a series circuit is equal to the total p.d. across the supply.

◯ EXTENDED That the current from the source is the sum of the currents in the separate branches of a parallel circuit.

End of topic questions

Note: The marks given for these questions indicate the level of detail required in the answers. In the examination, the number of marks given to questions like these may be different.

1. Look at the following circuit diagrams. They show a number of ammeters and in some cases the readings on these ammeters. All the lamps are identical.

a) For circuit X, what readings would you expect on ammeters A_1 and A_2?

(2 marks)

b) For circuit Y, what readings would you expect on ammeters A_4 and A_5?

(4 marks)

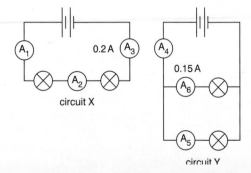

2. Look at the circuit diagram. It shows how three voltmeters have been added to the circuit. What reading would you expect on V_1?

(2 marks)

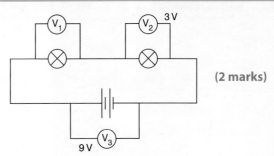

3. Look at the circuit diagram and answer the following questions:

a) What supplies the voltage in this circuit? **(1 mark)**

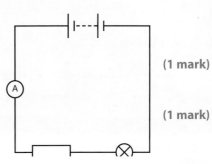

b) What will happen to the current if another cell is added? **(1 mark)**

c) If the resistance of the circuit is increased, what will happen to the current in the circuit? **(2 marks)**

d) If the ammeter shows 0.3 A and the resistor is 5 Ω, what is the voltage of the cell? **(2 marks)**

4. Copy and complete the following table.

Potential difference/V	Current/A	Resistance/Ω
	0.15	2
6	0.2	
	0.5	12
12	3	
240		18.5

(5 marks)

5. **EXTENDED** In the circuits shown here, what you would expect to read on each ammeter and voltmeter? All lamps are the same, and each cell produces 1.5 V. What *could* happen to these values if the bulbs had different resistances?

(12 marks)

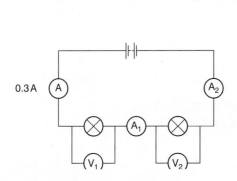

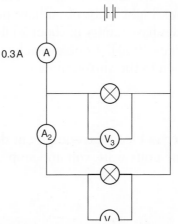

Electrical energy

△ Fig. 6.13 Electric lights have transformed the world.

INTRODUCTION

Electric circuits are particularly useful because they transfer energy. We have three different ways to calculate how much energy is involved – all three methods are useful in different situations. First, we may simply want to know the total energy transferred, which we would measure in joules. Second, we might want to know how much energy is transferred by each unit of charge, for which we would use p.d. or e.m.f. In this section we will meet a third method, how quickly the energy is transferred – the concept of power.

KNOWLEDGE CHECK

✓ Be able to measure currents and potential differences in circuits.
✓ Know that electric circuits can transfer energy.
✓ Be able to describe the energy transfers in a range of circuits.

LEARNING OBJECTIVES

✓ **EXTENDED** Be able to calculate the power transfer in electric circuits using $P = IV$.

✓ **EXTENDED** Be able to calculate the energy transferred in electric circuits using $E = IVt$.

ELECTRICAL ENERGY

All electrical equipment has a **power rating**, which indicates how many joules of energy are supplied each second. The unit of power used is the **watt** (W). Light bulbs often have power ratings of 60 W or 100 W. Electric kettles have ratings of about 2 kilowatts (2 kW = 2000 W). A 2 kW kettle supplies 2000 J of energy each second to the circuit components and then to the surroundings.

EXTENDED

The power of a piece of electrical equipment depends on the voltage and the current. The units watt, volt and amp are defined as follows:

- 1 watt = 1 J/s
- 1 volt = 1 J/C
- 1 amp = 1 C/s

From these definitions, we can see that 1 watt = 1 volt × 1 amp.

In other words: power = current × voltage

$$P = I \times V$$

where:
P = power in watts (W)

I = current in amps (A)

V = potential difference in volts (V)

You can use the triangle in Fig. 6.14 to help you to rearrange this equation.

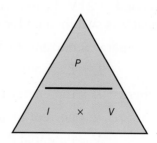

△ Fig. 6.14 Equation triangle for power, current and voltage to help with rearranging the equation.

WORKED EXAMPLES

1. What is the power of an appliance when a current of 7A is obtained from a 230V supply?

Write down the formula in terms of P: $P = V \times I$

Substitute the values: $P = 230 \times 7$

Work out the answer and write down the unit: $P = 1610$

2. An electric oven has a power rating of 2 kW. What will the current be when the oven is used with a 230V supply?

Write down the formula in terms of I: $I = P/V$

Substitute the values: $I = 2000/230$

Work out the answer and write down the unit $I = 8.7$ A

QUESTIONS

1. An appliance has 2A of current in it and operates at 110V. What is its power rating?

2. A 60W lamp has a current of 5A in it. At what voltage is it operating?

3. A 25W lamp is designed to be used with a voltage of 230V. Calculate the current in it.

Energy, current, voltage and time

If you switch on an electric kettle for 1 minute or a room heater for 5 minutes, then you can measure the temperature increase of the water in the kettle or of the room. The temperature will increase because energy has been given to the water or to the room. Energy is measured in **joules** (J). A heater rated at 1 watt will give out 1 joule of heat each second. A 2 kW heater will give out 2000 joules per second, and if the heater is switched on for 4 s, it will give out 8000 J.

energy = current × voltage × time

$$E = I \times V \times t$$

where: E = the energy transferred in joules (J)

I = current in amperes (A)

V = potential difference in volts (V)

t = time in seconds (s)

WORKED EXAMPLE

Calculate the energy transferred when a 12 V motor, running at a current of 0.5 A, is left on for 5 minutes.

Write down the formula: $E = I \times V \times t$

Substitute the values: $E = 0.5 \times 12 \times 300$

(Remember that the time *must* be in seconds.)

Work out the answer and write down the unit: $E = 1800 \text{ J}$

QUESTIONS

1. A laptop charger is designed for a country where the mains voltage is 230 V. The owner takes the charger to a country such as the USA, where the mains voltage is 110 V. Will they be able to charge their laptop?

2. How much energy is transferred when a current of 3 A flows in a circuit with a voltage of 12 V for 1 minute?

3. A heater which runs on 12 V transfers 4800 J of energy in 2 minutes. What is the current in the heater?

4. A bulb runs on a voltage of 110 V and has a current of 0.1 A in it. It transfers 2400 J of energy in a given time. What is this time?

5. A lamp transfers 24 J of energy and draws a current of 2 A for 1 s. What voltage is it operating at?

END OF EXTENDED

End of topic checklist

Key terms

joule, power rating, watt

During your study of this topic you should have learned:

○ How to draw and interpret circuit diagrams containing sources, switches, resistors (fixed and variable), lamps, ammeters, voltmeters, magnetising coils, transformers, bells, fuses and relays.

○ That the current at every point in a series circuit is the same.

○ How to give the combined resistance of two or more resistors in series.

○ That, for a parallel circuit, the current from the source is larger than the current in each branch.

○ That the combined resistance of two resistors in parallel is less than that of either resistor by itself.

○ About the advantages of connecting lamps in parallel in a lighting circuit.

○ **EXTENDED** How to use the fact that the sum of the potential differences across the components in a series circuit is equal to the total p.d. across the supply.

○ **EXTENDED** How to use the fact that the current from the source is the sum of the currents in the separate branches of a parallel circuit.

○ **EXTENDED** How to calculate the effective resistance of two resistors in parallel.

○ **EXTENDED** How to use the equations $P = IV$ and $E = IVt$.

End of topic questions

Note: The marks given for these questions indicate the level of detail required in the answers. In the examination, the number of marks given to questions like these may be different.

1. **EXTENDED** An appliance has a power rating of 1400 W. The potential difference of the mains is 230 V. Calculate the approximate current. **(3 marks)**

2. **EXTENDED** A lamp has a power rating of 11 W and runs from a supply of 230 V. What is the current in the lamp? **(2 marks)**

3. **EXTENDED** An appliance runs from a 110 V supply. It has a current of 3.2 A in it. What is its power rating? **(2 marks)**

4. **EXTENDED** An appliance has a current of 2.7 A in it and has a power rating of 300 W. What is the voltage of the supply? **(2 marks)**

Dangers of electricity

INTRODUCTION

Electricity is a clean and effective method of generating heat and movement. When a domestic appliance (such as a washing machine or a fridge) is switched on, a circuit is completed between the local substation and the appliance. Electrical energy travels from the substation to the appliance through the 'live' and 'neutral' wires. Some appliances have a third wire, the 'earth' wire. You will also meet the American word 'ground' instead of 'earth'. This wire does not normally carry any current, but it is there for safety.

△ Fig. 6.15 Electricity travels from power stations to our homes along the wires on pylons like these.

KNOWLEDGE CHECK

✓ Know some advantages of using mains electricity to transfer energy.
✓ Know some safety precautions to take when dealing with mains electricity.

LEARNING OBJECTIVES

✓ State the hazards of damaged insulation, overheating of cables and damp conditions.
✓ State that a fuse protects a circuit.
✓ Show an understanding of the use of fuses and choose the appropriate fuse ratings.

ELECTRICAL HAZARDS

Electricity can cause hazards in domestic situations. Table 6.2 gives some examples.

Hazard	Possible consequences
Frayed cables	Wiring can become exposed
Long trailing cables	These might cause a trip or a fall
Damaged plugs	Wiring can become exposed
Water around sockets	Water conducts electricity, so can connect a person into the mains supply
Pushing metal objects into sockets	This connects the holder to the mains supply and is likely to be lethal
Overloading of sockets	Causes too high a current, which might melt the insulation and cause a fire
Long, coiled cable to an electric heater	Cable can heat up because of the coiling and start a fire

△ Table 6.2 Examples of domestic electrical hazards.

If there is a fault in an electrical appliance, it could take too much electrical current. This might make the appliance itself dangerous, or it could cause the flex between the appliance and the wall to become too hot and start a fire.

Insulation and fuses

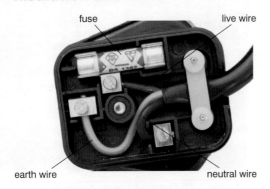

△ Fig. 6.16 A three-pin plug has a built-in safety device, the fuse.

There are several ways to make appliances safer to use and protect the user if a fault should develop. In some countries, a **fuse** is fitted into the plug of the appliance. The fuse fits between the brown live wire and the pin. The brown live wire and the blue neutral wire carry the current. The green and yellow striped **earth wire** is needed to make metal appliances safer.

The laws for the safe use of electricity are constantly being improved by governments, and electricians learn to work to the latest standards. The most important aids to the safe use of electricity are **insulation** and fuses.

Insulation these days is generally a plastic, such as PVC, which is used to cover the copper wires. This prevents them from touching each other, and also prevents the operator from touching them. In parts of appliances where the temperature goes above 100 °C, other plastics, glass or ceramic are used.

The electric current usually has to pass through a fuse before it reaches the appliance. If there is a sudden surge in the current, the wire in the fuse will heat up and melt – it 'blows'. This breaks the circuit and stops any further current flowing in.

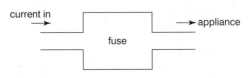

△ Fig. 6.17 How a fuse protects an appliance.

REMEMBER

Many students misunderstand the role of a fuse. It does not 'provide' current, nor does it 'allow' a certain amount of current to go through. It is just a wire that melts if the current gets too high.

In all houses with mains electricity, there is a distribution box (sometimes called a consumer unit) that takes all of the electricity for the house and sends it to the different rooms. In old houses this box may still contain fuses, but in modern installations the box has miniature circuit-breakers, often known as MCBs.

Where a fuse is fitted to a plug, it must have a higher current rating than the appliance needs, but should have the smallest current rating available above this. The most common ratings for plug fuses are 3A, 5A and 13A. Any electrical appliance with a heating element in it should be fitted with a 13A fuse. An appliance working at 3.5A should have a 5A fuse.

Metal-cased appliances, such as washing machines or electric cookers, must have an earth wire as well as a fuse. If the live wire works loose and comes into contact with the metal casing, the casing will become live and the user could be electrocuted.

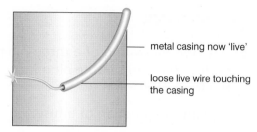

△ Fig. 6.18 A loose live wire can be dangerous if the metal casing of an appliance is not 'earthed'.

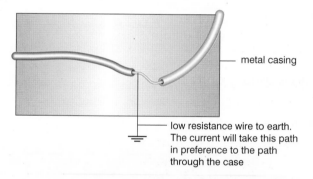

△ Fig. 6.19 An earth wire provides a path for current to flow to ground.

This low resistance means that a large current passes from the live wire to earth, causing the fuse to melt and break the circuit. If the earth wire is not fitted correctly, or if it has broken, the appliance will be extremely dangerous! If there is any doubt about the earthing of the appliance, or of the whole house, it must be checked by an electrician.

Appliances that are made with a plastic casing such as kettles do not need an earth wire. The plastic is an insulator and so can never become live. Appliances like this are said to be **double insulated**.

The earth wire provides a very low resistance route to the 0V earth. This low resistance means that a large current passes from the live wire to earth, causing the fuse to melt and break the circuit. This disconnects the appliance from the live connection, making it safe to touch (Fig. 6.20).

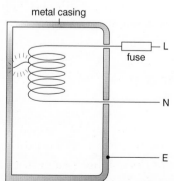

△ Fig. 6.20 The earth wire and fuse work together to make sure that the metal outer casing of this appliance can never become live and electrocute someone.

In some situations people may be unexpectedly exposed to electricity: for example, using an electric drill, especially drilling into a wall with hidden power cables, or using power tools out of doors, perhaps in wet conditions. In these cases, a special type of circuit breaker called a residual current **circuit breaker** (RCCB) must be used in the power socket on the wall. If any of the electricity starts to leak, through a **short circuit** (for example, because the device has got wet), the RCCB will turn off the power in 30 ms or less. The RCCB cannot be guaranteed to save the user's life, but it gives them a much better chance of surviving.

QUESTIONS

1. Describe the wiring of a three-pin plug. You should explain what each of the wires in the plug is connected to and the colour of the insulation.

2. Explain the function of a fuse.

3. A student wants to run an appliance that requires a current of 6 A. He chooses a fuse of 5 A 'because it's the nearest available'. Explain why this is not a good choice.

4. The earth wire connection to the ground is usually quite a thick piece of copper wire. Explain why.

5. Explain why appliances with plastic casing do not need to be earthed.

End of topic checklist

Key terms

double insulated, circuit breaker, earth wire, fuse, insulator, short circuit

During your study of this topic you should have learned:

○ About the hazards of:

- damaged insulation
- overheating of cables
- damp conditions.

○ That a fuse protects a circuit.

○ About the use of fuses and choose appropriate fuse ratings.

End of topic questions

Note: The marks given for these questions indicate the level of detail required in the answers. In the examination, the number of marks given to questions like these may be different.

1. **a)** A hairdryer works on 230 V mains electricity and takes a current of 4 A. Calculate the power of the hairdryer. **(2 marks)**

   b) In some countries it is illegal to have power sockets in a bathroom to stop you using electrical devices such as hairdryers near the wash basin or bath. Why would it be foolish to use a hairdryer near to water? **(2 marks)**

2. Why should a fuse always be connected in series with the live wire? **(1 mark)**

3. An appliance has a power rating of 1400 W. The potential difference of the mains is 230 V. Calculate the approximate current. Explain what size standard fuse you would use. **(3 marks)**

4. What rating of fuse would you use in a microwave of power rating 800 W with:

   a) a 240 V mains supply **(2 marks)**

   b) a 120 V mains supply? **(2 marks)**

5. What potential problems might there be with the fuse in question 4? **(2 marks)**

Doing well in examinations

INTRODUCTION

Examinations will test how good your understanding of scientific ideas is, how well you can apply your understanding to new situations and how well you can analyse and interpret information you have been given. The assessments are opportunities to show how well you can do these things.

To be successful in exams you need to:

✓ have a good knowledge and understanding of science

✓ be able to apply this knowledge and understanding to familiar and new situations

✓ be able to interpret and evaluate evidence that you have just been given.

You need to be able to do these things under exam conditions.

OVERVIEW

Ensure you are familiar with the structure of the examinations you are taking. Consult the relevant syllabus of the year you are entering your examinations for details of the different papers and the weighting of each, including the papers to test practical skills. Your teacher will advise you of which papers you will be taking.

You will be required to perform calculations, draw graphs and describe, explain and interpret ideas and information about physics. In some of the questions the content may be unfamiliar to you; these questions are designed to assess data-handling skills and the ability to apply physical principles and ideas in unfamiliar situations.

ASSESSMENT OBJECTIVES AND WEIGHTINGS

For the Cambridge IGCSE Combined Science examination, the assessment objectives and weightings are as follows:

✓ A01: Knowledge with understanding (50%)

✓ A02: Handling information and problem solving (30%)

✓ A03: Experimental skills and investigations (20%).

The types of question in your assessment fit the three assessment objectives shown in the table.

Assessment objective	Your answer should show that you can ...
AO1 Knowledge with understanding	Recall, select and communicate your knowledge and understanding of science.
AO2 Handling information and problem solving	Apply skills, including evaluation and analysis, knowledge and understanding of scientific contexts.
AO3 Experimental skills and investigations	Use the skills of planning, observation, analysis and evaluation in practical situations.

EXAMINATION TECHNIQUES

To help you to work to your best abilities in exams, there are a few simple steps to follow.

Check your understanding of the question

✓ **Read the introduction to each question carefully before moving on to the question itself**.

✓ Look in detail at any **diagrams, graphs** or **tables**.

✓ Underline or circle the **key words** in the question.

✓ **Make sure you answer the question that is being asked** rather than the one you wish had been asked!

✓ Make sure that you understand the meaning of the '**command words**' in the questions.

EXAMPLE

✓ '**Give**', '**state**' and '**name**' are used when recall of knowledge is required. For example, you could be asked to give a definition, or provide the best answers from a list of options.

✓ '**Describe**' is used when you have to give the main feature(s) of, for example, a biological process or structure.

✓ '**Explain**' is used when you have to give reasons, for example, for some experimental results or a scientific fact or observation. You will often be asked to 'explain your answer', that is, give reasons for it.

✓ '**Suggest**' is used when you have to come up with an idea to explain the information you're given – there may be more than one possible answer, no definitive answer from the information given, or it may be that you will not have learned the answer but have to use the knowledge you do have to come up with a sensible one.

✓ '**Calculate**' means that you have to work out an answer in figures.

✓ '**Plot**' and '**Draw a graph**' are used when you have to use the data provided to produce graphs and charts.

Check the number of marks for each question

✓ Look at the **number of marks** allocated to each question.

✓ Look at the **space provided** to guide you as to the length of your answer.

✓ Make sure you include at least as many points in your answer as there are marks.

✓ Do not use any more space than the space that has been provided in the examination paper.

REMEMBER

Beware of continually writing too much because it probably means you are not really answering the questions. Do not repeat the question in your answer.

Use your time effectively

✓ Don't spend so long on some questions that you don't have time to finish the paper.

✓ Regularly check how much time you have left.

✓ If you are really stuck on a question, leave it, finish the rest of the paper and come back to it at the end.

✓ Even if you eventually have to guess at an answer, you stand a better chance of gaining some marks than if you leave it blank.

ANSWERING QUESTIONS
Multiple choice questions

✓ Select your answer by placing a cross (not a tick) in the box. In the final exam multiple choice questions will be answered on a separate answer sheet using a pencil.

Short-answer and structured questions

✓ In short-answer questions, **don't write more than you are asked for**.

✓ You may not gain any marks, even if the first part of your answer is correct, if you've written down something incorrect later on or something that contradicts what you've said earlier. This might give the impression that you haven't really understood the question or are guessing.

✓ Always use **clear scientific language**. In some short answer questions one or two words may be sufficient, but in longer questions you should aim to use full sentences.

✓ Present the information in a logical sequence.

✓ Don't be afraid to use **labelled diagrams** or **flow charts** if it helps you to show your answer more clearly.

Questions with calculations

✓ **In calculations, always show your working**.

✓ Even if your final answer is incorrect you may still gain some marks if part of your attempt is correct.

✓ If you just write down the final answer and it is incorrect, you will get no marks at all.

✓ Write down your answers to as many **significant figures** as are used in the numbers in the question (and no more). If the question doesn't state how many significant figures, then a good general rule is to quote 3 significant figures.

✓ Don't round off too early in calculations with many steps.

✓ You may also lose marks if you don't use the correct **units**. In some questions, the units will be mentioned (for example, calculate the mass in grams) or the units may also be given on the answer line. If numbers you are working with are very large, you may need to make a conversion (for example, convert joules into kilojoules, or kilograms into tonnes).

Finishing your exam

✓ When you've finished your exam, **check through** your paper to make sure you've answered all the questions.

✓ Check that you haven't missed any questions at the end of the paper or turned over two pages at once and missed questions.

✓ Cover over your answers and read through the questions again and check that your answers are as good as you can make them.

You will be asked questions on investigative work. It is important that you understand the methods used by scientists when carrying out investigative work.

More information on carrying out practical work and developing your investigative skills are given in the next section.

Exam-style questions

Note: The questions, sample answers and marks in this section have been written by the authors as a guide only. The marks given for these questions indicate the level of detail required in the answers. In the examination, the number of marks given to questions like these may be different.

Sample student answer

Question 1

The diagram shows the structure of a part of an organism called *Mucor*.

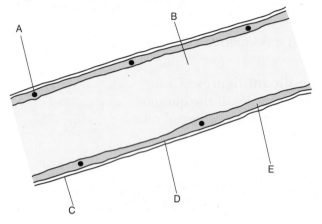

a) i) Name each of the parts A–E. Use the words in the box below. (5)

cytoplasm	membrane	nucleus
starch grain	vacuole	wall

A	nucleus	✓	①
B	vacuole	✓	①
C	membrane	✗	
D	wall	✗	
E	cytoplasm	✓	①

TEACHER'S COMMENTS

a) i) It is important to know the features of different groups of organisms and be able to label these.

A Correct – nucleus.

B Correct – vacuole.

C Incorrect – the outer layer of the hypha is the wall.

D Incorrect – the membrane is the next layer within the wall, pushed up against the wall.

E Correct – cytoplasm.

ii) Correct – the hyphae of moulds such as *Mucor* have a large central vacuole.

The answer is correct in that the hyphae have a wall, but this cannot be described as a 'cell wall' as the hyphae are not divided into cells. To be completely correct, the student simply had to repeat this information from part a) i).

iii) Correct. The student could have chosen from a selection of features but was only asked to give one for the mark.

b) The answer is correct, but the student could have added that food is digested outside the mould, and that this process is called saprotrophic nutrition.

c) Correct – the mycelium of *Mucor* has many nuclei distributed through the cytoplasm, with no cell boundaries; yeast is made up of single cells.

ii) State **two** features the organism has in common with plants. (2)

Mucor has a large central vacuole. ✔ ①

Mucor has a wall. ✔ ①

iii) **EXTENDED** Give one feature that tells you that *Mucor* is a fungus. (1)

It has many nuclei lying in the cytoplasm. ✔ ①

b) EXTENDED Describe how moulds such as *Mucor* feed. (4)

Mucor lives on its food, e.g. bread, and secretes enzymes into it. ✔ ①

The food is absorbed over the surface of the fungus. ✔ ①

c) EXTENDED Yeast is another type of fungus. State one major difference between *Mucor* and yeast. (1)

Yeast is single-celled. ✔ ①

(Total 13 marks)

Question 2

This question is about the characteristics of living things.

a) Copy and complete the sentences by writing the most appropriate word in each space.

Use only words from the box.

detect	development	energy	gravity
growth	light	location	nutrition
position	respiration	respond	sensitivity

.... is the taking in of substances needed for energy and for .... and
....

.... is a series of reactions that take place in living cells to release ....
from nutrient molecules so that cells can use this to keep them alive.

.... is the ability to .... and .... to changes in external and
internal conditions.

Movement causes a change in .... of the organism. In animals, this involves
their entire bodies. Plants often move parts of their body in response to external
stimuli, such as .... and ....   **(12)**

b) Define the term excretion.   **(3)**

c) Which of the characteristics shown by living things are shown by a motor car?   **(4)**

(Total 19 marks)

EXTENDED **Question 3**

Not everyone agrees that viruses should be called living things. Use your knowledge of
viruses and the characteristics of living things to discuss whether or not viruses should
be classed as living.   **(4)**

(Total 4 marks)

Question 4

Most living organisms are made up of cells.

a) Copy and complete the sentences by writing the most appropriate word in
each space.

The .... holds the cell together and controls substances entering
and leaving the cell. The .... is the jelly-like substance contained
within the cell. It is where many different chemical processes occur.

The .... is the control centre of the cell and contains genetic
material as .... These control how a cell grows and works.

Plant cells also have features that are not found in animal cells. These include
the .... , made of .... , which gives the cell extra
support and defines its shape.

Many plant cells have a large central, permanent .... that contains
cell sap. It is used for storage of some chemicals, and to support the shape of the cell.

Plant cells exposed to the light contain .... These contain the
green pigment .... , which absorbs the light energy that plants use
for the process of ....   **(10)**

b) The levels of organisation in multicellular organisms include cells, tissues, organs and systems.

State whether each of the following structures is a cell, tissue or organ.

Structure	Cell	Tissue	Organ
Blood			
Brain			
Liver			
Muscle			
Ovum			
Skin			
Sperm			

(8)

(Total 18 marks)

Question 5

The light micrograph shows a cell from the liver of a human.

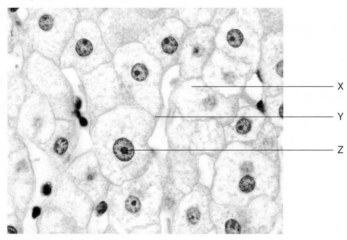

a) Identify the three structures labelled X, Y and Z. (3)

b) State the functions of the three structures. (6)

(Total 9 marks)

EXTENDED **Question 6**

A Biology student set up an investigation in which three cubes of different sizes were cut from an agar jelly block. The agar jelly contained a red indicator that turns blue in the presence of alkali.

The cubes were placed in an alkali. The student measured the time taken for the cubes to turn completely blue.

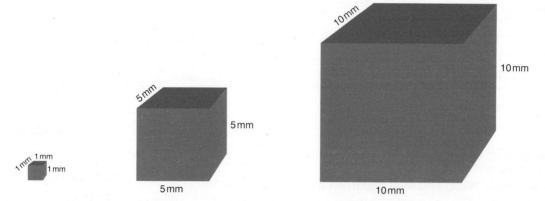

a) What is the name of the process that causes the alkali to penetrate the agar jelly? (1)

b) EXTENDED The student used cubes of three dimensions:

Dimensions of cube, mm
$1 \times 1 \times 1$
$5 \times 5 \times 5$
$10 \times 10 \times 10$

For each cube, calculate its:

  i) surface area

  ii) volume

iii) surface area : volume ratio. (9)

Dimensions of cube/mm	i) Surface area of cube/mm²	ii) Volume of cube/mm³	iii) Surface area : volume ratio
$1 \times 1 \times 1$			
$5 \times 5 \times 5$			
$10 \times 10 \times 10$			

c) EXTENDED Explain the relationship between surface area and volume as the cube increases in size. (1)

d) EXTENDED During the biology lesson, the alkali penetrated the two smaller cubes, but by the end of the lesson, the 10 mm × 10 mm × 10 mm cube had still not turned completely blue.

   i) Explain these results.   (3)

   ii) Explain what implications this has for organisms of increasing size.   (3)

                                                   (Total 17 marks)

Question 7

A student cut a number of cylinders from a potato and weighed them. These were placed in sucrose solutions of different concentrations.

After one hour, the cylinders were removed, blotted dry and reweighed. The student calculated the percentage change in mass for each cylinder. The results are shown in the table.

Concentration of sucrose/g per cm³	Percentage change in mass of potato cylinders				Average percentage change in mass
	Experiment 1	Experiment 2	Experiment 3	Experiment 4	
0.0	+31.4	+33.7	+31.2	+32.5	
0.2	+20.9	+22.2	+22.8	+21.3	
0.4	−2.7	−1.8	−1.9	−2.4	
0.6	−13.9	−12.8	−13.7	−13.6	
0.8	−20.2	−19.7	−19.3	−20.4	
1.0	−19.9	−20.3	−21.1	−20.3	

a) Calculate the average percentage changes in mass for each of the sucrose concentrations.   (6)

b) **i)** Draw a graph of these results. Join the points with a line of best fit.   (4)

   ii) At what concentration of sucrose was there no net movement of water?   (2)

   iii) Describe the changes in mass over the range of sucrose concentrations.   (3)

   iv) State the process involved in these changes in the potato cylinders.   (1)

                                                   (Total 16 marks)

Question 8

The table below lists some molecules that are important biologically.

a) Give the units that each one of the following biological molecules is made up of.

Biological molecule	Units that make up the molecule	
Glycogen		(1)
Fats		(2)
Proteins		(1)
Starch		(1)

b) Describe a test that can be carried out in the laboratory for the following carbohydrates:

i) glucose (5)

ii) starch. (4)

(Total 14 marks)

EXTENDED Question 9

A company that produces enzymes publishes information sheets of their performance. The graphs below show the performance of an enzyme, pectinase, at different pHs and temperatures.

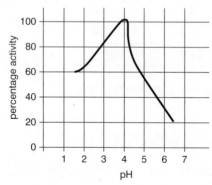
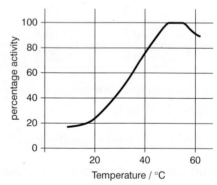

a) Describe and explain the effect of:

i) pH (6)

ii) temperature, on the activity of the enzyme pectinase. (5)

b) A student finds an information sheet on the effect of pH on the activity of a protease called papain, from the papaya plant.

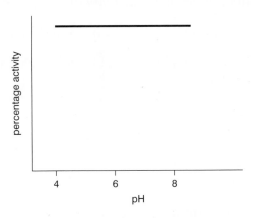

i) **EXTENDED** State two proteases produced by the human gut. (2)

ii) A student says that the graph shows that papain is unaffected by pH. Is the student correct? Explain your answer. (2)

(Total 15 marks)

EXTENDED Question 10

a) In humans, proteases are produced by the stomach, pancreas and small intestine. Copy the diagram and show the location of these organs. (3)

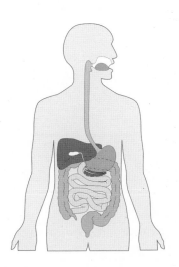

b) The graphs show the effect of pH on the activity of two proteases that break down proteins.

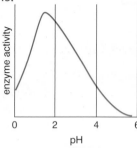

stomach protease

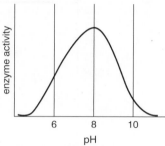

small intestine protease

i) EXTENDED What are the optimum pHs of stomach and small intestine proteases? (2)

ii) Describe an experiment to investigate the effect of temperature on the breakdown of a named food molecule. (5)

(Total 10 marks)

a) i) Correct. As light intensity increases, the rate of photosynthesis increases, because light energy supplies energy for the process of photosynthesis. At a certain point, the graph levels off, so any further increase in light intensity after that point will result in no further increase in photosynthesis. At this point, some other factor must be limiting, for example, carbon dioxide, and preventing any further increase.

ii) One mark has been given for the sketch given. The student has not responded to the second part of the question – explain the shape of the graph you have drawn. To gain a second mark, the student needed to explain that as carbon dioxide was a factor that was limiting the rate of photosynthesis where the graph levels off, in a higher concentration of carbon dioxide, the graph will continue to a higher point (i.e. a higher rate of photosynthesis) until it again levels off. The third mark would have been given for stating that at this point, with light and carbon dioxide being available, another factor (e.g. temperature) must

Exam-style questions
Sample student answer

EXTENDED **Question 1**

Plants respond to the light available.

a) The graph below shows the effect of light intensity on photosynthesis in a single-celled plant.

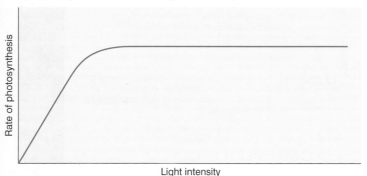

Light intensity

i) Describe and explain the effect of light intensity on the plant. **(4)**

> Increasing light intensity increases the rate of photosynthesis up to a certain point. ✓ ①
>
> This is because light energy is needed for photosynthesis. ✓ ①
>
> The graph stops getting steeper because increasing the light intensity can't increase the rate of photosynthesis any more. ✓ ①
>
> This is because another factor is limiting the rate of photosynthesis. ✓ ①

The investigation was also carried out in a high concentration of carbon dioxide.

ii) Sketch a graph of what you would expect so you can compare it with the graph above. Explain the shape of the graph you have drawn. **(3)**

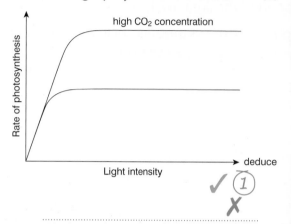

be preventing any further increase in the rate of photosynthesis.

b) The student has correctly identified that there is no significant difference between the amount of auxin in the plants in the light, the dark or those illuminated on one side. A further mark is gained by correctly stating that this shows that light has no effect on the production of auxin.

To gain further marks, the student needed to provide a more in-depth explanation using the results, such as: 'This shows that light has no effect on the production of auxin. In the plant illuminated from one side, about 71% of the auxin in the plant is on the dark side, so as the total auxin was unaffected by light, the auxin must have been redistributed from the light to dark side.'

b) A scientist investigating the response of plants to light placed:

- one group in the light, given even illumination
- one group of plants in the dark, and
- one group exposed to light from one side.

The plants were in an atmosphere of radioactive carbon dioxide, and after five hours, the amount of radioactive auxin in the area below the shoot tip was measured. The scientist's results are shown below.

	Plants in the light	Plants in the dark	Plants exposed to light from one side	
			Dark side	Lighted side
Total radioactive auxin/ counts per minute	2985	3004	2173	878

Explain fully what these results show about the effect of light on auxin in the plants. **(4)**

There is not much difference between the amount of auxin in the plants in the dark, the plants in the light or the plant that was half in the light and half in the dark. ✓ ①

This shows that altering the levels of light does not make any difference to the production of auxin. ✓ ①

(Total 11 marks)

EXTENDED Question 2

The leaf is the main organ of photosynthesis.

a) Write a word and symbol equation for photosynthesis. (5)

b) EXTENDED Explain how the leaf is adapted to exchanging gases required for photosynthesis. (5)

c) Chemical substances in a plant are transported in the xylem and phloem.

Copy and complete the table below.

	Phloem	Xylem
Substances transported	(2)	(2)
Substances are transported:		
from	(1)	(1)
to	(1)	(1)

(8)

(Total 18 marks)

Question 3

The diagram shows a section through a leaf.

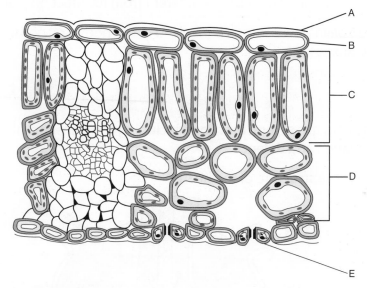

a) Name parts A–E shown in the diagram. (5)

b) The leaf is the main organ of photosynthesis. Write a word equation for photosynthesis. (3)

c) Define the term *transpiration*. (3)

d) Describe a technique used to investigate the effect of temperature on transpiration rate. (7)

(Total 18 marks)

EXTENDED Question 4

a) The diagram below shows a section of a plant root surrounded by soil particles.

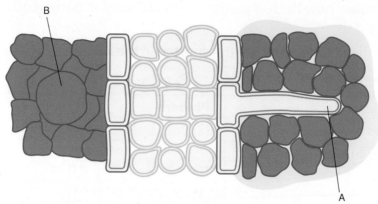

   i) Identify the parts of the plant root, A and B. (2)

 ii) EXTENDED Explain the importance of water potential in the uptake and transport of water by plant roots. (7)

b) The effect of osmosis on animal cells is different from its effect on plant cells.

   i) EXTENDED Explain how osmosis is involved in the body's response to a cholera infection. (5)

(Total 14 marks)

Question 5

The circulatory system has several functions, including the transport of substances, temperature regulation and defence.

The diagram shows the structure of the heart.

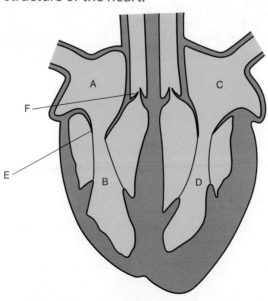

a) Name the chambers of the heart, A, B, C and D. (4)

b) Copy and complete the table for each of the different components of the blood.

Component of blood	Function
Red blood cells	
White blood cells	
Platelets	
Plasma	

(8)

(Total 12 marks)

Question 6

a) Describe how the back flow of blood in the heart is prevented. (4)

b) EXTENDED Explain why the wall of the left ventricle is four times as thick as the wall of the right ventricle. (4)

(Total 8 marks)

Question 7

a) Yeast, a fungus, is used in the production of biofuel.

   i) Give the word equation for the production of ethanol by yeast. (2)

   ii) Describe how yeast cells are involved in the process. (1)

b) Compare the energy produced by yeast undergoing anaerobic respiration with yeast undergoing aerobic respiration. (1)

(Total 4 marks)

Question 8

This question is about plants' responses towards stimuli.

Copy and complete the sentences by writing the most appropriate word in each space.

Growth in response to the direction of light is called ....
If the growth is towards light, it is called .... ,
as shown by plant ....

Growth in response to gravity is called .... Plant roots are
.... This response helps the plant
.... to grow .... , so the plant can
obtain the .... it needs. (Total 8 marks)

Question 9

a) The passage below describes the process of sexual reproduction.

Use suitable words to complete the sentences in the passage.

Sexual reproduction is the most common method of reproduction for the majority of larger organisms, including almost all animals and plants. To produce a new organism, two .... fuse. This process is known as .... **(2)**

Usually, sexual reproduction involves .... parent organisms of the same species. The .... formed is genetically different from each of the parents. **(2)**

b) **EXTENDED** Give one advantage and one disadvantage of:

   i) asexual reproduction. **(2)**

   ii) sexual reproduction. **(2)**

c) **i)** Label the diagram of the human female reproductive system. **(4)**

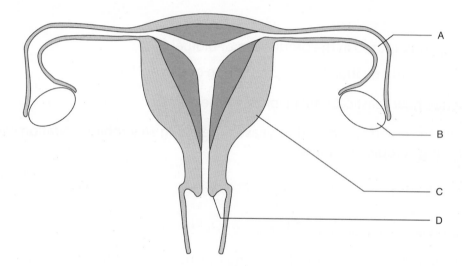

(Total 12 marks)

Question 10

Flowers are adapted to be pollinated by insects or by the wind.

a) Name the structures of an insect-pollinated flower shown in the diagram below.   **(10)**

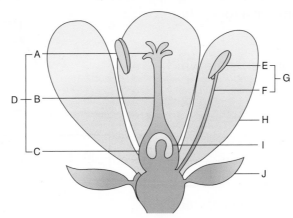

b) Explain how each of the structures is adapted for pollination in wind-pollinated flowers:

   i) petals (2)

 ii) stigma (2)

iii) stamens (2)

 iv) pollen grains. (2)

(Total 18 marks)

EXTENDED Question 11

This question is about reproduction.

a) The diagram shows a human sperm.

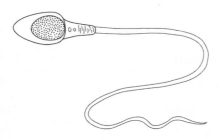

Explain how the sperm is adapted to fertilising a human egg. (4)

b) Explain the roles of hormones in controlling the menstrual cycle and preparing the uterus for a fertilised egg. (3)

(Total 7 marks)

Question 12

The food web below shows the relationship of some of the organisms on a rocky shore.

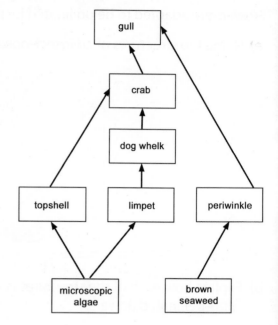

a) EXTENDED In the food web, state which organisms are:

  i) producers   (2)

  ii) primary consumers   (3)

  iii) secondary consumers.   (3)

b) In an ecosystem, the numbers of crabs is severely reduced.

  i) State **one** reason for the reduction of an organism in an ecosystem.   (3)

  ii) Describe the impact on dog whelks, limpets and gulls in the food web.   (5)

c) A student investigated the distribution of a species of brown seaweed and periwinkles down a rocky shore. Her results are shown here.

Distance below high water on seashore/ metres	Distribution of organisms	
	Observed distribution of brown seaweed	**Density of periwinkles/ mean number of periwinkles per m²**
0	Absent	0
10	Rare	16
20	Occasional	52
30	Abundant	156
40	Abundant	128
50	Occasional	44
60	Rare	12

Suggest two reasons for the distribution of the periwinkles.   (2)

(Total 18 marks)

Question 13

Nutrients are cycled in nature.

a) The passage below describes the stages in the carbon cycle.

Use suitable words to complete the sentences in the passage. (9)

.... from the atmosphere is converted to complex carbon compounds in .... by the process of .... This is often called carbon ....

Plants are then often eaten by ...., which build up their own complex carbon compounds.

The process of .... in both plants and animals, returns some of this carbon back to the atmosphere as ....

When organisms die, their bodies decay as they are worked on by .... Some of the complex carbon compounds are taken into the bodies of these organisms, where some may be converted to carbon dioxide during their ....

b) EXTENDED By what other process does carbon dioxide enter the air? (1)

(Total 10 marks)

EXTENDED Question 14

In an ecosystem, the following measurements were made.

Organism	Number in ecosystem	Biomass of organisms/g
Oak trees	1	500 000
Aphids	100 000	100
Ladybirds	200	10

a) Draw a food chain to illustrate the feeding relationships of the three organisms. (3)

b) EXTENDED In observations of the ecosystem, ladybirds were seen to be fed on by spiders, and spiders fed on by blackbirds.

  i) Draw a food chain to illustrate these feeding relationships. (1)

  ii) Explain fully why food chains longer than this are rare. (5)

(Total 9 marks)

Question 15

The table gives information on how the land area covered by forest has changed from 1990 to 2005.

Country	Area covered by forest/millions of hectares		Area of forest lost from 1990 to 2005/%
	1990	**2005**	
Bolivia	109.9	58.7	46.6
Brazil	851.5	477.7	
Colombia	113.9	60.7	
French Guiana	9.0	8.1	
Peru	125.5	68.7	
Suriname	16.3	14.8	
Venezuela	91.2	47.7	

Data from http://rainforests.mongabay.com/deforestation_alpha.html.

a) EXTENDED Calculate the area of forest lost for each country, as a percentage of the area in 1990. The first one has been done for you. (6)

b) During the time period 1990 to 2005, in which country is there:

　i) the greatest deforestation? (1)

　ii) the least deforestation? (1)

c) Suggest two reasons for deforestation. (2)

d) List the effects of deforestation. (5)

(Total 15 marks)

TEACHER'S COMMENTS

a) **i)** Correct.

ii) The answers given for carbon dioxide and nitrous oxide are detailed and correct.

For methane, the student has not appreciated that the graph has been drawn to the scale on the right hand axis, which ranges from 0–2000 ppb. This has meant that although the trends have been described, the values of methane concentration are incorrect.

It is important to check scales carefully when reading data from graphs. The answer should therefore be:

'The concentration of methane has shown a very slow, slight upward trend from 0 to around 1750, ranging from 650 ppb to around 750 ppb.'

'Then after a slight dip, a steep increase to around 1925 ppb in 2005.'

EXTENDED Question 1

This question is about the greenhouse effect and global warming.

a) The graph shows the concentration of greenhouse gases in the air from the Year 0 to 2000.

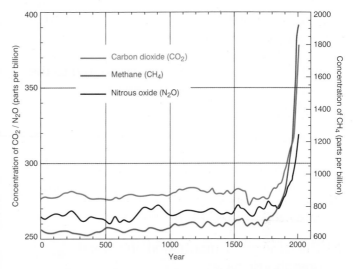

i) Which greenhouse gas was present in the highest concentration in the air in 2000? **(1)**

Methane ✓ ①

ii) Describe the trends in the changes of the concentration of each greenhouse gas from 0 to 2005. **(6)**

The concentration of carbon dioxide has been fairly stable at around 280 parts per billion from 0 to 1600. ✓ ①

Then, after a dip, it shows a steep increase to 380 ppb in 2005. ✓ ①

The concentration of nitrous oxide has shown a little fluctuation from 0 to around 1800, ranging from 265–275 ppb. ✓ ①

Exam-style questions continued

But there has been a steep increase to around 320 ppb in 2005. ✓ ①

The concentration of methane has shown a very slow, slight upward trend from 0 to around 1750, ranging from 255 ppb to around 260 ppb. ✗

But then a steep increase to around 390 ppb in 2005. ✗

iii) How has human activity contributed to the change in the concentration of carbon dioxide in the air? **(2)**

Carbon dioxide production has increased from the burning of fossil fuels in transport, heating and cooling, and in manufacture. ✓ ①

b) The table gives information on several greenhouse gases.

iii) The student has written a good answer for the contribution of the burning of fossil fuels to the increase in carbon dioxide. These all refer to the burning of fossil fuels, however, and the student could have picked up the second mark by referring to deforestation.

b) **i)** The student has picked up two marks, but for the third mark, has not mentioned the fact that sulfur hexafluoride has the longest lifetime – a greenhouse gas that's around for a shorter time will make less of a contribution to the greenhouse effect.

Gas	Chemical formula	Lifetime (years)	Global Warming Potential*
Carbon dioxide	CO_2	Variable	1
Methane	CH_4	12	21
Nitrous oxide	N_2O	114	310
CFC-11	CCl_3F	45	3800
CFC-12	CCl_2F_2	100	8100
Sulfur hexafluoride	SF_6	3200	23 900

*The **Global Warming Potential (GWP)** is a measure of how much heat a greenhouse gas traps in the atmosphere relative to that trapped by the same mass of carbon dioxide. A GWP is calculated over a time interval. The values in the table are for a 100-year time scale.

From: IPCC/TEAP (2005) *Special Report on Safeguarding the Ozone Layer and the Global Climate System: Issues Related to Hydrofluorocarbons and Perfluorocarbons* [Metz, B., et al. (eds.)]. Cambridge University Press.

i) Which greenhouse gas contributes most to global warming?
   Explain your answer. (3)

Sulfur hexafluoride ✓ ①

It has the highest GWP. ✓ ①

ii) Explain how greenhouse gases result in the greenhouse effect and global warming. (6)

Shortwave radiation from the Sun passes through the Earth's atmosphere and warms the ground. ✓ ①

The warmed Earth gives off longer wave radiation that is prevented from leaving the Earth by greenhouse gases in the atmosphere. ✓ ①

The trapping of the radiation leads to the Earth warming up, which is called the greenhouse effect. ✓ ①

The greenhouse effect is important, because without it, the temperature on the Earth would be 33 °C lower - the Earth would be uninhabitable. ✓ ①

But increases in greenhouse gases as a result in human activity is leading to a significant warming of the Earth called global warming. ✓ ①

(Total 18 marks)

EXTENDED **Question 2**

Modern technology is used to increase the yields of crop plants.

a) One method used to increase crop yields is to apply fertilisers. List the effects, in sequence, when nitrates are washed from the fields and pollute water. **(4)**

b) Pesticides are often applied to crops.

 i) Explain why pesticides are applied to crops. **(5)**

 ii) Describe two negative impacts of pollution by one type of pesticide. **(2)**

(Total 11 marks)

EXTENDED **Question 3**

This question is about eutrophication. The effect of sewage into a river was monitored over a number of years. The results are shown in the graphs.

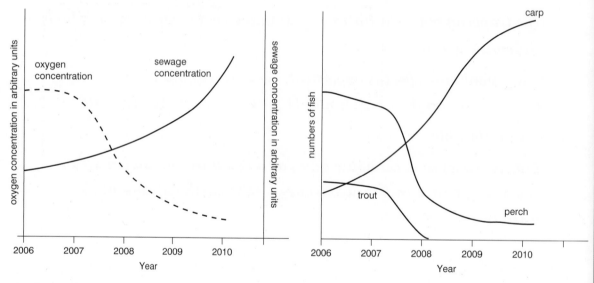

a) i) Describe the trends in sewage and oxygen concentration between 2006 and 2010. **(2)**

 ii) Explain why sewage had this effect on the oxygen concentration in the water. **(3)**

b) Describe and explain the effects of the changing oxygen concentration on fish populations. **(6)**

(Total 11 marks)

a) It is important to identify the states of matter:

A = gas, B = liquid, C = solid.

i) Correct – evaporation process.

ii) Correct – solidifying.

iii) Incorrect – should be 'AC' order because ethene is a gas, poly(ethene) a solid.

iv) Correct – equation shows solid → gases (sublimation).

Exam-style questions

Note: The questions, sample answers and marks in this section have been written by the authors as a guide only. The marks given for these questions indicate the level of detail required in the answers. In the examination, the number of marks given to questions like these may be different.

Sample student answer

Question 1

a) The diagrams show the arrangement of particles in the three states of matter.

Each circle represents a particle.

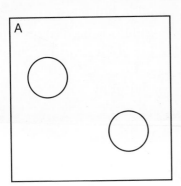 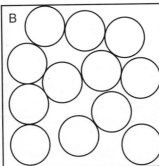 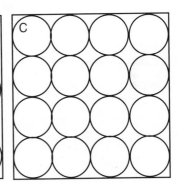

Use the letters A, B, and C to give the starting and finishing states of matter for each of the changes in the table. For the mark, both the starting state and the finishing state need to be correct.

Change	Starting state	Finishing state		
i) The formation of water vapour from a puddle of water on a hot day	B	A	✓	①
ii) The formation of solid iron from molten iron	B	C	✓	①
iii) The manufacture of poly(ethene) from ethene	B	A	✗	
iv) The reaction whose equation is ammonium hydrogen chloride(s) → ammonia(g) + hydrogen chloride(g)	C	A	✓	①

(4)

Exam-style questions continued

b) Which state of matter is the *least* common for the elements of the Periodic Table at room temperature?

gases ✗ .... (1)

c) The manufacture of sulfuric acid can be summarised by the equation:

$$2S(s) + 3O_2(g) + 2H_2O(l) \rightarrow 2H_2SO_4(l)$$

Tick one box in each line to show whether the formulae in the table represents a compound, an element or a mixture.

	Compound	Element	Mixture	
i) $2S(s)$		✓		✓ ①
ii) $2S(s) + 3O_2(g)$		✓		✗
iii) $3O_2(g) + 2H_2O(l)$			✓	✓ ①
iv) $2H_2SO_4(l)$	✓			✓ ①

(4)

(Total 9 marks)

⑥/⑨

Question 2

This question is about atoms.

a) Choose words from the box to label the diagram of an atom. (3)

proton	neutron	electron	ion

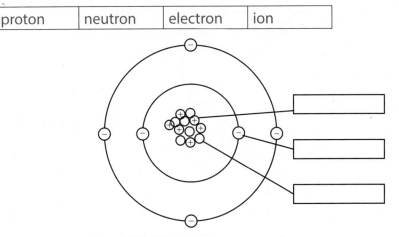

b) What is the proton number of this atom? (1)

c) What is the nucleon number of this atom? (1)

(Total 5 marks)

Question 3

a) Some elements combine together to form ionic compounds. Use words from the box to complete the sentences.

Each word may be used once, more than once or not at all.

gained	high	lost	low
medium	metals	non-metals	shared

Ionic compounds are formed between .... and ....

Electrons are .... by atoms of one element and .... by atoms of the other element.

The ionic compound formed has a .... melting point and a .... boiling point. (6)

b) Two elements react to form an ionic compound with the formula $MgCl_2$. (proton number of Mg = 12; proton number of Cl = 17)

   i) Give the electronic configurations of the two elements in this compound *before* the reaction. (2)

   ii) Give the electronic configurations of the two elements in this compound *after* the reaction. (2)

(Total 10 marks)

Question 4

The structures of some substances are shown here:

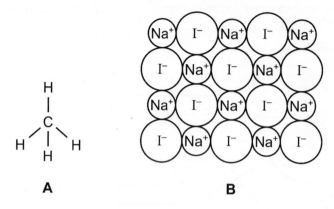

A   **B**   **C**

H — Br

D   **E**

Answer these questions using the letters **A**, **B**, **C**, **D** or **E**.

a) Which structure is methane?   (1)

b) Which two structures are giant structures?   (1)

c) Which two structures are hydrocarbons?   (1)

d) Which structure contains ions?   (1)

e) Which two structures have very high melting points?   (1)

(Total 5 marks)

Question 5

Strontium and sulfur chlorides both have a formula of the type XCl_2 but they have different properties.

Property	Strontium chloride	Sulfur chloride
Appearance	White crystalline solid	Red liquid
Melting point/°C	873	−80
Particles present	Ions	Molecules
Electrical conductivity of solid	Poor	Poor
Electrical conductivity of liquid	Good	Poor

a) The formulae of the chlorides are similar because both elements have a valency of 2. Explain why Group II and Group VI elements both have a valency of 2. (2)

b) Draw a dot-and-cross diagram of one covalent molecule of sulfur chloride. Use x to represent an electron from a sulfur atom. Use o to represent an electron from a chlorine atom. (3)

c) Explain the difference in electrical conductivity between the following:

   i) solid and liquid strontium chloride (1)

   ii) liquid strontium chloride and liquid sulfur chloride. (1)

(Total 7 marks)

Exam-style questions

The questions, sample answers and marks in this section have been written by the authors as a guide only. The marks given for these questions indicate the level of detail required in the answers. In the examination, the number of marks given to questions like these may be different.

Sample student answer

Question 1

Solutions of silver nitrate and sodium chloride react together to make the insoluble substance silver chloride.

The equation for the reaction is

$$AgNO_3(aq) + NaCl(aq) \rightarrow NaNO_3(aq) + AgCl(s)$$

An investigation was carried out to find out how much precipitate formed with different volumes of silver nitrate solution.

A student measured out 15 cm³ of sodium chloride solution using a measuring cylinder.

He poured this solution into a clean boiling tube.

Using a clean measuring cylinder, he measured out 2 cm³ of silver nitrate solution (of the same concentration as the sodium chloride solution). He added this to the sodium chloride solution.

A cloudy white mixture formed and the precipitate was left to settle.

The student then measured the height (in cm) of the precipitate using a ruler.

The student repeated the experiment using different volumes of silver nitrate. The graph shows the results obtained.

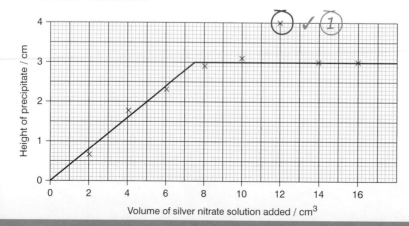

TEACHER'S COMMENTS

a) i) Correct point marked.

ii) Correct explanation. Also correct would be tube not being vertical when being set up so precipitate not level.

iii) Correct response.

b) Correct response – also correct is 'silver nitrate in excess'.

c) i) Correct reading of ruler.

ii) Answer is 3.9 cm³ – the student has misread the horizontal axis scale.

d) i) Correct response.

ii) Correct – this is the purpose of the experiment.

e) 2 marks have been lost here because the filtered-off precipitate needs to be washed (1) and 'dried' (1) before being weighed.

a) i) On the graph, circle the point that seems to be anomalous. (1)

ii) Explain two things that the student may have done in the experiment to give this anomalous result.

Precipitate not settled ✓ ①   *Because not left long enough* ✓ ① (2)

iii) Why must the graph line go through (0, 0)?

Cannot have a precipitate if no silver nitrate added yet. ✓ ① (1)

b) Suggest a reason why the height of the precipitate stops increasing.

No more sodium chloride left to react. ✓ ① (1)

c) i) How much precipitate has been made in the tube drawn on the right?

1.5 cm ✓ ① (1)

ii) Use the graph to find the volume of silver nitrate solution needed to make this amount of precipitate.

2.9 cm ✗ (1)

solution of soluble salts

precipitate of solid silver chloride iodide

d) After he had plotted the graph, the student decided he should obtain some more results.

i) Suggest what volumes of silver nitrate solution he should use.

Between 6 cm³ and 10 cm³ ✓ ① (1)

ii) Explain why he should use these volumes.

Need to know exactly where the graph levels off ✓ ① (1)

e) Suggest a different method for measuring the amount of precipitate formed.

Filter ✓ ①   *off each precipitate and weigh it* ✓ ① (4)

(Total 13 marks)

⑩/⑬

Question 2

Dilute nitric acid reacts with marble chips to produce carbon dioxide. The equation is given below:

$$2HNO_3(aq) + CaCO_3(s) \rightarrow Ca(NO_3)_2(aq) + H_2O(l) + CO_2(g)$$

Some students investigated the effect of changing the temperature of the nitric acid on the rate of the reaction. The method is:

- Use a measuring cylinder to pour $50\,cm^3$ of dilute nitric acid into a conical flask.
- Heat the acid to the required temperature.
- Put the flask on the balance.
- Add 15 g (an excess) of marble chips to the flask.
- Time how long it takes for the mass to decrease by 1.00 g.
- Repeat the experiment at different temperatures.

The students' results are shown in the table.

Temperature of acid (°C)	Time to lose 1.00 g (s)
20	93
33	68
44	66
55	40
67	30
76	25

a) **i)** Draw a graph of the results. (3)

   ii) One of the points is inaccurate. Circle this point on your graph. (1)

   iii) Suggest a possible cause for this inaccurate result. (1)

b) Use the graph to find the times taken to lose 1.00 g at 40 °C and 60 °C. (2)

c) The rate of the reaction can be found using the equation:

$$\text{rate of reaction} = \frac{\text{mass lost}}{\text{time taken to lose mass}}$$

   i) Use this equation and your results from **b)** to calculate the rates of reaction at 40 °C and 60 °C. (2)

ii) What will be the unit for these rates? (1)

iii) State how the rate of reaction changes when the temperature increases. (1)

iv) Explain in terms of particles and collisions why the rate changes when the temperature increases. (1)

d) Describe how the method could be changed to obtain a result at 5 °C. (1)

(Total 13 marks)

Question 3

The diagram shows the apparatus used to electrolyse lead(II) bromide.

a) The wires connected to the electrodes are made of copper.

Explain why copper conducts electricity. (1)

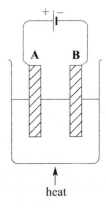

b) Explain why electrolysis does not occur unless the lead(II) bromide is molten. (2)

(Total 3 marks)

Question 4

Read the following instructions for the preparation of hydrated copper(II) sulfate ($CuSO_4.5H_2O$), then answer the questions that follow.

1. Put 25 cm^3 of dilute sulfuric acid in a beaker.

2. Heat the sulfuric acid until it is just boiling and then add a small amount of copper(II) carbonate.

3. When the copper carbonate has dissolved, stop heating, then add a little more copper carbonate. Continue in this way until copper carbonate is in excess.

4. Filter the hot mixture into a clean beaker.

5. Make the hydrated copper(II) sulfate crystals from the copper(II) sulfate solution.

The equation for the reaction is

$$CuCO_3(s) + H_2SO_4(aq) \rightarrow CuSO_4(aq) + CO_2(g) + H_2O(l)$$

a) What piece of apparatus would you use to measure out 25 cm^3 of sulfuric acid? **(1)**

b) Why is the copper(II) carbonate added in excess? **(1)**

c) When copper(II) carbonate is added to sulfuric acid, there is fizzing. Explain why. **(1)**

d) Draw a diagram to describe step 4. You must label your diagram. **(3)**

e) After filtration, which one of the following describes the copper(II) sulfate in the beaker?

Select the correct answer.

crystals   filtrate   precipitate   water   **(1)**

f) Explain how you would obtain pure dry crystals of hydrated copper(II) sulfate from the solution of copper(II) sulfate. **(2)**

(Total 9 marks)

TEACHER'S COMMENTS

a) The mark would have been given for stating that all the elements have the same number of electrons in their outer shell. The student has gone further and correctly stated that they all have one electron in the outer shell.

b) i) The correct answer has been given.

ii) This answer lacks precision. The mark would be given for either explaining that the reaction produced heat or that the reaction was exothermic.

iii) The student has not scored both marks. Apart from stating that the reaction would be more rapid than that with lithium, observations have not been given. The products have been correctly named but these are not what you would *observe*. Marks would be given for: fizzing/effervescence/bubbles (of gas), the sodium floats/moves around on the water, forms a ball/disappears.

iv) The correct answer and explanation have been given.

Exam-style questions

The questions, sample answers and marks in this section have been written by the authors as a guide only. The marks given for these questions indicate the level of detail required in the answers. In the examination, the number of marks given to questions like these may be different.

Sample student answer

Question 1

Lithium (Li), sodium (Na) and potassium (K) are in Group I of the Periodic Table.

a) These elements have similar chemical properties

Explain why, using ideas about electronic structures.

All the elements have one electron in their outer shell. ✓ ① **(1)**

b) Lithium reacts with water to form a solution of lithium hydroxide and a colourless gas. During this reaction the temperature of the water increases.

i) What is the name of the colourless gas produced?

hydrogen ✓ ① **(1)**

ii) Why does the temperature of the water increase?

The reaction between lithium and water is rapid. ✗ **(1)**

iii) Describe what you would observe if a small piece of sodium is added to water.

The sodium forms sodium hydroxide and hydrogen gas is given off. ✗

The reaction would be more rapid than the lithium reaction. ✓ ① **(2)**

Exam-style questions continued

iv) Caesium (Cs) is another Group I metal.

Is caesium more or less reactive than lithium? Give a reason for your answer.

More reactive, because reactivity in Group I increases down the group.

(1)

(Total 6 marks)

Question 2

Use the Periodic Table to help you answer this question.

a) State the symbol of the element with proton number 14. (1)

b) State the symbol of the element that has a relative atomic mass of 32. (1)

c) State the number of the group that contains the alkali metals. (1)

d) Which group contains elements whose atoms form ions with a 2+ charge? (1)

e) Which group contains elements whose atoms form ions with a 1– charge? (1)

(Total 5 marks)

Question 3

Three of the elements in Group VII of the Periodic Table are chlorine, bromine and iodine.

a) Chlorine has a proton number of 17. What is the electron configuration of chlorine? (1)

b) How many electrons will be in the outer shell of a bromine atom? (1)

c) Bromine reacts with hydrogen to form hydrogen bromide. The equation for the reaction is:

$Br_2(g) + H_2(g) \rightarrow 2HBr(g)$

What is the colour change during the reaction? (1)

(Total 3 marks)

Question 4

The reactivity of metals can be compared by comparing their reactions with dilute sulfuric acid. Pieces of zinc, iron and magnesium of identical size are added to separate test tubes containing this acid.

a) What order of reactivity would you expect? Put the most reactive metal first. (1)

b) Write a word equation for the reaction between magnesium and dilute sulfuric acid. (1)

c) Write a balanced equation for the reaction in **b)**. (1)

d) Name a metal that does not react with dilute sulfuric acid. (1)

e) What other reaction could be used to compare the reactivity of metals? (1)

(Total 5 marks)

Question 5

Look at the list of five elements below:

argon, bromine, chlorine, iodine, potassium.

a) Put these five elements in order of increasing proton number. (1)

b) Put these five elements in order of increasing relative atomic mass. (1)

c) The orders of proton number and relative atomic mass for these five elements are different. Which one of the following is the most likely explanation for this?

A The proton number of a particular element may vary.

B The presence of neutrons.

C The atoms easily gain or lose electrons.

D The number of protons must always equal the number of neutrons. (1)

d) Which of the five elements in the list are in the same group of the Periodic Table? (1)

e) i) From the list, choose one element that has one electron in its outer shell. (1)

ii) From the list, choose one element that has a full outer shell of electrons. (1)

f) Which two of the following statements about argon are correct?

A   Argon is a noble gas.

B   Argon reacts readily with potassium.

C   Argon is used to fill weather balloons.

D   Argon is used in light bulbs.   (2)

(Total 8 marks)

Question 6

The table gives some information about the elements in Group I of the Periodic Table.

Element	Boiling point (°C)	Density (g/cm³)	Radius of atom in the metal (nm)	Reactivity with water
Lithium	1342	0.53	0.157	
Sodium	883	0.97	0.191	Rapid
Potassium	760	0.86	0.235	Very rapid
Rubidium		1.53	0.250	Extremely rapid
Caesium	669	1.88		Explosive

a) How does the density of the Group I elements change going down the group?   (2)

b) Suggest a value for the boiling point of rubidium.   (1)

c) Suggest a value for the radius of a caesium atom.   (1)

d) Use the information in the table to suggest how fast lithium reacts with water compared with the other Group I metals.   (1)

e) State three properties shown by all metals.   (3)

f) When sodium reacts with water, hydrogen is given off:

$$2Na(s) + 2H_2O(l) \rightarrow 2NaOH(aq) + H_2(g)$$

i) State the name of the other product formed in this reaction.   (1)

ii) Describe a test for hydrogen.   (2)

(Total 11 marks)

b) Correct answer, alkenes are unsaturated hydrocarbons. A and C are alkanes which are saturated hydrocarbons.

c) Correct answer, as this is a test for an unsaturated hydrocarbon.

d) Carbon dioxide is correct but the other product is water and not hydrogen.

e) Correct answer, as the general formula is C_nH_{2n}.

f) i) Correct answer, the process for breaking down larger hydrocarbons into smaller ones is cracking.

ii) High temperature is correct but a catalyst is also needed.

Exam-style questions

The questions, sample answers and marks in this section have been written by the authors as a guide only. The marks given for these questions indicate the level of detail required in the answers. In the examination, the number of marks given to questions like these may be different.

Sample student answer

Question 1

This question is about the following organic compounds:

A C_4H_{10} **B** C_3H_6 **C** $C_{10}H_{22}$

a) Which compound belongs to the alkene homologous series?

B ✓ ①   **(1)**

b) Which compound is an unsaturated hydrocarbon?

B ✓ ①   **(1)**

c) Which compound will decolourise bromine water?

B ✓ ①   **(1)**

d) Name two products that are formed when compound A burns in a plentiful supply of air.

hydrogen ✗ and *carbon dioxide* ✓ ① **(2)**

e) Write the formula of another hydrocarbon that is in the same homologous series as compound B.

C_2H_4 ✓ ①   **(1)**

f) Compound C undergoes the following reaction:

$$C_{10}H_{22}(g) \rightarrow C_4H_{10}(g) + 2C_3H_6(g)$$

i) What is the name of this process?

Cracking ✓ ①   **(1)**

ii) What conditions are needed for this reaction?

high temperature ✓ ①   **(2)**

Total ⑦/9

Question 2

The alkanes are a homologous series of saturated hydrocarbons.

a) Say whether each of the following statements about the members of the alkane homologous series is TRUE or FALSE.

   i) They have similar chemical properties. (1)

   ii) They have the same displayed formula. (1)

   iii) They have the same general formula. (1)

   iv) They have the same physical properties. (1)

   v) They have the same relative formula mass. (1)

b) Define the following terms:

   i) hydrocarbon (1)

   ii) saturated. (1)

c) The third member of the alkane homologous series is propane.

   i) What is the molecular formula of propane? (1)

   ii) Draw the displayed formula of propane. (2)

(Total 10 marks)

Question 3

Many useful substances are produced by the fractional distillation of crude oil.

a) Bitumen, fuel oil and gasoline are three fractions obtained from crude oil.

Name the fractions that have the following properties:

   i) the highest boiling point (1)

   ii) molecules with the fewest carbon atoms (1)

   iii) the darkest colour. (1)

b) Some long-chain hydrocarbons can be broken down into more useful products.

What is the name of this process and how is it carried out? (3)

c) Methane is used as a fuel. When methane is burned in a plentiful supply of air, carbon dioxide is formed.

   i) Write a balanced equation for this reaction. (2)

   ii) Carbon dioxide is a greenhouse gas. How may it contribute to climate change? (2)

(Total 10 marks)

Question 4

Poly(ethene) is a plastic that is made by polymerising ethene, C_2H_4.

a) Which one of the following best describes the ethene molecules in this reaction?
alcohols, alkanes, monomers, polymers, products (1)

b) The structure of ethane is shown below.

Explain, by referring to its bonding, why ethane cannot be polymerised. (1)

c) Draw the structure of ethene, showing all its atoms and bonds. (1)

d) Ethene is obtained by cracking alkanes.

   i) Explain the meaning of the term *cracking*. (1)

   ii) What condition is needed to crack alkanes? (1)

   iii) Copy and complete the equation for cracking decane, $C_{10}H_{22}$.

      $C_{10}H_{22} \rightarrow C_2H_4 + $ _____ (1)

e) Some oil companies crack the ethane produced when petroleum is distilled.

   i) Copy and complete the equation for this reaction.

      $C_2H_6 \rightarrow C_2H_4 + $ _____ (1)

   ii) Describe the process of fractional distillation, which is used to separate the different fractions in petroleum. (2)

   iii) State a use for the following petroleum fractions:

      gasoline fraction (1)
      lubricating fraction. (1)

(Total 11 marks)

Exam-style questions

Sample student answers

Question 1

The diagram shows a car travelling to the right.

The arrows represent four forces on the car as it moves.

The arrows are **not** drawn to scale.

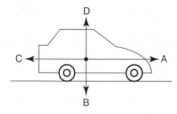

a) i) Which arrow represents the weight of the car?

B ✔ ①   (1)

ii) Which arrow represents the driving force acting on the car?

A ✔ ①   (1)

b) The horizontal forces on the car are unbalanced.

i) State the equation linking unbalanced force, mass and acceleration.

$F = ma$ ✔ ①   (1)

The car accelerates at $2\,\text{m/s}^2$ and it has a mass of $1500\,\text{kg}$.

ii) Calculate the magnitude of the unbalanced force on the car.

$1500 \times 2 = 3000$

Force ✔ ① $= 3000\,\text{N}$ ✔ ①   (2)

TEACHER'S COMMENTS

a) i) and **ii)** The first two questions are straightforward, recalling and applying simple ideas in a familiar situation. Two marks are given.

b) i) and **ii)** The student has handled the mathematical parts of the question correctly.

iii) The student almost certainly knew which direction the force was acting in (forward) but loses the mark because 'horizontally' has two possibilities – forward and backward. This is an easy slip to make, and the student might have noticed the mistake if they had time to check through their answers.

iv) The student has missed the point of the question. It is still in part b) and that is the clue that the answer is still related to $F=ma$. The question asks about forces so the answer should also refer to forces, the calculated value is a resultant force made up from the engine force – the friction.

Instead, the student talks about 'speeding up' and 'energy losses' which are too vague to gain credit here.

c) 'Bigger' is too vague to gain credit. The student needed to mention 'larger mass', the question is still linked to $F=ma$. The student does then go on to link force to acceleration correctly so gains the second mark. The third mark is also gained because the idea of more air resistance is a sensible suggestion (and it is a 'suggest' question). It also makes it clearer what the student meant by writing 'bigger' at the start of the sentence. Overall, however, the student needs be clearer so that the ideas don't have to be linked together by the reader.

iii) Which direction does this force act in?

Horizontal ✗ (1)

iv) The force provided by the car engine is larger than the value you have calculated. Explain why.

The car is speeding up and energy is lost to the surroundings, so the engine has to give a bigger force. ✗ ✗ (2)

c) A truck tries to keep up with the car as it accelerates.

The truck falls behind, even though the engine in the truck provides a bigger force than the engine in the car.

Suggest why the truck falls behind.

The truck is bigger, so it has to have a bigger force to get the same acceleration because there will be more air resistance. ✗ ✓ ✓ ①① (3)

(Total 11 marks)

7/11

Question 2

In 2009, Usain Bolt set the world record for 100 m at 9.58 s.

a) **i)** State the equation linking average speed, distance moved and time taken. (1)

   ii) Calculate the average speed for Usain Bolt's run.

   Give your answer to an appropriate number of significant figures. (3)

   iii) At some part of the race, Usain Bolt must have run faster than this.

   Explain why. (3)

b) At the start of the race, it is counted as a false start when a runner moves before 0.10 s *after* the starter has fired the starting gun.

Suggest why. (3)

(Total 10 marks)

Question 3

The graph shows how the velocity of a cyclist varies.

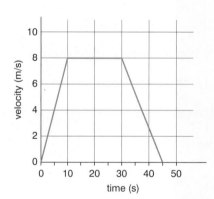

a) Between what times was the cyclist travelling at a constant speed? (1)

b) How can you tell from the graph? (1)

   Calculate the acceleration of the cyclist during the first 10 s. (3)

   Use the graph to find the total distance travelled by the cyclist. (3)

(Total 8 marks)

Question 4

A student is planning to investigate air resistance.

He makes a number of paper parachutes of different sizes and attaches them one at a time to a small mass.

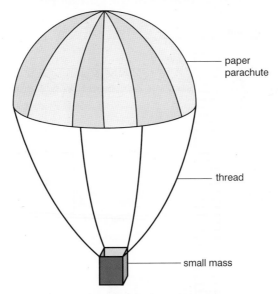

paper parachute

thread

small mass

a) Describe how the student should carry out his investigation.
You should include:

   i) what other equipment the student should use

   ii) what measurements the student should make

  iii) how the student should use his measurements to draw a conclusion.   (5)

b) Use ideas about the forces acting on falling objects to explain how a falling object can reach a terminal velocity.   (5)

(Total 10 marks)

Question 5

A crane at a building site is lifting a container.

The point labelled C is the centre of gravity of the container.

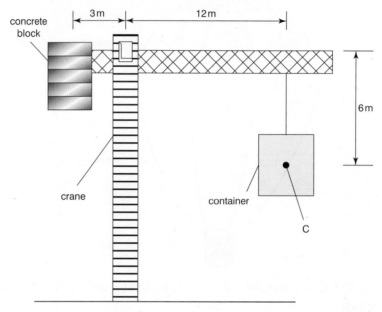

The container has a mass of 2000 kg.

a) State the equation linking weight, mass and g. (1)

b) Calculate the weight of the container. (2)

c) Use ideas about moments to explain the purpose of the concrete block. (3)

d) In the diagram, the crane is balanced.

Calculate a suitable value for the mass of the concrete block. (4)

(Total 10 marks)

Question 6

A student investigates stretching a spring.

The student hangs 100 g masses onto the spring one at a time and measures the length of the spring.

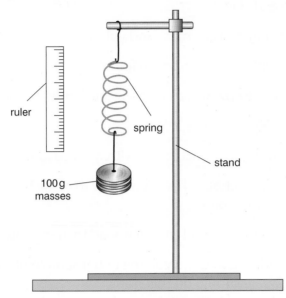

a) Describe how the student should use a ruler to obtain accurate measurements for the length of the spring. **(3)**

b) Suggest one safety precaution the student should take during the investigation. **(1)**

The student calculates the extension in the spring for each mass attached.

The table shows the results.

Weight/N	Extension/cm
0	0.0
1	3.1
2	6.0
3	8.8
4	13.7
5	14.9

i) Draw a graph of these results. **(5)**

ii) Is there any evidence that the student made a mistake in the measurements?

Explain your answer. **(2)**

iii) Does the evidence from this experiment support Hooke's law?

Explain your answer. **(3)**

(Total 14 marks)

Question 7

Two students want to find the density of clay.

They each have a sample of clay.

a) State the equation linking density, mass and volume. (1)

The students suggest different methods to find the volumes of their samples of clay.

b) One student shapes his sample of clay into a regular cube shape.

Then he measures the length of the sides.

He finds the volume by doing a calculation with his measurements.

  i) How should the student choose his equipment to make his measurement of volume as precise as possible? (2)

  ii) Describe a feature of this method that may lead to inaccurate results. (1)

c) The other student decides to find the volume of her sample of clay using a measuring cylinder. Describe how she should do this. (4)

d) Describe how the students can use an electronic balance to find the masses of their samples of clay. (1)

  i) If the electronic balance is incorrectly calibrated, how will this affect their measurements? (1)

  ii) How could the students check the calibration of the electronic balance? (2)

(Total 12 marks)

TEACHER'S COMMENTS

a) i) This is a common mistake – 0 marks. Any energy source cannot be used again – once the fuel is burnt it cannot be burnt again! However, renewable sources can be replaced over reasonably short time scales.

ii) 1 mark.

b) This will probably score 1 mark, but it would be better if the student specified 'Natural Gas'. Lots of materials are 'gas', for example oxygen, but they are not energy sources.

c) This is likely to score 3 marks. An advantage and disadvantage is given for both energy sources, but simply saying 'makes pollution' is too vague and 'gives you lots of energy' really needs to include a reference to how quickly the energy is delivered.

d) These are sensible responses, given in enough detail – 2 marks.

Exam-style questions

Note: The questions, sample answers and marks in this section have been written by the authors as a guide only. The marks given for these questions indicate the level of detail required in the answers. In the examination, the number of marks given to questions like these may be different.

Sample student answers

Question 1

Different energy resources are used to generate electricity.

a) Some energy sources are renewable.

   i) What does 'renewable' mean when it refers to energy sources? (1)

     It can be used again. ✗

   ii) Give one example of a renewable energy source. (1)

     The wind. ✓ ①

b) Fossil fuels are non-renewable energy sources.

   Give one example of a fossil fuel. (1)

   Gas ✓ ①

c) Use your examples from **a)** and **b)** to describe advantages and disadvantages of using renewable and non-renewable energy sources to generate electricity. (4)

   Renewable energy sources are free once the wind turbine is made, but it is not always windy so you can't always make electricity. An advantage of using gas is that it gives you lots of energy, but it also makes pollution. ✓ ✓ ✓ ③

d) Less than 10% of the electricity generated in the UK is from renewable energy sources.

Suggest two reasons, apart from cost, why the UK does not generate more electricity using renewable energy sources. (2)

1. *The UK uses very large amounts of electricity which is more than renewable sources can provide.* ✔ ①

2. *The UK is not sunny or windy enough to generate enough electricity using these methods. The climate is wrong.* ✔ ①

(Total 9 marks)

Question 2

A car is travelling at 20 m/s and has a mass of 1500 kg.

a) Calculate the kinetic energy of the car in joules when it is travelling at 20 m/s. (2)

b) Write down the work done stopping the car and give the correct unit. (2)

c) If the brakes provide a force of 10 000 N, calculate the distance required to stop the car.

d) What happens to the kinetic energy of the car when it stops at the traffic lights? (2)

(Total 6 marks)

Question 3

A child climbs to the top of a slide at a playground.

The child has a mass of 30 kg and the top of the slide is 3.0 m above the ground.

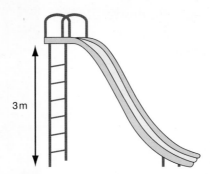

a) State the equation linking gravitational potential energy, mass, *g* and height.   (1)

b) Calculate the gravitational potential energy gained by the child when he climbs to the top of the slide.   (2)

c) State the link between the gravitational potential energy gained and the work done by the child.   (1)

d) To calculate the power of the child as he climbs to the top of the slide, what other measurement would be needed?   (1)

e) The child rides down the slide. Assuming that there are no energy losses, calculate the speed of the child at the bottom of the slide.   (4)

(Total 9 marks)

Exam-style questions

Note: The questions, sample answers and marks in this section have been written by the authors as a guide only. The marks given for these questions indicate the level of detail required in the answers. In the examination, the number of marks given to questions like these may be different.

Sample student answers

Question 1

This question is about particles.

a) i) Use ideas about particles to explain why liquids can flow but solids keep the same shape.

> *This is because the particles in a solid are close together but in a liquid they are more spread out.* ✗ **(2)**

ii) Use ideas about particles to explain why a gas fills its container but solids keep the same size.

> *In a gas the particles are moving about so they can go to all parts of the container. In a solid the particles are joined together in a fixed pattern.* ✓ ✓ ✗ ② **(3)**

a) i) 0 marks. Although the particles are close together in a solid, this does not answer the question about why they cannot flow. The student needed to say that the particles were held in fixed positions. The student then says that the particles are more spread out in a liquid – this is correct in itself, but still makes no reference to the particles being able to 'slide' over each other, which is the key point when explaining why a liquid can flow.

ii) 2 marks – but only just! The description of a gas would be stronger if it referred to the spacing of the particles and their random motion. However, it does score 1 mark for saying that the particles can go to all parts of the container. The description of the particles in a solid scores one mark.

When you are preparing for exams, make sure you cover every point in the syllabus. Your teachers will use it as well, but that is not an excuse for you to ignore it!

b) i) These parts of the question are handled well and score full marks.

ii) The student made a good use of the relevant equations to make their argument. This is a powerful way to answer questions of this type and can save a lot of descriptive writing.

b) The photo shows bubbles rising in water.

The bubbles increase in volume as they move towards the surface of the water.

i) Explain why the bubbles rise to the surface.

The bubbles are less dense than the water ✔ ①
.so the forces on them are unbalanced and they float upwards. ✔ ①

(2)

ii) Explain why the volume of the bubbles increases as they rise.

The pressure = depth × density ×g, ✔ ① *so the pressure gets less as the bubbles move upwards.* ✔ ① *Also, pressure × volume is constant, so if the pressure decreases then the volume increases.* ✔ ① *So as the pressure on the bubbles decreases their volume will increase and the bubbles will get bigger.* ✔ ①

(4)

(Total 11 marks)

Question 2

A student investigates the effect of insulation on cooling.

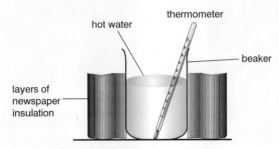

The student puts some hot water into a beaker and measures the temperature drop in 20 minutes.

He repeats the experiment using layers of paper as insulation.

His results are shown in the table.

Number of layers of insulation	Temperature drop in 20 minutes/°C
0	21
5	20
10	18
15	17
20	18

a) Draw a graph of these results. (5)

b) The student concludes that the graph shows that thicker insulation reduces heat loss.

Is this a correct conclusion from this data?

Explain your answer. (2)

(Total 7 marks)

Question 3

Two experiments are carried out to investigate energy transfer in water.

In Experiment 1, cold water is gently heated at the top of a glass boiling tube. A block of ice trapped at the bottom remains solid even when the water at the top begins to boil.

In Experiment 2, cold water is gently heated at the bottom of the tube. Ice at the top of the tube melts before the water boils.

a) What is the process by which thermal (heat) energy travels through the glass?

(1)

b) i) What is the principal process in Experiment 2, which takes the energy from the water at the bottom to the ice at the top? (1)

ii) Describe how the process in **b) i)** occurs. (2)

c) Suggest two reasons why the ice in Experiment 1 does not melt, even when the water at the top begins to boil. (2)

(Total 6 marks)

Exam-style questions

Note: The questions, sample answers and marks in this section have been written by the authors as a guide only. The marks given for these questions indicate the level of detail required in the answers. In the examination, the number of marks given to questions like these may be different.

Sample student answers

Question 1

The table shows some of the regions of the electromagnetic spectrum.

Gamma	X-rays	A	Visible	Infrared	B	Radio

a) i) Complete the chart by writing the names of the missing regions (A and B) of the spectrum.

A: UV ✗ (1)

B: microwaves ✓ ① (1)

ii) The table lists the spectrum in what order?

A. increasing amplitude

B. increasing density

C. increasing frequency

D. increasing wavelength

D ✓ ① (1)

b) i) Describe one situation in which X-rays are useful.

in a hospital ✗ (1)

ii) Explain why X-rays are useful in the example you have given.

X-rays can show broken bones

because they will pass through the skin

and flesh ✓ ① but there will be a shadow where the (1)

bones are. ✓ ① (1)

c) i) Describe one situation in which X-rays can be harmful.

When they get into your body. ✗ (1)

TEACHER'S COMMENTS

a) i) The question clearly asks for the name of the missing regions. Writing 'UV' is not a name, nor even a recognised standard symbol, so it loses a mark. The student must take care to read the question carefully and answer it in the way it is asked.

ii) The student should know the different ways to describe the order of the spectrum. Here the answer is given in terms of increasing wavelength, but they should also be able to describe the sequence in order of frequency and possibly in order of energy.

b) i) Stating 'in a hospital' is too vague to be a 'situation' as the question asks.

ii) Clearly, the student was aware of the correct situation – imaging – so should have written this in their answer.

c) i) Again, the student has not described a situation, so cannot gain the first mark.

ii) The student has two valid points, but there is only one mark available. The second was for describing how the risks can be reduced and the student has failed to give any response to this. Always read the question carefully and answer each point required.

d) 'Not strong enough' is too vague. The correct term 'amplitude' is required here. The second part of the answer correctly links to energy and the student might have thought a little further about this, possibly then getting the link from 'strong' to 'amplitude'.

This part of the question tests basic definitions and these should be learned thoroughly.

ii) Describe how X-rays can be harmful and how the risks can be reduced.

They can damage body cells ✓ ① **(1)**

and cause some cells to become cancer cells. ✗ **(1)**

d) A remote control for a television uses infrared signals.

The human body detects infrared radiation as heat.

Explain why the infrared signal from a television remote control does not make your skin feel hot.

The signal from the remote is not strong enough to make you feel hot. ✗

There is not enough energy to burn you. ✓ ① **(2)**

 6/11

(Total 11 marks)

Question 2

The diagram shows some waves on the surface of water in a ripple tank.

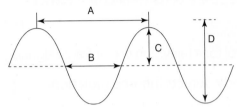

a) **i)** Which letter represents the wavelength of the wave? (1)

   ii) Which letter represents the amplitude of the waves? (1)

b) The waves in the water are transverse waves.

   i) Give another example of a transverse wave. (1)

   ii) Describe the difference between transverse waves and longitudinal waves. (2)

c) A student makes some observation about the waves in the ripple tank.

   number of waves passing by in 6 seconds – 10

   distance between wave crests – 5 cm

   i) State the equation linking wave speed, frequency and wavelength. (1)

   ii) Calculate the wave speed of these waves in m/s. (2)

(Total 8 marks)

Exam-style questions

Note: The questions, sample answers and marks in this section have been written by the authors as a guide only. The marks given for these questions indicate the level of detail required in the answers. In the examination, the number of marks given to questions like these may be different.

Sample student answers

Question 1

The table shows the maximum current that can safely pass through copper wires of different cross-sectional areas.

Cross-sectional area in mm²	Maximum safe current in amps
1.0	11.5
2.5	20.0
4.0	27.0
6.0	34.0

a) i) The student's answer needs to use correct terminology as given in the question and state clearly that as the cross-sectional area increases the maximum current increases.

Both sets of numbers are getting bigger ✗ (1)

ii) To get the 2nd mark the student's answer needs to show they understand that more electrons flowing per second means a larger current.

A wire with a larger diameter allows more electrons to move through ✓ ① (2)

b) To get both marks the student needs to use the terminology and data given in the question and state that the oven takes a bigger current than the maximum safe current of 20 A for the 2.5 mm² wire.

A cooker uses a much bigger electric current ✓ ① (2)

c) The student's answer correctly gives the working, the numerical answer and the unit.

$R = \dfrac{V}{I} = \dfrac{230}{20.0} = 11.5 \ ohms$ ✓ ① (2)

d) The student's answer needs more detail to get both marks and so should also refer to the wire overheating.

It will melt ✓ ① (2)

Exam-style questions

Note: The questions, sample answers and marks in this section have been written by the authors as a guide only. The marks given for these questions indicate the level of detail required in the answers. In the examination, the number of marks given to questions like these may be different.

Sample student answers

Question 1

a) i) and **ii)** The student has chosen the correct fuse but the explanation lacks detail and should also refer to the value of the fuse chosen being the smallest value above the 1.5 A working current..

i) Which fuse should be fitted? Choose from 1 A, 3 A, 5 A, 13 A.   (1)

3 A ✓ ①

ii) Explain your answer.   (2)

smallest value for the lamp to work ✓ ①   (2)

b) To get all 5 marks the student's answer should include a link to the safety of the lamp, as mentioned in the question. It should include a reference to the lamp switching off when the fuse melts.   (5)

The earth wire has a very low resistance so if the casing becomes live and a

big current flows through it the fuse will melt ✓ ①

Developing experimental skills

INTRODUCTION

As part your Combined Science course, you will develop practical skills and have to carry out investigative work in science. This will either be done as coursework, a practical test or a written paper, depending on your school.

This section provides guidance on carrying out an investigation.

The experimental and investigative skills are divided into four parts as follows (each of the four sections carry equal weighting):

1. Using and organising techniques, apparatus and materials

2. Observing, measuring and recording

3. Handling experimental observations and data

4. Planning and evaluating investigations

1. USING AND ORGANISING TECHNIQUES, APPARATUS AND MATERIALS

Learning objective: to demonstrate and describe appropriate experimental and investigative methods, including safe and skilful practical techniques.

Questions to ask:

How shall I use the equipment safely to minimise the risks – what are my safety precautions?

✓ When writing a Risk Assessment, investigators need to be careful to check that they've matched the hazard with the technique used and with the concentration of a chemical used. Many acids, for instance, are corrosive in higher concentrations, but are likely to be irritants or of low hazard in the concentration used when working in biology or chemistry experiments.

✓ Don't forget to consider the hazards associated with all the chemicals and biological materials, even if these are very low.

✓ In the exam, you may be asked to describe the precautions taken when carrying out an investigation.

How much detail should I give in my description?

✓ You need to give enough detail so that someone else who has not done the experiment would be able to carry it out to reproduce your results.

How should I use the equipment to give me the precision I need?

✓ You should know how to read the scales on the measuring equipment you are using.

✓ You need to show that you are aware of the precision needed.

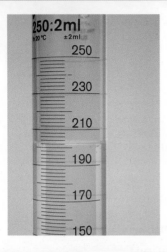

◁ Fig. 7.1 The volume of liquid in a measuring cylinder must be read to the bottom of the meniscus. The volume in this measuring cylinder is 202 cm³ (ml), not 204 cm³.

EXAMPLE 1

This is an extract from a student's notebook. It describes how she carried out an experiment to investigate the production of carbon dioxide by yeast at different temperatures.

What are my safety precautions?

a) Chemicals.

I have looked up the hazards associated with the

chemical I am using:

Glucose (solutions from 0.05 to 0.25M): LOW HAZARD

Although it is only a low hazard, it is still best to wear eye

protection when using the solutions, especially as some of

the liquids will be hot. It is also important to handle all

chemicals carefully, and wipe up any spills of liquid.

COMMENT

The student has used a data source to look up the chemical hazards.

EXAMPLE 2

This is an extract from a student's notebook. It describes how she investigated the motion of an object for which the acceleration is not constant.

Experimental detail

The student's method is given below.

1 *The tube was marked every 10 cm using tape.*

2 *The ball was released carefully from the surface of the oil.*

3 *At the same time, a stopclock was started.*

4 *As the ball passed each mark, the time was noted.*

5 *Since the marks are 10 cm apart, the speed of the ball in each section of the tube can be calculated.*

Precision and accuracy

An example from the notebook is:

The speed measured to the nearest 0.1 cm/s.

COMMENT

The method is well written and detailed. Point 1 could have been improved if the student had noted the width of the tape used. The student has appreciated the accuracy that can be achieved using this method.

---

2. OBSERVING, MEASURING AND RECORDING

Learning objective: to make observations and measurements with appropriate precision, record these methodically, and present them in a suitable form.

Questions to ask:

How many different measurements or observations do I need to take?

✓ Sufficient readings have been taken to ensure that the data are consistent.

✓ It is usual to repeat an experiment to take more than one measurement. If an investigator takes just one measurement, this may not be typical of what would normally happen when the experiment was carried out.

✓ When repeat readings are consistent, they are said to be **repeatable**.

Do I need to repeat any measurements or observations that are anomalous?

✓ An **anomalous result** or **outlier** is a result that is not consistent with other results.

✓ You want to be sure that a single result is accurate (as in Example 3). So you will need to repeat the experiment until you get close agreement in the results you obtain.

✓ If an investigator has made repeat measurements, they would normally use these to calculate the arithmetical mean (or just mean or average) of these data to give a more accurate result. You calculate the mean by adding together all the measurements, and dividing by the number of measurements. Be careful, though: anomalous results should not be included when taking averages.

✓ Anomalous results might be the consequence of an error made in measurement. But sometimes outliers are genuine results. If you think an outlier has been introduced by careless practical work, you should omit it when calculating the mean. But you should examine possible reasons carefully before just leaving it out.

✓ You are taking a number of readings in order to see a changing pattern. For example, measuring the speed every 10 cm for 60 cm (so six different readings). It is likely that you will plot your results onto a graph and then draw a **line of best fit**.

✓ You can often pick an anomalous reading out from a results table (or a graph if all the data points have been plotted, as well as the mean, to show the range of data). It may be a good idea to repeat this part of the practical again, but it's not necessary if the results show good consistency.

✓ If you are confident that you can draw a line of best fit through most of the points, it is not necessary to repeat any measurements that are obviously inaccurate. If, however, the pattern is not clear enough to draw a graph then readings will need to be repeated.

How should I record my measurements or observations – is a table the best way? What headings and units should I use?

✓ A table is often the best way to record results.

✓ Headings should be clear.

✓ If a table contains numerical data, do not forget to include units; data are meaningless without them.

✓ The units should be the same as those that are on the measuring equipment you are using.

✓ Sometimes you are recording observations that are not quantities. Putting observations in a table with headings is a good way of presenting this information.

EXAMPLE 3

How many different measurements or observations do I need to take?

A student cut a number of cylinders of tissue from a potato and weighed them, and recorded the mass of each cylinder. Six dishes were set up with each dish containing a different concentration of sucrose. Four potato cylinders were placed into each dish. After one hour, the cylinders were removed, blotted dry and reweighed. The student then calculated the percentage change in mass for each cylinder. The results are shown on the next page.

Concentration of sucrose/M	Percentage change in mass of potato cylinders/g				Average percentage change in mass	Texture of potato cylinders (qualitative)
	Experiment 1	Experiment 2	Experiment 3	Experiment 4		
0.0	+31.4	+33.7	+31.2	+32.5	+42.9	Firm
0.2	+20.9	+33.4	+22.8	+21.3	+21.7	Firm
0.4	−2.7	−1.8	−1.9	−2.4	−2.2	Slightly soft
0.6	−13.9	−12.8	−13.7	−13.6	−13.5	Soft
0.8	−20.2	−19.7	−19.3	−20.4	−19.9	Floppy
1.0	−19.9	−20.3	−21.1	−20.3	−20.4	Very floppy

△ Table 7.1 Results for Example 3.

EXAMPLE 4

In an experiment to investigate the efficiency of a small motor, the student has sensibly recorded her results in a table. Notice each column has a heading *and* units.

Mass lifted/g	Distance lifted/m	Useful work done/J	Voltage of motor/V	Current in motor/A	Time to lift the mass/s	Electrical energy supplied/J
0.01	1.0		2.4	0.20	22.0	
0.03	1.0		2.4	0.22	24.4	
0.05	1.0		2.4	0.25	26.5	
0.07	1.0		2.3	0.28	27.6	
0.09	1.0		2.3	0.29	28.7	

△ Table 7.2 Table of results.

EXAMPLE 5

In another experiment, a student has recorded his results obtained in an experiment to investigate the strength of an electromagnet as the current in the coil varies.

variable resistor

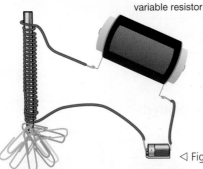

◁ Fig. 7.2 Apparatus for experiment.

Current/A	Number of paper clips held
0	0
0.3	2
0.5	5
0.7	6
0.9	9
1.0	9

△ Table 7.3 Results of experiment.

COMMENT

In this table of results:

✓ the description of each measurement is clear

✓ the units are given in each case.

3. HANDLING EXPERIMENTAL OBSERVATIONS AND DATA

Learning objective: to analyse and interpret data to draw conclusions from experimental activities that are consistent with the evidence, using scientific knowledge and understanding, and to communicate these findings using appropriate specialist vocabulary, relevant calculations and graphs.

Questions to ask:

What is the best way to show the pattern in my results? Should I use a bar chart, line graph or scatter graph?

✓ Graphs are usually the best way of demonstrating trends in data.

✓ A bar chart or bar graph is used when one of the variables is a **categorical variable**; for example, when the melting points of the oxides of the Group 2 elements are shown for each oxide, the names are categorical and not continuous variables.

✓ A line graph is used when both variables are continuous, for example, time and temperature, time and volume.

✓ Scatter graphs can be used to show the intensity of a relationship, or degree of *correlation*, between two variables.

✓ Sometimes a line of best fit is added to a scatter graph, but usually the points are left without a line.

When drawing bar charts or line graphs:

✓ Choose scales that take up most of the graph paper.

✓ Make sure that the axes are linear and allow points to be plotted accurately. Each square on an axis should represent the same quantity. For example, one big square = 5 or 10 units; not 3 units.

✓ Label the axes with the variables (ideally with the independent variable on the x-axis).

✓ Make sure the axes have units.

✓ If more than one set of data is plotted use a key to distinguish the different data sets.

If I use a line graph should I join the points with a straight line or a smooth curve?

✓ When you draw a line, do not just join the dots!

✓ Remember, there may be some points that don't fall on the curve – these may be incorrect or anomalous results.

✓ A graph will often make it obvious which results are anomalous and so it would not be necessary to repeat the experiment (see Example 10).

✓ In biology, if following the biological rhythms of an organism over a period of time, you should join the data points, point-to-point.

Do I have to calculate anything from my results?

✓ It is usual to calculate means from the data.

✓ Sometimes it is helpful make other calculations, before plotting a graph; for example, you might calculate 1/time for a rate of reaction experiment in physics or the energy content of food *per gram* when burning a sample of food in biology.

✓ Sometimes you will have to make some calculations before you can draw any conclusions.

✓ Investigators also look for numerical trends in data, for example, the doubling of a reaction rate every 10 °C; the doubling of numbers of microorganisms every 20 minutes.

Can I draw a conclusion from my analysis of the results, and what scientific knowledge and understanding can be used to explain the conclusion?

✓ You need to use your scientific knowledge and understanding to explain your conclusion.

✓ It is important to be able to add some explanation which refers to relevant scientific ideas in order to justify your conclusion.

EXAMPLE 6

What is the best way to show the pattern in my results?

A student did an experiment to compare the loss of water from leaves of three different species of tree – apple, hazel and oak.

He measured the mass of 10 leaves of similar size and hung the leaves on a line. After three hours, he removed the leaves and measured the masses of the leaves again and calculated the average loss of water in grams per hour.

Species	Average loss of water/g per hour
Apple	0.30
Hazel	0.05
Oak	0.01

△ Table 7.4 Results.

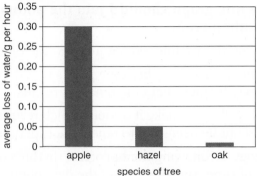

△ Fig. 7.4 Bar chart showing water loss from different leaves.

A bar chart or bar graph is used to display the data in this instance, as the type of leaf is a categoric variable.

EXAMPLE 7

A student carried out an experiment to find out how the rate of a reaction changes during the reaction. She added some hydrochloric acid to marble chips and measured the volume of carbon dioxide produced in a gas syringe. She took a reading of the volume of gas in the syringe every 10 seconds for 1.5 minutes.

The apparatus she used and the results obtained are shown in Figs 7.5 and 7.6:

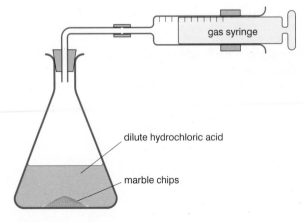

△ Fig. 7.5

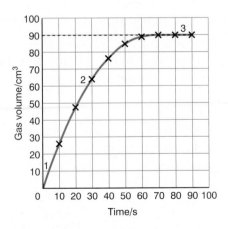

△ Fig. 7.6 Graph of experimental results.

What is the best way to show the pattern in my results?

✓ If the experiment involves **continuous variables**, a line graph is needed.

Straight line or a smooth curve?

✓ The results obtained will either require a smooth curve or a straight line of best fit. The shape of the results should show you which is needed.

Do I have to calculate anything from my results?

✓ If you have to calculate a quantity you will often be able to do this by looking at the change in steepness/gradient of the curve.

Can I draw a conclusion from my analysis of the results?

✓ You need to write a sentence summarising what you have learned from your investigation.

✓ Make sure that you write a clear statement. You might refer, for example, to 'the gradient of the line' at points 1, 2 and 3 to make your conclusion even more precise.

COMMENT

A good conclusion will make direct links to scientific knowledge in relation to the topic.

4. PLANNING AND EVALUATING INVESTIGATIONS

4a Planning

Learning objective: to devise and plan investigations, drawing on scientific knowledge and understanding in selecting appropriate techniques.

Questions to ask:

What do I already know about the area of science I am investigating and how can I use this knowledge and understanding to help me with my plan?

✓ Think about what you have already learned and any investigations you have already done that are relevant to this investigation.

✓ List the factors that might affect the process you are investigating.

What is the best method or technique to use?

✓ Think about whether you can use or adapt a method that you have already used.

✓ A method, and the measuring instruments, must be able to produce **valid** measurements. A measurement is valid if it measures what it is supposed to be measuring.

You will make a decision as to which technique to use based on:

✓ The accuracy and precision of the results required; investigators might require results that are as accurate and precise as possible but if you are making a quick comparison, or a preliminary test to check a range over which results should be collected, a high level of accuracy and precision may not be required.

✓ The simplicity or difficulty of the techniques available, or the equipment required; is this expensive, for instance?

✓ the scale, for example, using standard laboratory equipment or on a micro-scale, which may give results in a shorter time period.

✓ The time available to do the investigation.

✓ Health and safety considerations.

What am I going to measure?

✓ The factor you are investigating is called the **independent variable**. A **dependent variable** is affected or changed by the independent variable that you select.

✓ You need to choose a range of measurements that will be enough to allow you to plot a graph of your results and so find out the pattern in your results.

✓ You might be asked to explain why you have chosen your range rather than a lower or higher range.

How am I going to control the other variables?

✓ These are **control variables**. Some of these may be difficult to control.

✓ You must decide how you are going to control any other variables in the investigation and so ensure that you are using a fair test and that any conclusions you draw are valid.

✓ You may also need to decide on the concentration or combination of reactants.

What equipment is suitable and will give me the accuracy and precision I need?

✓ The **accuracy** of a measurement is how close it is to its true value.

✓ **Precision** is related to the smallest scale division on the measuring instrument that you are using; for example, when measuring a distance, a ruler marked in millimetres will give greater precision than one divided into centimetres only.

✓ A set of precise measurements also refers to measurements that have very little spread about the mean value.

✓ You need to be sensible about selecting your devices and make a judgement about the degree of precision. Think about what is the least precise variable you are measuring and choose suitable measuring devices. There is no point having instruments that are much more precise than the precision you can measure the variable to.

What are the potential hazards of the equipment and technique I will be using and how can I reduce the risks associated with these hazards?

✓ In the exam, be prepared to suggest safety precautions when presented with details of an investigation.

✓ You can find out about hazards associated with reactants using CLEAPSS Student Safety Sheets or a similar resource. Visit www.cleapss.org.uk.

EXAMPLE 8

You have been asked to design and plan an investigation to explore the motion of a trolley down a ramp. In a previous investigation, you have investigated such motion using ticker tape so you are familiar with what happens and the measurements you need to take.

What do I already know?

Previously, you have investigated the motion of a trolley down a ramp. You know that you can use ticker tape to measure the distance the trolley travels in a given time.

What is the best method or technique to use?

The technique you used in your previous investigation can be re-used. You set up the apparatus as shown in the diagram.

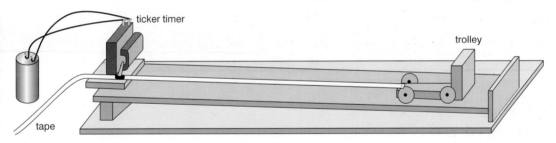

△ Fig. 7.7 Trolley and ramp apparatus for the investigation.

What am I going to measure?

You are investigating the motion of a trolley down a ramp. You will measure the length of each five-dot strip of ticket tape with a ruler.

How am I going to control the other variables?

It is important that you decide on the angle at which to set the ramp at the start. As you have carried out this investigation before, you can look back and see what angle you used previously and decide whether you will use the same angle, or increase or decrease it.

What equipment is suitable and will give me the accuracy and precision I need?

You now know what you will need to measure and so can decide on your measuring devices.

Measurement	Quantity	Device
Length of ticker tape	Five-dot strips	Ruler to measure to nearest centimetre

△ Table 7.5 Suitable equipment for experiment.

Choosing a ruler that can measure to the nearest millimetre would not be appropriate, as the width of the ticker tape dots is of the order of millimetres.

What are the potential hazards and how can I reduce the risks?

The hazards are as follows:

✓ trolley and ramp.

These indicate that there are no specific hazards you need to be aware of.

In terms of the equipment and technique, the major hazard will be the trolley rolling off the end of the ramp. You can limit this hazard by putting a buffer at the end of the ramp, as shown in Fig. 7.7.

EXAMPLE 9

You have been asked to design and plan an investigation to find out the effect of temperature on the rate of reaction between sodium thiosulfate and hydrochloric acid. In a previous investigation you have used this reaction so you are familiar with what happens and how the rate of the reaction can be measured.

4b Evaluating

Learning objective: to evaluate data and methods.

Questions to ask:

Do any of my results stand out as being inaccurate or anomalous?

✓ You need to look for any anomalous results or outliers that do not fit the pattern.

✓ You can often pick this out from a results table (or a graph if all the data points have been plotted, as well as the mean, to show the range of data).

What reasons can I give for any inaccurate results?

✓ When answering questions like this it is important to be specific. Answers such as 'experimental error' will not score any marks.

✓ It is often possible to look at the practical technique and suggest explanations for anomalous results.

✓ When you carry out the experiment you will have a better idea of which possible sources of error are more likely.

✓ Try to give a specific source of error and avoid statements such as 'the measurements must have been wrong'.

Your conclusion will be based on your findings, but must take into consideration any uncertainty in these introduced by any possible sources of error. You should discuss where these have come from in your evaluation.

Error is a difference between a measurement you make, and its true value.

The two types of error are:

✓ random error

✓ systematic error.

With **random error**, measurements vary in an unpredictable way. This can occur when the instrument you're using to measure lacks sufficient precision to indicate differences in readings. Random errors can also occur when it is difficult to make a measurement.

With **systematic error**, readings vary in a controlled way. They are either consistently too high or too low. One reason could be the way you are making a reading, for example, measuring with a measuring cylinder at the wrong point on the meniscus, or not being directly in front of an instrument when reading from it.

What an investigator *should not* discuss in an evaluation are problems introduced by using faulty equipment, or by using the equipment inappropriately. These errors can, or could have been, eliminated, by:

✓ checking equipment

✓ practising techniques before the investigation, and taking care and patience when carrying out the practical.

Overall was the method or technique I used good enough?

✓ If your results were good enough to provide a confident answer to the problem you were investigating, the method probably was good enough.

✓ If you realise your results are not accurate when you compare your conclusion with the standard result, it may be that you have a **systematic error** (an error that has been made in obtaining all the results). A systematic error would indicate an overall problem with the experimental method.

✓ If your results do not show a convincing pattern, it is fair to assume that your method or technique was not precise enough and there may have been a **random error** (that is, measurements vary in an unpredictable way).

If I were to repeat the investigation what would I change or improve?

✓ Having identified possible errors, it is important to say how these could be overcome. Again you should try and be absolutely precise.

✓ When suggesting improvements, do not just say 'do it more accurately next time' or 'measure the volumes more accurately next time'.

✓ For example, if you were measuring small lengths, you could improve the method by using a vernier scale to measure the lengths rather than a ruler.

✓ Investigations can also often be improved by extending the range (temperature, time, pH, and so on) over which it is carried out.

EXAMPLE 10

A student was measuring how current varies with voltage. He used the circuit shown in Fig. 7.8.

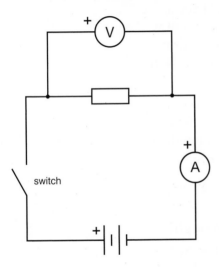

△ Fig. 7.8 Circuit used to measure how current varies with voltage.

Do any of my results stand out as being inaccurate or anomalous?

The student plotted his results on a graph, as shown in Fig. 7.9. An inaccurate result stands out from the rest, as shown by the circle on the graph. Given the pattern obtained with the other results, there is no real need to repeat the result – you could be very confident that the result should have followed the pattern set by the others. A result like this is referred to as an anomalous result. It was an error but not a systematic error.

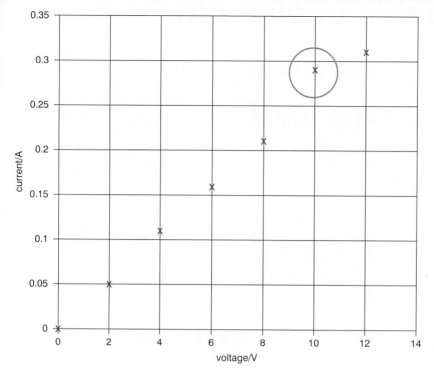

△ Fig. 7.9 Graph of results.

What reasons can I give for any inaccurate results?

The possible source of the error is that one of the variables was noted incorrectly.

Was the method or technique I used precise enough?

You can be reasonably confident that using digital meters for the current and voltage readings will give you precise measurements.

How can I improve the investigation?

For example, you could take readings after the component has heated up so that the steady state resistance is noted.

EXAMPLE 11

A student was measuring the height of a precipitate produced in a test tube when different volumes of lead(II) nitrate solution were added to separate 15 cm³ samples of potassium iodide solution:

Periodic Table of elements

Group

Key

| atomic number |
| atomic symbol |
| name |
| relative atomic mass |

	1
	H
	hydrogen
	1

I	II	III	IV	V	VI	VII	VIII
							2
							He
							helium
							4

3	4											5	6	7	8	9	10
Li	Be											B	C	N	O	F	Ne
lithium	beryllium											boron	carbon	nitrogen	oxygen	fluorine	neon
7	9											11	12	14	16	19	20

11	12											13	14	15	16	17	18
Na	Mg											Al	Si	P	S	Cl	Ar
sodium	magnesium											aluminium	silicon	phosphorus	sulfur	chlorine	argon
23	24											27	28	31	32	35.5	40

19	20	21	22	23	24	25	26	27	28	29	30	31	32	33	34	35	36
K	Ca	Sc	Ti	V	Cr	Mn	Fe	Co	Ni	Cu	Zn	Ga	Ge	As	Se	Br	Kr
potassium	calcium	scandium	titanium	vanadium	chromium	manganese	iron	cobalt	nickel	copper	zinc	gallium	germanium	arsenic	selenium	bromine	krypton
39	40	45	48	51	52	55	56	59	59	64	65	70	73	75	79	80	84

37	38	39	40	41	42	43	44	45	46	47	48	49	50	51	52	53	54
Rb	Sr	Y	Zr	Nb	Mo	Tc	Ru	Rh	Pd	Ag	Cd	In	Sn	Sb	Te	I	Xe
rubidium	strontium	yttrium	zirconium	niobium	molybdenum	technetium	ruthenium	rhodium	palladium	silver	cadmium	indium	tin	antimony	tellurium	iodine	xenon
85	88	89	91	93	96	–	101	103	106	108	112	115	119	122	128	127	131

55	56	57–71	72	73	74	75	76	77	78	79	80	81	82	83	84	85	86
Cs	Ba	lanthanoids	Hf	Ta	W	Re	Os	Ir	Pt	Au	Hg	Tl	Pb	Bi	Po	At	Rn
caesium	barium		hafnium	tantalum	tungsten	rhenium	osmium	iridium	platinum	gold	mercury	thallium	lead	bismuth	polonium	astatine	radon
133	137		178	181	184	186	190	192	195	197	201	204	207	209	–	–	–

87	88	89–103	104	105	106	107	108	109	110	111	112		114		116		
Fr	Ra	actinoids	Rf	Db	Sg	Bh	Hs	Mt	Ds	Rg	Cn		Fl		Lv		
francium	radium		rutherfordium	dubnium	seaborgium	bohrium	hassium	meitnerium	darmstadtium	roentgenium	copernicium		flerovium		livermorium		
–	–		–	–	–	–	–	–	–	–	–		–		–		

lanthanoids

57	58	59	60	61	62	63	64	65	66	67	68	69	70	71
La	Ce	Pr	Nd	Pm	Sm	Eu	Gd	Tb	Dy	Ho	Er	Tm	Yb	Lu
lanthanum	cerium	praseodymium	neodymium	promethium	samarium	europium	gadolinium	terbium	dysprosium	holmium	erbium	thulium	ytterbium	lutetium
139	140	141	144	–	150	152	157	159	163	165	167	169	173	175

actinoids

89	90	91	92	93	94	95	96	97	98	99	100	101	102	103
Ac	Th	Pa	U	Np	Pu	Am	Cm	Bk	Cf	Es	Fm	Md	No	Lr
actinium	thorium	protactinium	uranium	neptunium	plutonium	americium	curium	berkelium	californium	einsteinium	fermium	mendelevium	nobelium	lawrencium
–	232	231	238	–	–	–	–	–	–	–	–	–	–	–

The volume of one mole of any gas is 24 dm³ at room temperature and pressure (r.t.p.)

Glossary

Biology

acrosome A bag of enzymes at the front of a sperm cell that makes it possible for the sperm nucleus to enter the egg cell.

active site The space in an enzyme into which the substrate molecule fits.

active transport The movement of molecules across a cell membrane using energy from respiration; movement is often against a concentration gradient.

aerobic respiration Respiration (the breakdown of glucose to release energy) using oxygen.

alimentary canal The tubular part of the digestive system, from mouth to anus.

alveoli The tiny bulges of the air sacs in lungs where gases diffuse between the air in the lungs and the blood (singular: alveolus).

amino acid The basic unit of a protein.

amniotic fluid Fluid surrounding the developing fetus in the uterus.

amniotic sac The tough membrane surrounding the developing fetus and amniotic fluid in the mother's uterus.

amylase An enzyme that digests starch.

anther The male part of flower that produces pollen.

aorta The largest artery, which receives blood from the left ventricle of the heart.

artery Blood vessel that carries blood away from the heart.

asexual reproduction Production of young without fertilisation.

atrium (plural *atria*) One of two chambers of the heart that receive blood from veins and pump it into the ventricles.

auxin A plant hormone that controls growth of roots and shoots.

B

balanced diet The intake of food that supplies all the protein, fat, carbohydrate, vitamins and minerals that the body needs in the right amounts.

Benedict's reagent Solution that changes colour in the presence of reducing sugars, used to test for their presence in a food sample.

biodiversity The range of variation in species in an area.

biological catalyst A catalyst of reactions inside living organisms.

biomass The mass of a living organism.

biuret test The test used to indicate the presence of protein in a food sample.

bronchioles The tiny tubes in the lungs that carry air to the alveoli.

bronchitis A disease of the lungs that produces a hacking cough as a result of damage to the cilia by smoking or infection.

bronchus The division of the trachea as it joins to the lungs.

C

capillary Smallest blood vessel, found within every tissue, which exchanges substances with the cells.

carbohydrate A large molecule, such as starch or glycogen, made of many simple sugars.

carbon cycle How the element carbon cycles in different forms between living organisms and the environment.

carcinogenic Causes cells to produce cancers, such as some of the chemicals in tobacco smoke.

carnivore An animal that eats animals.

carpel The female structure in flowers that contains one or more ovaries and their stigmas and styles.

catalyst A substance that increases the rate of a chemical reaction, such as an enzyme.

cell membrane The structure surrounding cells that controls what enters and leaves the cell.

cell wall A layer of cellulose that surrounds plant cells, giving them support and shape. Bacterial cells also have a cell wall outside the cell membrane.

cellular respiration See *respiration*.

chemical digestion The breakdown of large molecules into smaller ones using enzymes.

chlorophyll The green chemical in chloroplasts that captures light energy for photosynthesis.

chloroplast An organelle found in plant cells and some protoctist cells that can capture energy from light for use in photosynthesis.

chromosome A long DNA molecule that is found in a cell nucleus.

cilia Tiny hairs that project from a cell surface and that move things across the cell surface.

ciliated cell A cell that has cilia on its surface.

circulatory system The organ system that transports substances around the body.

combustion Burning, such as that of fossil fuels.

community All the organisms that live in the same habitat.

complementary Being similar but opposite, like the shape of the space in an enzyme and the shape of its substrate.

concentration gradient The difference in the amount of a substance between two areas; diffusion is usually down the concentration gradient: molecules move from the area of high concentration to the area of low concentration.

constipation The slow movement of digested food through the intestines and rectum as a result of there being too little fibre in the food for peristalsis to be effective.

consumer An organism that gets its food by eating other organisms; an animal.

COPD (Chronic Obstructive Pulmonary Diseases) Diseases of the lung that include bronchitis and emphysema.

coronary heart disease Diseases such as angina, high blood pressure or heart attack, caused by the partial or complete blockage of a coronary artery, such as by cholesterol.

cuticle A waxy layer that covers leaves, particularly the upper surface, to reduce water loss from the leaf.

cytoplasm The jelly-like liquid inside the cell that contains the organelles and where many chemical reactions take place.

D

decomposer An organism that feeds on dead plants or animals, or animal waste.

deforestation The destruction of large areas of forest or woodland.

denature When an enzyme stops working, which may be the result of high temperature or pH change.

deoxygenated Lacking in oxygen.

deoxyribonucleic acid See *DNA*.

diffusion The net movement of molecules along a concentration gradient from a region of higher concentration to a region of lower concentration; it is a passive process (it does not require energy).

digestion The breakdown of large food molecules into smaller molecules.

digestive enzyme An enzyme found in the digestive system.

double circulation A circulatory system as in humans where the blood passes through the heart to be pumped first to the lungs, then returned to the heart to be pumped to the rest of the body.

E

effector An organ that responds to the nervous system, such as a muscle or a gland.

egestion The removal of undigested material from the body (faeces) (compare with *excretion*).

egg cell The female sex cell.

embryo Developing young, in which cell division and differentiation are taking place rapidly. The stage before a fetus.

emphysema A disease of the lungs caused by damage to alveoli, which reduces the area for gas exchange in the lungs.

endocrine gland An organ that produces a hormone.

enzyme A protein that acts as a biological catalyst, changing the rate of reactions in the body.

epidermis The layer of cells on the outer surface of a body or organ, such as a leaf.

eutrophication The addition of nutrients to water, which may lead to water pollution.

evaporation When particles in a liquid (e.g. water) gain enough energy to move fast enough and become a gas (as in water vapour).

excess More than is needed.

F

faeces The undigested material that remains after digestion of food in humans.

famine Starvation of many people in an area as a result of extreme shortage of food.

fat A solid lipid.

fatty acid One of the basic units of a lipid, along with glycerol.

fertilisation Joining of male and female sex cells.

fetus The name given to the developing baby in the uterus.

fibre Plant material that is difficult to digest and keeps the food in the alimentary canal soft and bulky, aiding peristalsis. Also called roughage.

food chain diagram A diagram that shows the transfer of energy from organism by ingestion.

food web diagram A diagram showing the feeding relationships between organisms in the same community.

fossil fuel Fuel formed from organic material, such as peat, coal and oil.

G

gamete A sex cell.

gas exchange The exchange of gases between the air and the body across a gas exchange surface such as the lungs or a leaf.

germination The start of plant growth from a seed, which only occurs when there is the right amount of oxygen and water and an appropriate temperature.

glycerol One of the basic units of a lipid, along with fatty acids.

gravitropism A plant's growth in response to gravity.

growth The permanent increase in body size and dry mass of an organism, usually from an increase in cell number or cell size (or both).

H

haemoglobin The red chemical in red blood cells that combines reversibly with oxygen.

heart The organ of the circulatory system that pumps blood.

heart rate The number of heart beats in a given time, e.g. beats per minute.

herbivore An animal that eats plants.

hormonal system The chemical response system of the body to changes in the environment.

hormone A chemical substance produced by a gland resulting in a change in another part of the body.

humidity A measure of the concentration of water molecules in the air.

hypha A single thread of fungal mycelium (plural *hyphae*).

I

immune system The system of the body that protects the body against infection; includes white blood cells.

implantation When the embryo settles into the thickened uterus lining.

ingestion The taking of food into the alimentary canal.

insoluble A substance that does not dissolve.

K

kinetic energy The energy carried by moving molecules.

L

larynx The 'voice box' at the top of the trachea, which produces sounds when air moves through it, e.g. when speaking.

leaching The loss of dissolved mineral nutrients in soil water as it soaks deep into the ground beyond the reach of plant roots.

lipase An enzyme that digests lipids (fats and oils).

lipid Molecule made from fatty acids and glycerol.

lungs Organs in the human body where gas exchange takes place.

M

magnification The amount by which a microscope increases the observed size of a structure compared with its actual size: calculated by multiplying the eyepiece magnification with the objective magnification.

malnutrition Not getting the right amounts and balance of nutrients and other essential substances in the diet, including a diet that has too much or too little of any of these.

mechanical digestion The breaking up of food into smaller pieces through biting and chewing by teeth.

menstrual cycle The continuous sequence of events in a woman's reproductive organs; each cycle of ovulation (ripening and release of an egg) and menstruation (shedding of the unwanted uterus lining) takes about 28 days and is controlled by the hormones oestrogen and progesterone.

metabolism All the reactions that occur inside the body that keep an organism alive.

mineral (mineral ion) Nutrients that plants and animals need in small amounts, such as nitrates that are needed for making amino acids.

movement The ability to change the position of all or some of the body.

mucus Slimy liquid that is produced by cells lining the trachea, bronchi and bronchioles.

multicellular An organism that has a body which contains many cells.

mycelium A mass of hyphae that forms the body of a fungus.

N

nerve A bundle of nerve cells.

net movement The sum of all the movements in different directions, e.g. the movement of all the particles being considered in diffusion or osmosis.

nucleus The organelle in plant and animal cells that contains the genetic material.

nutrient cycle How a nutrient cycles between living organisms and the environment.

nutrition The taking in of nutrients to the body from the environment.

O

obesity A condition of the body that has large amounts of fat.

oil A liquid lipid.

omnivore An animal that eats both plants and animals.

optimum pH The pH at which an enzyme works best.

optimum temperature The temperature at which an enzyme works best.

osmosis The net movement of water molecules through a partially permeable membrane, from a solution that has a higher concentration of water molecules (a dilute solution) to one that has a lower concentration of water molecules (a more concentrated solution).

ovary (in plants and humans) A structure that contains egg cells.

oviduct A tube that carries the egg released from the ovary to the uterus; where fertilisation occurs.

ovulation When an egg is released by an ovary.

ovule Female structure in a flower that contains one egg cell.

oxygenated Containing a lot of oxygen.

P

palisade cell Cells in the upper part of a leaf that contain the most chloroplasts and carry out most of the photosynthesis in a leaf.

partially permeable The condition of a membrane that lets some substances pass through but not others.

passive The opposite of *active*, happening without the need for additional energy.

pathogen A disease-causing organism.

peristalsis The rhythmic muscular contractions of the alimentary canal that moves food from mouth to anus.

phagocytosis To flow and engulf, as when phagocytes engulf pathogens.

phloem The plant tissue that carries sucrose through the veins of a plant.

photosynthesis The process carried out in plant cells that makes sugars by combining carbon dioxide and water molecules using energy from light.

phototropism Growth of a plant in response to light.

physical digestion The breaking up of food particles, such as fats and oils, into smaller pieces (droplets) by substances in the alimentary canal, such as bile.

placenta A structure formed by the developing fetus that attaches to the wall of the mother's uterus and across which substances are exchanged between the mother and fetus.

platelet Fragment of a much larger cell that causes blood clots to form at sites of damage in blood vessels.

pollination The process in which pollen from one flower is transferred to another flower, before fertilisation can take place.

producer An organism that produces its own food; for example, plants using energy transferred from light in photosynthesis to produce glucose.

product A molecule that is formed during a reaction.

protease An enzyme that digests proteins.

protein A large molecule that is made of many amino acids joined together.

R

receptor organ An organ that receives information about the environment (such as eye or ear) and responds by stimulating a neurone.

reproduction The process of creating new members of a species.

respiration The chemical process in which glucose is broken down inside the mitochondria in cells, releasing energy and producing carbon dioxide and water.

root hair cell A cell in the epidermis of roots that has a long extension of cytoplasm, where uptake of substances from soil water occurs.

S

secretion The release of chemicals that have been made inside the cell into the fluid outside the cell.

seed The structure formed from an ovule that contains the plant embryo and food stores.

sensitivity The detection of changes (stimuli) in the surroundings by a living organism, and its responses to those changes.

sexual reproduction Production of new individuals by the fusion of a male and a female gamete.

simple sugar A basic sugar unit (e.g. glucose) that can join together with other sugar units to make large carbohydrates such as starch and glycogen.

soil erosion The washing away of soil as a result of wind and rainfall when there is little vegetation to hold on to the soil.

soluble Dissolves easily in a solvent, such as water.

specialisation When a cell develops special features that help it work in a particular way.

specific Limited, usually to one or a few. For example, enzymes are specific because they only work with one or a few similar substrates.

sperm cell Male gamete in animals.

spongy mesophyll The layer of cells in the lower part of the leaf in which there are many air spaces, so increasing the internal surface area to volume ratio.

stamen The male structure in flowers that contains the anther.

starch A complex carbohydrate made from many glucose units.

starvation Eating too little food to supply the body with its need for energy and nutrients.

stigma The female structure in flowers to which pollen grains attach in pollination.

stimulus A change in the internal or external environment that produces a response by an organism.

stomata Tiny holes in the surface of a leaf (mostly the lower epidermis), which allow gases to diffuse into and out of the leaf.

style The structure that supports the stigma in a flower.

substrate A molecule that fits into an enzyme molecule at the start of a reaction.

sucrose Common sugar that is formed of pairs of glucose units.

synthesis The building of larger molecules from smaller ones, such as the formation of proteins from amino acids.

T

target organ An organ that is affected by a hormone.

testis The site of production of sperm in men.

toxic Poisonous.

trachea The tube leading from the mouth to the bronchi, sometimes called the windpipe.

translocation The movement of dissolved substances, such as sucrose and amino acids, through the phloem tissue of a plant.

transpiration Evaporation of water vapour from the surface of a plant.

trophic level A feeding level in a food chain or food web, e.g. producer, primary consumer.

tropism A growth response of a plant as a result of the environment.

U

uterus Where a baby develops inside a mother.

V

vacuole A large sac found in the middle of many plant cells, containing cell sap.

vagina An elastic muscular tube where sperm is received from the penis during sexual intercourse.

valve Flaps in the heart, and in veins, that prevent the flow of blood in the wrong direction.

vascular bundle Tissue that forms the veins in plant roots, stems and leaves, containing xylem vessels and phloem cells.

vein (*animal*) A blood vessel that carries blood towards the heart. (*plant*) See *vascular bundle*.

vena cava The largest human vein that delivers blood from the body to the right aorta.

ventilation Moving air into and out of the lungs.

ventricle One of two chambers of the heart that receive blood from the atria and pump it out through arteries.

vitamin A nutrient needed by the body in tiny amounts to remain healthy, such as vitamins A, C and D.

W

waste product A product of a chemical reaction that is not needed, such as oxygen in photosynthesis.

water potential The potential for a solution to take up more water molecules; it is 0 for pure water and has a negative value for solutions.

water potential gradient The difference in water potential between two regions, e.g. in a plant.

X

xylem vessel A tube formed from dead cells in the vascular bundles of a plant, which carries water and dissolved substances from the roots to the leaves and other parts of the plant.

Z

zygote A fertilised egg, formed from the fusion of a male gamete and female gamete.

Chemistry

A

acid A substance that contains replaceable hydrogen atoms which form H^+ ions when the acid is dissolved in water. It has a pH less than 7.

activation energy The minimum energy that must be provided before a reaction can take place.

addition polymer A polymer that is made when molecules of a single monomer join together in large numbers.

alkali metal A Group I element.

alkane A hydrocarbon in which the carbon atoms are bonded together by single bonds only.

alkene A hydrocarbon that contains a carbon–carbon double bond.

alloy A mixture of a metal and one or more other elements.

anhydrous Literally means 'without water' – a compound, usually a salt, with no water of crystallisation.

anion A negatively charged ion.

anode A positively charged electrode in electrolysis.

atom The smallest particle of an element. Atoms are made of protons, electrons and neutrons.

B

boiling The change of state from liquid to gas.

boiling point The temperature of a boiling liquid – the highest temperature that the liquid can reach and the lowest temperature that the gas can reach.

burning The reaction of a substance with oxygen in a flame.

C

calorimetry A method for determining energy changes in reactions or when substances are mixed together.

carbonate A salt formed by the reaction of carbon dioxide with alkalis in solution.

catalyst A chemical that is added to speed up a reaction, but remains unchanged at the end.

catalytic cracking The process by which long-chain alkanes are broken down to form more useful short-chain alkanes and alkenes, using high temperatures and a catalyst.

cathode A negatively charged electrode in electrolysis.

cation A positive ion.

chemical change A change that is not easily reversed because new substances are made.

chemical formula The combination of element symbols that represents a compound or molecule.

chemical reaction A chemical change that produces new substances and which is not usually easily reversed.

chemical symbol A unique symbol that represents a particular chemical element.

chromatogram A visible record (usually a coloured chart or graph) showing the separation of a mixture using chromatography.

chromatography The process for separating dissolved solids using a solvent and filter paper (in the school laboratory).

collision theory A theory used to explain differences in the rates of reactions as a result of the frequency and energy associated with the collisions between the reacting particles.

combustion The reaction that occurs when a substance (usually a fuel) burns in oxygen.

compound A pure substance formed when elements react together.

condensation The change of state from gas to liquid.

control variable Something that is fixed and is unchanged in an investigation.

covalent bond A bond that forms when electrons are shared between the atoms of two non-metals.

D

dependent variable A variable that changes as a result of changes made to value of the independent variable.

desalination The separation of salt from sea water by evaporation of the water.

diatomic Two atoms combined together (for example, in a molecule).

displacement reaction A reaction in which one element takes the place of another in a compound, removing (displacing) it from the compound.

dissociation The splitting of a molecule to form smaller molecules or, in the presence of water, ions.

distillation The process for separating a liquid from a solid (usually when the solid is dissolved in the liquid) or a liquid from a mixture of liquids.

E

effective collision A collision between particles with enough energy to cause a chemical reaction.

electrode The carbon or metal material that delivers electric charge in electrolysis reactions.

electrolysis The breaking down of a compound by passing an electric current through it.

electrolyte A substance that allows electric current to pass through it when it is molten or dissolved in water.

electron Negatively charged particle with a negligible mass that forms the outer part of all atoms.

electronic configuration The arrangement of electrons in an atom, molecule or ion.

element A substance that cannot be broken down into other substances by any chemical change.

endothermic A type of reaction in which energy is taken in from the surroundings.

enthalpy change (ΔH) The heat energy change when the reactants shown in a chemical equation react together.

enzyme A chemical that speeds up certain reactions in biological systems, such as digestive enzymes that speed up the chemical digestion of food.

evaporation When liquid changes to gas at a temperature lower than its boiling point.

exothermic A type of reaction in which energy is transferred out to the surroundings.

F

filtrate The clear solution produced by filtering a mixture.

fossil fuel Fuel made from the remains of decayed animal and plant matter compressed over millions of years.

fraction A collection of hydrocarbons that have similar molecular masses and boil at similar temperatures.

fractional distillation A process for separating liquids with different boiling points.

freezing Changing a liquid to a solid.

freezing point The temperature at which a liquid changes to a solid.

functional group A part of an organic molecule which is responsible for the characteristic reactions of the molecule.

G

gas The state of matter in which the substance has no volume or shape.

global warming The rise in the average temperature of the Earth's atmosphere and oceans.

greenhouse effect The trapping of long-wave radiation emitted from the Earth's surface by gases in the atmosphere.

greenhouse gas A gas that can trap long-wave radiation emitted from the Earth's surface.

group A vertical column of elements in the Periodic Table.

H

halogens The Group VII elements (F, Cl, Br, I, At).

homologous series A group of organic compounds with the same general formula, similar chemical properties and physical properties that change gradually from one member of the series to the next.

hydrocarbon A compound containing only hydrogen atoms and carbon atoms.

I

independent variable A variable that is deliberately changed in an investigation and, as a result, causes changes to the dependent variable.

indicator A substance that changes colour in either an acid or alkali and so can be used to identify acids or alkalis.

intermolecular force The force of attraction or repulsion between molecules.

intramolecular bond A bond within a molecule.

ion A charged atom or molecule.

ionic bond A bond that involves the transfer of electrons to produce electrically charged ions.

ionic compound A compound formed by the reaction between a metal and one or more non-metals.

ionic equation A chemical equation showing how the ions involved react together.

L

liquid The state of matter in which a substance has a fixed volume but no definite shape.

litmus An indicator that has different colours in acids (red) and alkalis (blue).

M

mass number The number of protons and neutrons in an atom (also known as the nucleon number).

melting Changing a solid into a liquid at its melting point.

melting point The temperature at which a solid changes to a liquid.

metal An element with particular properties (usually hard, shiny and a good conductor of heat and electricity).

mineral A solid inorganic substance that occurs naturally.

mixture Two or more substances combined without a chemical reaction – they can be separated easily.

molecule A group of two or more atoms covalently bonded together.

monatomic An element composed of separate atoms.

monomer Small molecules that can be joined in a chain to make a polymer.

N

neutralisation A reaction in which an acid reacts with a base or alkali to form a salt and water.

neutron Particle in the nucleus of atoms that have mass but no charge.

noble gas Group 0 elements (He, Ne, Ar, Kr, Xe, Rn). They have full outer electron shells.

non-metal An element with particular properties (usually a gas or soft solid and a poor conductor of heat and electricity).

non-renewable A fuel that cannot be made again in a short time span.

nucleus, atomic The tiny centre of an atom, typically made up of protons and neutrons.

nucleon number (mass number) The total number of protons and neutrons in an atom.

O

ore A mineral from which a metal may be extracted.

oxidation state The degree of oxidation of an element.

oxidation The addition of oxygen in a chemical reaction.

P

particle theory The theory describing the movement of particles in solids, liquid and gases.

period A row in the Periodic Table, from an alkali metal to a noble gas.

Periodic Table The modern arrangement of the chemical elements in groups and periods.

periodicity The gradual change in properties of the elements across each row (period) of the Periodic Table.

pH scale A scale measuring the acidity (lower than 7) or alkalinity of a solution (higher than 7). It is a measure of the concentration of hydrogen ions in a solution.

photosynthesis A reaction that plants carry out to make food.

physical change A change in a substance that is easily reversed and does not involve the making of new chemical bonds.

polymer A large molecule made up of linked smaller molecules (monomers). Polythene is a polymer made from ethene.

polymerisation Making polymers from monomers.

pressure A physical characteristic of a substance caused by particles of the substance colliding with the walls of the container it is in.

products The substances that are produced in a reaction.

proton Positively charged particles in the nucleus of atoms.

proton number (atomic number) the number of protons in an atom.

R

radical An element, molecule or ion that is highly reactive.

rate of reaction How fast a reaction goes in a given interval of time.

reactant The substances taking part in a chemical reaction. They change into the products.

reactivity series A list of elements showing their relative reactivities. More reactive elements will displace less reactive ones from their compounds.

redox A reaction involving both oxidation and reduction.

reduction When a substance loses oxygen.

renewable energy Energy from a source that will not run out, such as wind, water or solar energy.

retention factor (R_f) The distance travelled by a substance in a chromatography experiment (through the stationary phase) compared to the distance travelled by the solvent in the same time (expressed as a number between 0 and 1).

reversible reaction A reaction in which reactants form products and products form reactants.

rusting The chemical reaction in which iron is oxidised to iron(III) oxide in the presence of air (oxygen) and water.

S

salt A compound formed when the replaceable hydrogen atom(s) of an acid is (are) replaced by a metal.

saturated Describes an organic compound that contains only single bonds (C–C).

soap A cleaning agent made from fats or oils using sodium hydroxide.

shell A grouping of electrons around a nucleus. The first shell in an atom can hold up to 2 electrons, the next can hold up to 8.

solid The state of matter in which a substance has a fixed volume and a definite shape.

soluble A substance that dissolves in a solvent to form a solution.

solute A substance that dissolves in a solvent producing a solution.

solution This is formed when a substance dissolves in a liquid. Aqueous solutions are formed when the solvent used is water.

solvent The liquid in which solutes are dissolved.

spectator ions Ions that play no part and are unchanged in a chemical reaction.

state symbols These denote whether a substance is a solid (s), liquid (l), gas (g) or is dissolved in aqueous solution (aq).

surface area The total area of the outside of an object.

T

thermal decomposition The breaking down of a compound by heat.

titration An accurate method for calculating the concentration of an acid or alkali solution in a neutralisation reaction.

transition metal Elements found between Group II and III in the Periodic Table. Often used as catalysts and often make compounds that have coloured solutions.

U

universal indicator Indicating solution that turns a specific colour at each pH value.

unsaturated Describes carbon compounds that contain carbon-to-carbon double bonds.

V

valency electrons The outermost electrons of an atom that are involved when the atom reacts with other atoms or compounds.

vapour Another term for gas.

variable A factor that can either be changed in an investigation or changes as a result of other factors changing.

viscous The description of a liquid which does not flow very easily (for example, does not flow as readily as water).

volatile Easily turns to a gas.

W

water cycle The processes that cause the movement of water between the Earth's surface and the atmosphere.

water of crystallisation Water that occurs in crystals.

Physics

A

acceleration A change in speed divided by the time taken to change.

ammeter An instrument that measures electrical current in amperes.

ampere A unit of current measuring the electric charge that flows during one second.

amplitude The maximum change of the medium from normal in a wave. For example, the height of a water wave above the level of calm water.

angle of incidence The angle between the incident light ray and the normal to the surface.

angle of reflection The angle between the reflected light ray and the normal to the surface of the material.

atom A particle of matter. The smallest particle in a chemical element.

average speed The distance an object has moved, divided by the time taken.

B

bimetallic strip A metal bar made from two different metals which have different expansion rates.

C

charge A fundamental property of matter that produces all electrical effects. It is equal to current × time.

chemical energy Energy stored in molecules. Batteries and fuels contain stored chemical energy.

circuit breaker A device that breaks a circuit when there is an increase in current.

compression The squashing together of particles in a particular region.

conduction The transfer of heat energy through a material.

conductors, electricity Substances that conduct electricity well.

conductors, heat Substances that conduct heat very well.

convection Heat transfer in a liquid or gas – when particles in a warmer region gain energy and move into cooler regions carrying this energy with them.

coulomb The unit of electric charge.

D

decibel A unit of sound intensity.

density The mass, in kilograms, of a one metre cube of a substance: mass divided by volume.

diffraction Waves spreading into the shadow when they pass an edge.

distance–time graph A visual representation of how distance travelled varies with time.

double-insulated When a device has a casing that is made of an insulator and does not need an earth wire.

E

earth wire A wire connecting the case of an electrical appliance, through the earth pin on a three-pin plug, to earth.

echo A reflected sound wave.

elastic strain energy A form of stored energy from stretching or compressing an object like a spring.

electric field A region in which any electrical charges will feel a force.

electromagnetic spectrum The 'family' of electromagnetic radiations (from longest to shortest wavelength): radio, microwave, infrared, visible light, ultraviolet, X-rays, gamma rays. In order of frequency, the order is reversed. They all travel at the same speed in a vacuum.

electromagnetic wave A wave that transfers energy – it can travel through a vacuum and travel at the speed of light.

electromotive force The energy per coulomb transferred to the charge carriers by a source of electrical energy

electrostatic forces Forces due to charged particles.

evaporation The change of state from a liquid to a gas.

extension The increase in length when something is stretched.

F

fiducial mark A marker used to judge a timing point, for example, to decide when a pendulum has completed a full swing.

force Change in momentum divided by time taken.

fossil fuel Non-renewable energy resource such as coal, oil or natural gas.

frequency The number of vibrations per second or number of peaks or troughs that pass a point each second, measured in hertz (Hz). It is equal to 1/time period.

friction The force that resists when you try to move something. It can cause insulators to become charged.

fuse A special wire that protects an electric circuit. If the current gets too large, the fuse melts and stops the current.

G

gamma ray Ionising electromagnetic radiation – radioactive and dangerous to human health.

gradient The slope of a curve.

gravitational field strength The force of gravity on a mass of one kilogram. The unit is the newton per kilogram, and it is different on different planets.

gravitational potential energy A form of stored energy given by mass × g × height.

H

Hooke's law The extension of a spring is in direct proportion to the force applied to it, as long as the force is smaller than the material's elastic limit.

hydrometer An instrument used to measure the density of liquids.

I

infrared The part of the electromagnetic spectrum that has a slightly longer wavelength than the visible spectrum.

insulators (of electricity) Substances that do not conduct electricity.

insulators (of heat) Substances that do not conduct heat very well.

internal energy The energy inside an object.

inversely proportional The relationship between two quantities if one doubles when the other halves.

J

joule The unit of energy. One joule is the energy needed to push an object through one metre with a one newton force.

K

kinetic energy The energy of moving objects, equal to $\frac{1}{2}$ × mass × (speed)^2.

kinetic molecular model The theory describing the movement of particles in solids, liquid and gases.

kinetic theory of gases The idea that all matter is made of tiny particles. Kinetic describes how the particles interact to produce the behaviour of gases that we can measure.

L

limit of proportionality The extension up to which Hooke's law applies when stretching an object, for example, a spring.

longitudinal wave A wave in which the change of the medium is parallel to the direction of the wave. Sound is an example.

M

mass The amount of material in an object, measured in kilograms.

microwave Part of the electromagnetic spectrum with a lower frequency than infrared.

molecule A small particle of matter, typically composed of a small number of atoms bonded together.

N

newton A unit of force.

Newton's first law of motion For a body to change the way it is moving, a resultant force needs to act on it.

non-renewable (resources) An energy resource that will run out, such as oil or natural gas.

normal A construction line drawn at 90° to a boundary where a ray of light hits the boundary.

O

Ohm's law The current flowing through a component is proportional to the potential difference between its ends, providing temperature is constant.

oscilloscope An instrument used to display waveforms and measure voltages.

P

parallel Describes a circuit in which the current.

pitch Whether a note sounds high or low to your ear.

potential difference (p.d.) The energy transferred from one coulomb of charge between two points. Measured in volts. Often called the 'voltage'.

potential energy A form of stored energy.

power The amount of energy transferred every second, equal to work done/time taken. Power can be transferred from somewhere (e.g. a power station) or to somewhere (e.g. electric kettle).

power rating The energy transferred per second by an appliance.

pressure The effect of a force spread out over an area. Pressure is equal to force/area.

proportional behaviour Where the two key variables are linked by a simple multiplying factor.

R

radiation Energy, such as electromagnetic rays, that travels in straight lines.

radio wave The part of the electromagnetic spectrum that has a long wavelength and is used for communications.

rarefaction Region where particles are stretched further apart than normal.

reflection When waves bounce off a mirror. The angle of incidence is the same size as the angle of reflection.

refraction When waves change direction because they have gone into a different medium. They change direction because their speed changes.

renewable (resource) An energy resource that is constantly available or can be replaced as it is used, such as solar power or wind power.

resistance The property of an electrical conductor that limits how easily an electric current flows through it. Measured in ohms.

resistor A circuit component used to restrict the flow of current.

resultant force A single imaginary force that is equivalent to all the forces acting on an object, equal to mass × acceleration.

S

series Describes a circuit in which the current travels along one path through every component.

short circuit The unwanted branch of an electrical circuit that bypasses other parts of the circuit and causes a large current to flow.

speed A measure of how far something moves every second. Average speed = distance travelled/time taken.

static charge A charge caused by an excess or deficiency of, usually, electrons – the charged particles are unable to flow and cause a current.

T

thermal transfer The transfer of heat energy.

transverse wave A wave in which the change of the medium is at 90 degrees to the direction of the wave. Light is an example.

turbine A machine that rotates. It is pushed by the movement of a fluid such as air or water.

U

ultrasound Sound with a frequency above 20 000 Hz, above the range of human hearing.

ultraviolet The part of the electromagnetic spectrum that has a slightly longer wavelength than the visible spectrum.

V

variable resistor A component with a resistance that can be manually altered.

visible spectrum The part of the electromagnetic spectrum that can be detected by cells in the eye.

virtual image An image that cannot be projected onto a screen.

volt A unit of voltage. The energy carried by one coulomb of electric charge.

voltage A measure of the energy carried by an electric current.

volume A measure of the space filled by an object, usually in ml or m^3.

watt A unit of power. One watt is one joule transferred every second.

wavelength The distance between the same points of successive waves, for example, the distance from one crest to the next.

weight The force of gravity on a mass, equal to mass × gravitational field strength. The unit of weight is the newton.

work The energy transferred when a job is done, equal to force × distance moved in the direction of the force.

x-ray Part of the electromagnetic spectrum with a higher frequency than ultraviolet.

Answers: Biology

The answers given in this section have been written by the author and are not taken from examination mark schemes.

SECTION 1 CHARACTERISTICS OF LIVING ORGANISMS

Page 13

1. **a)** Any suitable answers for human, such as: movement – walking; respiration – combination of oxygen with glucose to release energy, carbon dioxide and water; sensitivity – vision; growth – increase in height; reproduction – having a baby; excretion – producing urine; nutrition – eating food.

   b) Any suitable answers for a specific animal, such as: ovement – crawling; respiration – combination of oxygen with glucose to release energy, carbon dioxide and water; sensitivity – smell; growth – increase in length; reproduction – producing young; excretion – losing carbon dioxide through respiratory surface; nutrition – eating food.

   c) Any suitable answers for a plant, such as: movement – growing towards light; respiration – combination of oxygen with glucose to release energy, carbon dioxide and water; sensitivity – detecting direction of light; growth – increase in height; reproduction – producing seeds; excretion – diffusion of waste products out of leaf for photosynthesis (oxygen) and respiration (carbon dioxide); nutrition – taking in nutrients from soil and making glucose by photosynthesis.

2. movement – to reach best place to get food or other conditions favourable for growth
   respiration – to release energy from food that can be used for all life processes
   sensitivity – to detect changes in the environment
   growth – to increase in size until large/mature enough for reproduction
   reproduction – to pass genes on to next generation
   excretion – to remove harmful substances from body
   nutrition – to take in substances needed by the body for growth and reproduction

SECTION 2 CELLS

Cell structure

Page 22

1. **a)** Drawing should be drawn with thin, clear pencil lines, no crossing out, to show the outline of the cell in the photograph and the central shape.

   b) Diagram should be labelled to show nucleus, cytoplasm and cell membrane.

2. cell wall, large vacuole, chloroplast

3. **a)** chloroplast

   b) large vacuole

   c) cell wall

Page 25

1. **a)** EXTENDED Lining some tubes in animal organs, such as the respiratory tract of humans; the cilia on the outside of the cells help move substances along inside the tubes.

   b) EXTENDED In blood; carry oxygen around attached to haemoglobin inside the cell.

   c) EXTENDED Near the tips of plant roots; have long cell extensions to increase surface area for absorption of materials into the root.

2. EXTENDED Sperm cells are small, and have a tail for movement. Mitochondria provide energy for movement of the tail, and the acrosome contains enzymes that digest the egg cell membrane so the sperm nucleus can enter the egg cell for fertilisation. Egg cells are large and contain a lot of cytoplasm to provide nutrients for the fertilised cell during the early stages of division.

Page 26

1. 2 mm

2. If you are not using a suitable magnification for the specimen you are looking at, you may not be able to see what you want to. (It is most useful to start by focusing at a lower magnification and then moving up to the magnification you want to use.)

3. actual size $= \dfrac{2.5\,mm}{100}$

Movement in and out of cells

Page 29

1. Any answer that means the same as the following:

   net movement – the sum of movement in all the different directions possible

   diffusion – the sum of the movement of particles from an area of high concentration to an area of lower concentration in a solution or across a partially permeable membrane

2. Only particles that are small enough to pass through the membrane can diffuse. Larger molecules cannot diffuse through the membrane.

Page 33

1. **EXTENDED** Any answer that means the same as the following: the net movement of water molecules from a region of their high concentration to a region of their lower concentration.

2. **a)** It is a passive movement of molecules as the result of a concentration gradient.

   b) Osmosis only considers the movement of water molecules; diffusion considers the solute molecules.

3. The strong cell wall prevents more water entering a plant cell than there is space for in the cell (i.e. when the cell is full of water). The cell wall gives cells that are full of water a specific shape, and this helps to support the plant, keeping it upright.

SECTION 3 BIOLOGICAL MOLECULES

Page 39

1. **a)** fatty acids and glycerol
   b) simple sugars
   c) amino acids

2. Protein is formed from amino acids, carbohydrates from simple sugars; carbohydrates are often made from one kind of simple sugar, proteins from many different kinds of amino acids.

Page 40

1. It is important because many substances dissolve in it.

Page 42

1. **a) i)** An orange–red precipitate would form, because glucose is a reducing sugar. **ii)** The solution wouldn't change colour as there is no starch present.

   b) i) There would be no change in colour because sucrose and the starch in wheat flour are not glucose (reducing sugar). **ii)** The solution would turn blue–black because of the starch in flour.

2. Crush the walnut using a mortar and pestle, then:

   a) mix part with ethanol, decant the liquid and add water, if the mixture turns cloudy, then fat is present

   b) mix part with water to form a solution, add a few drops of biuret solution – if protein present, a blue ring forms at the surface, which disappears to form a purple solution

SECTION 4 ENZYMES

Page 49

1. A substance that speeds up the rate of reaction but remains unchanged at the end of the reaction.

2. A chemical that is found in living organisms that acts as a catalyst.

3. Without enzymes, the metabolic reactions of a cell would happen too slowly for life processes to continue.

4. A substrate is a molecule that an enzyme joins with at the start of a reaction. Substrate molecules are changed to product molecules during a reaction.

Page 50

1. **EXTENDED** The sequence of amino acids in the amino acid chain determines the way the chain will fold up to make the three-dimensional structure of the protein.

2. **EXTENDED** The shape in an enzyme into which a substrate fits closely during a reaction.

3. **EXTENDED** Only a substrate with a shape that is complementary to the shape of the active site can fit into it. So, an enzyme can only work with a particular shape of substrate.

Page 54

1. As temperature increases, the rate of the reaction will increase, up to a maximum point (the optimum) after which it decreases rapidly as the enzyme is denatured.

2. The optimum pH for pepsin is around pH 2, which is very acidic, like the contents of the stomach. The optimum for trypsin is around pH 8, which is more alkaline, like the contents of the small intestine. Each enzyme has an optimum pH that matches the environment in which they work, so that they act most efficiently there.

Page 55

1. **a)** **EXTENDED** The cooler molecules are, the slower they move. So, the longer it takes for the enzymes and substrate molecules to bump into each other and the substrate to fit into the active site. Therefore, the cooler the temperature, the slower the rate of reaction.

   b) **EXTENDED** As temperature increases, the atoms in the enzyme vibrate more. This changes the shape of the active site, making it more difficult for the substrate to fit into the active site and so slowing down the rate of reaction. Eventually, the atoms vibrate so much that the shape of the active site is destroyed and the enzyme is denatured.

2. **EXTENDED** At a pH above and below the optimum of pH 2, the shape of the active site is changed as the interactions between the amino acids in the enzyme are affected by the pH. This makes it more difficult for the substrate to fit into the active site, so the rate of reaction slows down.

SECTION 5 PLANT NUTRITION

Page 62

1.

carbon dioxide + water $\xrightarrow[\text{light energy}]{\text{chlorophyll}}$ glucose + oxygen

2. Without light, photosynthesis cannot take place in plant cells.

3. a) **EXTENDED** $6CO_2 + 6H_2O \xrightarrow[\text{light energy}]{\text{chlorophyll}} C_6H_{12}O_6 + 6O_2$

   b) **EXTENDED** Labels should show: CO_2 from air, H_2O from soil water, $C_6H_{12}O_6$ used in cells for respiration or converted to other chemicals for use in cells, O_2 released into air if not needed in respiration.

4. **EXTENDED** Most organisms other than plants get their energy in chemical form from the food that they eat. That energy was originally converted from light energy to chemical energy during photosynthesis in a plant cell and then transferred as chemical energy along the food chain.

Page 64

1. Test the leaf of a variegated plant for starch. Starch is only produced in the green parts of the leaf, where there is chlorophyll, so only the green parts of the leaf photosynthesise.

2. Heat in a water bath, keeping the ethanol away from open flames, such as from a Bunsen burner, because ethanol gives off flammable fumes.

Page 66

1. a) As light increases, so rate of photosynthesis increases.

   b) As temperature increases, the rate of photosynthesis increases up to a maximum, after which it decreases rapidly.

2. a) As light increases, more energy is supplied to drive the process of photosynthesis.

   b) As temperature increases, up to the maximum the particles in the reaction including enzymes are moving faster and bump into each other more. Above the maximum the rate of photosynthesis decreases because the enzymes that control the process start to become denatured.

Page 68

1. Any four from: cuticle, epidermis, spongy mesophyll, palisade mesophyll, xylem.

2. **EXTENDED** Thin broad leaves, chlorophyll in cells, veins containing xylem tissue that transports water and mineral ions to the leaves and phloem tissue that takes products of photosynthesis to other parts of the plant, transparent epidermal cells, palisade cells tightly packed in a single layer near top of leaf, stomata to allow gases into and out of leaf, spongy mesophyll layer with large internal surface.

3. **EXTENDED** A large surface area helps to maximise the rate of diffusion, in this case diffusion of carbon dioxide into cells for photosynthesis and oxygen out of cells so that it can be released into the air.

4. **EXTENDED** It allows as much light as possible to pass through the epidermal cells to reach the palisade cells below, where there are chloroplasts.

Page 69

1. Plants make their own foods and need to convert the carbohydrates made by photosynthesis into other substances, such as proteins, which contain additional elements.

2. a) Nitrogen is an essential element for making substances other than carbohydrates, such as proteins.

   b) Magnesium is needed to make chlorophyll, which is the green substance in plants.

3. a) **EXTENDED** Stunted growth: because without proteins, the plant cannot make new cells and will not grow well.

   b) **EXTENDED** Without enough magnesium the plant will not be able to make enough chlorophyll, so it will lose the green colour and become yellow. Any magnesium in the plant is transported to the new leaves, so that photosynthesis can continue there for making food for growth.

SECTION 6 ANIMAL NUTRITION

Diet

Page 78

1. Carbohydrates, proteins and fats.

2. Carbohydrates from pasta, rice, potato, bread, wheat flour; proteins from meat, pulses, milk products, nuts; fats from vegetable oils, butter, full-fat milk products, red meat.

3. Vitamins, minerals, water and fibre.

4. Vitamins and minerals are needed for maintaining the health of skin, blood, bones, etc. Water is needed to maintain the water potential of cells. Fibre is needed to help digested food to move easily through the alimentary canal.

1. Any answer along the lines of: different people need different amounts of energy every day; for example, active people need more than people who are seated for much of the day; men have a larger average body mass than women so will need more energy to support that extra tissue; some groups of people need more of a particular group of nutrients than others, e.g. pregnant women need additional iron.

2. Food that contains more energy than the body uses is converted into body fat, leading to obesity, which is associated with many health problems. A diet that is too low in energy leads to health problems as a result of low body weight.

3. a) Obesity is caused by a diet that contains too much energy, and is associated with many diseases.

   b) Starvation is a diet too low in energy and/or nutrients, leading to health problems from deficiency diseases or breakdown of muscle tissue for energy.

   c) Constipation is caused by too little fibre in the diet and may lead to diseases such as bowel cancer and diverticulitis.

Alimentary canal

Page 85

1. Sketch should show the following labels correctly attached to organs shown on the diagram:

   - mouth, where food is broken down by physical digestion (chewing) and amylase enzyme starts digestion of starch in food
   - oesophagus moves food from mouth to stomach by peristalsis
   - stomach, where churning mixes food with protease enzymes and acid to start digestion of protein molecules
   - small intestine, where alkaline bile neutralises the acid chyme and enzymes from pancreas complete digestion of proteins, lipids and carbohydrates, and where digested food molecules are absorbed into the body
   - large intestine, where water is absorbed from undigested food
   - rectum, where faeces are held until they are egested through the anus
   - liver, where bile is made
   - gall bladder, where bile is stored until needed
   - pancreas, where proteases, lipases and amylase which pass to the small intestine

2. Egestion is the removal of undigested food from the alimentary canal – food that has never crossed the intestine wall into the body. Excretion is the removal of waste substances that have been produced inside the body.

3. Peristalsis caused by contraction of the circular muscles of the alimentary canal, followed by relaxation as the longitudinal muscles contract.

Digestion

Page 86

1. Chemical digestion uses chemicals (enzymes) to help break down large food molecules into smaller ones. Mechanical/physical digestion is the chewing by the teeth to break large pieces of food into smaller ones before swallowing, or the breaking up of large fat droplets into smaller ones by bile.

Page 88

1. **EXTENDED** The digestive enzymes break down food molecules that are too large to cross the wall of the small intestine into smaller ones that can be absorbed across cell membranes and so enter the body. If we did not have enzymes, we would not be able to absorb many nutrients from our food.

2. a) **EXTENDED** amylase

   b) **EXTENDED** glucose

3. **EXTENDED** The acid increases stomach acidity, providing the right conditions for enzymes that digest food in the stomach.

SECTION 7 TRANSPORT

Transport in plants

Page 98

1. In vascular bundles that form veins throughout the roots, stems and leaves.

2. Xylem vessels are long continuous tubes formed from dead cells, which allow water and dissolved substances to pass easily through the plant.

3. Phloem cells link together to form continuous phloem tissue in the vascular bundles. They carry dissolved food materials, such as sucrose and amino acids, from the leaves where they are formed to other parts of the plant that use them for life processes or where they will be stored.

Page 100

1. It enters through the root hair cells, moves through the root cortical cells to the xylem in the centre of the root. It moves through the xylem up the stem and into the leaves. In the leaves, it moves out of the xylem into the spongy mesophyll cells.

2. Place a stem of a plant in water containing food colouring. The colour will travel through the xylem with the water, and show where the xylem is in the stem, leaves and flowers.

Page 103

1. Evaporation from the surfaces of a plant, particularly from the stomata of a leaf into the air.

2. Diagram should include annotations like the following, at the appropriate point: water molecules evaporate from surfaces of spongy mesophyll cells into air spaces; water molecules from air spaces move into and out through stomata into the air – diffusion (net movement) usually from inside leaf to outside; osmosis causes water molecules to move from xylem into neighbouring leaf cells, and then from cell to cell until they reach a photosynthesising cell or a spongy mesophyll cell; transpiration is the evaporation of water from a leaf.

3. Closing stomata reduces diffusion of water molecules out of the leaf. At night, oxygen is not needed for photosynthesis, so keeping stomata open would lose water unnecessarily.

4. a) When temperature is higher, particles move faster, so water molecules will diffuse out of the leaf more quickly.

   b) When air humidity is high, there is a high concentration of water molecules in the air. So, more water particles will move from the air through the stomata into the leaf while water particles are moving out of the leaf into the air. This means the rate of diffusion will be lower.

Transport in mammals

Page 105

1. To pump blood around the body

2. Valves in the heart and veins

3. **EXTENDED** Blood passes twice through the heart for every once round the body – there are effectively two separate circulations of blood from the heart.

Page 107

1. left atrium, right atrium, left ventricle, right ventricle

2. Arteries carry blood away from the heart; veins carry blood towards the heart.

3. **EXTENDED** vena cava, right atrium, right ventricle, pulmonary artery, pulmonary vein, left atrium, left ventricle, aorta

Page 109

1. **EXTENDED** Resting heart rate varies widely due to many factors, including age, health and fitness, so a single value for the average is too limited.

2. As level of activity increases, so heart rate increases.

3. **EXTENDED** Heart rate increases with exercise so the blood can circulate faster round the body, delivering oxygen and glucose to muscle cells for the increased rate of respiration to generate the energy needed for contraction. It also removes waste carbon dioxide from muscle tissue more rapidly to prevent it building up and affecting cells.

Page 111

1. **EXTENDED** To supply the oxygen and glucose needed for the heart muscle cells to respire and to remove waste carbon dioxide.

2. smoking, diet containing a lot of saturated fat, stress, genetic factors

3. Eat a diet that is relatively low in saturated fat, don't smoke and try to control stress and the effects that it has on behaviour.

Page 114

1. a) renal arteries b) aorta c) pulmonary veins

2. Arteries are large vessels with thick, elastic muscular walls; capillaries are tiny blood vessels with very thin walls that are often only one cell thick; veins are large vessels with a large lumen and valves to prevent backflow of blood.

3. **EXTENDED** The walls stretch as blood enters then slowly relax as the blood flows through, balancing out the pressure so that the change in pressure is reduced.

Page 116

1.

Blood component	Function
plasma	carries dissolved substances, such as carbon dioxide, glucose, urea and hormones; also transfers heat energy from warmer to cooler parts of the body
red blood cell	carries oxygen
white blood cell	protects against infection
platelet	causes blood clots to form when a blood vessel is damaged

2. **EXTENDED** The biconcave disc shape increases surface area to volume ratio, so rate of diffusion of oxygen into and out of a cell is maximised. Haemoglobin inside the cell binds with oxygen when oxygen concentration is high and releases oxygen when oxygen concentration is low. The cell has no nucleus, so there is as much room as possible for haemoglobin. The cell has a flexible shape so can squeeze through the smallest capillaries and reach all tissues.

SECTION 8 GAS EXCHANGE AND RESPIRATION

Gas exchange

Page 128

1. **EXTENDED** Exchange of gases between the body and the environment is by diffusion. Organisms need plenty of oxygen for respiration to provide energy for all life processes, and need to get rid of the waste carbon dioxide. So a rapid rate of diffusion supports a higher rate of respiration and all the other processes in the body.

2. The trachea carries air from the mouth down to the lungs; the bronchi (the two large divisions of the trachea as it reaches the lungs) are supported with rings of cartilage to prevent collapse during breathing; the bronchioles (the fine tubes in the lungs) carry the air to the alveoli; the alveoli (the bulges of the air sac) have a large surface area and are very thin for efficient diffusion of gases.

3. **EXTENDED** Sketch similar to Fig. 8.4, with annotations showing: thin lining of alveolar wall and wall of capillary allows rapid diffusion; high concentration gradients for gases between blood and air in alveolus due to continuous blood flow through capillary and ventilation of alveolus (lungs); large area of contact between capillary and alveolus, maximising area over which diffusion can occur.

4. **EXTENDED** The mucus traps particles and microorganisms that are in the air breathed in, and the cilia move the mucus and anything trapped in it up out of the lungs to the throat, where it can be swallowed. This protects the lungs from damage and infection.

Page 131

1. The percentage of oxygen is less in exhaled air than inhaled air. The percentage of carbon dioxide is greater in exhaled air than inhaled air. The percentage of water vapour is higher in exhaled air than inhaled air.

2. As level of exercise increases, rate and depth of breathing increase.

3. **EXTENDED** There is less oxygen in expired air than inspired air because oxygen in the body is used for respiration. There is more carbon dioxide in exhaled air than inhaled air because the body produces carbon dioxide in respiration. There is more water vapour in exhaled air than inhaled air because water molecules evaporate from the surface of the alveoli due to warmth of the body.

4. **EXTENDED** More exercise means more carbon dioxide is produced from an increased rate of respiration. Carbon dioxide is a soluble acidic gas so causes the body tissues and blood to become more acidic. A change in pH can affect many enzymes and so affect the rate at which life processes are carried out in the body. Slowing down the rate of life processes may harm the body.

Page 132

1. Bronchitis – hacking cough caused by build-up of mucus in tubes of lungs as a result of damage to cilia; emphysema – shortage of breath due to breakdown of walls of alveoli so there is a smaller surface area for gas exchange in the lungs.

2. Nicotine is addictive, which makes smoking difficult to give up; carbon monoxide replaces oxygen on haemoglobin, reducing the amount of oxygen that blood can carry.

Respiration

Page 135

1. **a)**, **b)** and **c)**:

   glucose (*from digested food from alimentary canal*) + oxygen (*from air via lungs*) ⟶ carbon dioxide (*excreted through lungs*) + water (*used in cells or excreted through kidneys*) (+ energy (*transferred to other chemicals in cell processes*))

   d) glucose replaced by fats from hump, and very little water excreted through kidneys

2. inside cells

3. Any three from: muscle cells for contraction, synthesis of new molecules, such as proteins, for growth, active transport across cell membranes, passage of nerve impulses, maintenance of core body temperature.

4. **EXTENDED** $C_6H_{12}O_6 + 6O_2 \longrightarrow 6CO_2 + 6H_2O$

SECTION 9 COORDINATION AND RESPONSE

Hormones in humans

Page 144

1. **a)** A chemical messenger in the body that produces a change in the way some cells work.

   b) A gland that secretes hormones.

c) An organ that contains cells that are affected by hormones.

2. When faced with attack, or when suddenly frightened.

3. It prepares the body for action by increasing the amount of oxygen and glucose delivered to muscle cells for rapid respiration, and improving vision.

Tropic responses

Page 147

1. A growth response of a plant to a stimulus.

2. a) Shoots grow towards light.

   b) Roots grow in the direction of the force of gravity.

3. EXTENDED Auxin is produced in the tip of the growing shoot and diffuses down the shoot. Auxin on the bright/light side of the shoot moves across the shoot to the darker side as it diffuses down the shoot. Cells on the dark side of the shoot elongate more than the cells on the light side of the shoot, so the shoot starts to bend as it grows so that the tip is pointing towards the light.

SECTION 10 REPRODUCTION

Asexual and sexual reproduction

Page 154

1. Reproduction without the fusion of gametes, using a cell from only one parent.

2. Binary fission is where the genetic material is copied and the cell splits in half. Only one cell is involved and there is no fusion of parent cells before division.

Page 154

1. a) The fusion of a male gamete and a female gamete to produce a zygote.

   b) The production of offspring from two parents as a result of fertilisation.

Sexual reproduction in plants

Page 156

1. Stigma, where pollen grains attach; style, which supports the stigma; ovary, which surrounds and protects the ovule, inside which is the female gamete.

2. Stamen, which includes an anther that contains pollen grains, inside which are the male gametes; filament, which holds the anther above the flower to help with shedding of pollen.

Page 160

1. Pollination is the transfer of pollen from a stamen to a stigma. Fertilisation is the fusion of the male gamete with the female gamete to form a zygote.

2. Any three from: wind-pollinated flowers usually small, no colour (white), make masses of lightweight pollen. Insect-pollinated plants usually large, may be brightly coloured, produce nectar and sometimes scent, make small amounts of larger pollen grains.

3. a) EXTENDED Can make less pollen; less waste of pollen as insects more likely to deliver pollen to flower than random distribution in wind.

   b) If the insect species die out, the plant will not get pollinated.

Page 163

1. When the embryo in a seed starts to grow, splitting the seed coat and increasing in size and complexity.

2. a) Seeds need a supply of oxygen for growth, although they may be able to start germination using anaerobic respiration.

   b) Seeds need water for germination and will not germinate in dry soil.

   c) Seeds need warmth for germination, although the amount of warmth they need may depend on where they naturally grow. Seeds from plants that live in colder areas may need a period of deep cold before they will germinate. Seeds from plants that live in areas prone to fire may not germinate until after a fire.

Sexual reproduction in humans

Page 165

1. Sketch should be similar to Fig. 10.17. Labels and annotations as follows:
   - testes, where sperm (male gametes) are produced
   - sperm duct, which carries sperm to urethra
   - prostate gland and seminal vesicles, which produce liquid in which sperm swim
   - penis, which when erect delivers sperm into vagina of female
   - urethra, the tube that carries sperm from sperm ducts to outside the body.

2. Sketch should be similar to Fig. 10.18. Labels and annotations as follows:
   - ovaries, where egg cells form
   - oviducts, which carry the eggs to the uterus and where fertilisation by sperm takes place

- uterus, where embryo implants into lining and fetus develops
- cervix, base of uterus where sperm are deposited during sexual intercourse
- vagina, where penis is inserted during sexual intercourse.

3. EXTENDED

	Egg cell	Sperm cell
size	very large, 0.2 mm diameter	very small, 45 µm long
numbers	thousands in ovary but usually only one released each month	>100 million produced each day
mobility	unable to move on its own	self-propelling with tail

Page 168

1. a) The cell produced by fusion of a male gamete and female gamete.

    b) Formed from the division of cells in the zygote – until distinctive structures are obvious, such as limbs, when it becomes a fetus.

    c) Developing baby in the uterus (womb), from about 3 months after fertilisation.

2. In an oviduct.

3. In early stages, rapid cell division, and differentiation of cells to produce the main structures; later, development of nervous system and movement; increase in size and weight.

4. EXTENDED Provides nutrients from mother's blood and carries waste to mother's blood to be excreted.

SECTION 11 ORGANISMS AND THEIR ENVIRONMENT

Page 180

1. Producer: an organism that produces its own food from simpler materials, e.g. plants making carbohydrates in photosynthesis.

    Consumer: an organism that gets its food from eating other organisms, e.g. animal.

    Herbivore: an animal that eats plants.

    Carnivore: an animal that eats other animals.

2. EXTENDED Decomposer: an organism that gets its food from dead plants and animals or waste material, such as some fungi and bacteria.

    Trophic level: the feeding level of an organism within a food chain or food web.

3. The Sun provides light energy, transferred as chemical energy to build plant tissue, which is then transferred as chemical energy through all other organisms in the ecosystem.

4. A food chain shows the relationship between one producer, one herbivore, the carnivore that eats the herbivore, and so on.

    A food web shows the feeding relationships between all the organisms living in an area.

5. a) Food webs help us to understand the relationship between organisms in an area, and can help us predict what might happen to the organisms as a result of a change to the ecosystem.

    b) It can be difficult to organise the information in a food web because some organisms feed at many trophic levels, and it may not be possible to include all organisms (e.g. decomposers) on a food web because of space for the drawing.

Page 183

1. EXTENDED Energy in light from Sun (gain) > some reflected, some passes straight through, some wrong wavelength (losses) > energy in light transferred to chemical substances during photosynthesis > energy transferred to environment from photosynthetic reactions and from respiration by heating (losses) > energy stored in plant biomass.

2. EXTENDED Energy stored in food (gain) > energy in undigested food lost, transferred to environment in faeces (loss) > energy stored in absorbed food molecules transferred to energy in waste products, such as urea in urine, and transferred to the environment (loss) > energy released in respiration transferred by heating to environment (loss) > energy stored in animal biomass.

3. EXTENDED Not all the energy gained is stored in new tissue in the organism. When the next trophic level feeds on the previous level, only the energy stored in the body tissue is available to it.

SECTION 12 HUMAN INFLUENCES ON ECOSYSTEMS

Page 193

1. a) Respiration releases carbon dioxide into the atmosphere from the breakdown of complex carbon compounds inside organisms.

    b) Photosynthesis fixes/converts carbon dioxide from the atmosphere into complex carbon compounds in plant tissue.

    c) Decomposition decays/breaks down dead plant and animal tissue by decomposers, releasing carbon dioxide into the atmosphere during respiration.

2. a) carbon dioxide **b)** complex carbon compounds, **c)** complex carbon compounds

3. EXTENDED Combustion increases the carbon dioxide concentration in the atmosphere more rapidly than natural processes such as respiration. Deforestation removes trees, so this reduces the amount of oxygen taken from the atmosphere for photosynthesis and increases the amount of carbon dioxide released if the forest is burnt. So this can rapidly change the oxygen/carbon dioxide balance in the atmosphere near the forest.

Page 197

1. a) EXTENDED Changes the amount of water transferred from soil to air through transpiration, so more water remains in soil and enters rivers.

    b) Soil washed away and nutrients leached from soil by increased water flow through ground, so decreasing soil fertility.

    c) Increases carbon dioxide concentration as less carbon dioxide taken from air through photosynthesis and stored as wood.

2. The addition of nutrients to water.

3. Run-off of fertiliser into water as a result of heavy rainfall; leaching of soluble nutrients in fertiliser through soil into water systems.

4. Eutrophication leads to the rapid growth of algae and other microorganisms, which remove large amounts of oxygen from the water for respiration. This does not leave enough oxygen in the water for the fish, so they die.

5. EXTENDED Sewage added to water > adds nutrients to water = eutrophication > plant and microorganism growth rate increases > respiration rate of microorganisms increases, removing dissolved oxygen from water > less dissolved oxygen for other organisms, which die = water pollution.

Answers: Chemistry

The answers given in this section have been written by the author and are not taken from examination mark schemes.

SECTION 1 PRINCIPLES OF CHEMISTRY

The particulate nature of matter

Page 205

1. (I)
2. Only the solid state has a fixed shape.
3. Fine sand will pour or flow like a liquid; it takes the shape of the container it is poured into (although under a microscope you would see gaps at the surface of the container).

Page 209

1. The particles in a solid vibrate about a fixed point.
2. The particles are held together the most strongly in solid water (ice).
3. Evaporation is the process that occurs when faster-moving particles in a liquid escape from the liquid surface.
4. Melting is the name of the process when a solid changes into a liquid.

Page 210

1. All elements contain atoms.
2. Methane contains one carbon atom and four hydrogen atoms.
3. This molecule is made up of 6 carbon atoms, 12 hydrogen atoms and 6 oxygen atoms.
4. An ion is a charged particle and an atom is neutral.

Experimental techniques

Page 216

1. A baseline drawn in pencil will not dissolve in the solvent.
2. If the solvent were above the baseline the substances would just dissolve and form a solution in the beaker.
3. The dye may be insoluble in the solvent.
4. **EXTENDED** $R_f = 1.7/10 = 0.17$

Page 219

1. A solvent is a liquid that will dissolve a substance (solute).
2. If a substance is soluble in a solvent it dissolves in that solvent.
3. Distillation
4. Boiling points

Atoms, elements and compounds

Page 224

1. In a compound the elements are chemically combined together. In a mixture, the elements or components are not chemically combined together.

Page 229

1. The electron has the smallest relative mass.
2. Atoms are neutral. The number of positive charges (protons) must equal the number of negative charges (electrons).
3. a) The nucleon number is 27.
   b) 14 neutrons

Page 230

1. a) Magnesium has two electrons in its outer electron shell.
   b) It is in Group II.
3. a) Aluminium

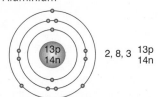

   2, 8, 3   13p 14n

   b) Calcium

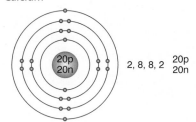

   2, 8, 8, 2   20p 20n

3. The noble gases have full outer electron shells or have eight electrons in their outer electron shells and so do not easily lose or gain electrons.

Ions and ionic bonds

Page 236

1.

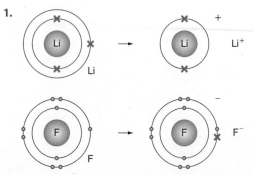

2. EXTENDED

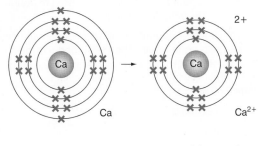

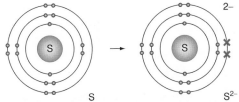

3. EXTENDED Both phosphorus and oxygen are non-metals. (A metal is needed to form an ionic bond.)

Page 239

1. EXTENDED The ions are held together strongly in a giant lattice structure. The ions can vibrate but cannot move around.

2. EXTENDED Sodium chloride is made up of singly charged ions, Na^+ and Cl^-, whereas the magnesium ion in magnesium oxide has a double charge, Mg^{2+}. The higher the charge on the positive ion, the stronger the attractive forces between the positive ion and the negative ion.

Molecules and covalent bonds

Page 245

1.

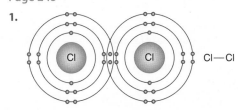

2. EXTENDED

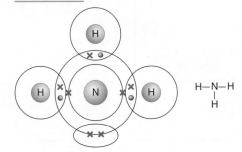

3. EXTENDED

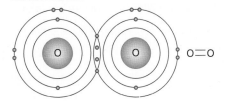

4. EXTENDED

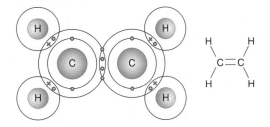

5. EXTENDED

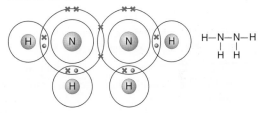

Page 247

1. EXTENDED The intermolecular forces of attraction between the molecules are weak.

2. No. There are no ions or delocalised electrons present.

Stoichiometry

Page 253

1. a) KBr
  b) CaO
  c) $AlCl_3$
  d) CH_4

2. a) $Cu(NO_3)_2$
  b) $Al(OH)_3$
  c) $(NH_4)_2SO_4$
  d) $Fe_2(CO_3)_3$

3. a) EXTENDED $ZnCl_2$
  b) EXTENDED Cr_2O_3
  c) EXTENDED $Fe(OH)_2$

Page 257

1. a) $2Ca + O_2(g) \rightarrow 2CaO(s)$

b) $2H_2S(g) + 3O_2(g) \rightarrow 2SO_2(g) + 2H_2O(l)$

c) $2Pb(NO_3)_2(s) \rightarrow 2PbO(s) + 4NO_2(g) + O_2(g)$

2. a) EXTENDED $S(s) + O_2(g) \rightarrow SO_2(g)$

b) EXTENDED $2Mg(s) + O_2(g) \rightarrow 2MgO(s)$

c) EXTENDED $CuO(s) + H_2(g) \rightarrow Cu(s) + H_2O(l)$

Page 259

1. a) EXTENDED $2C_5H_{10}(g) + 15O_2(g) \rightarrow 10CO_2(g) + 10H_2O(l)$

b) EXTENDED $Fe_2O_3(s) + 3CO(g) \rightarrow 2Fe(s) + 3CO_2(g)$

c) EXTENDED $2KMnO_4(s) + 16HCl(aq) \rightarrow 2KCl(s) + 2MnCl_2(s) + 8H_2O(l) + 5Cl_2(g)$

SECTION 2 PHYSICAL CHEMISTRY

Electricity and chemistry

Page 268

1. The breaking down (decomposition) of an ionic compound by the use of electricity.

2. The positive electrode is the anode.

3. The substance must contain ions and they must be free to move (in molten/liquid state or dissolved in water).

Page 272

1. a) An inert electrode is an unreactive electrode; it will not be changed during electrolysis.

b) Carbon is commonly used as an inert electrode (Platinum is another inert electrode).

2. a) Lead and chlorine.

b) Magnesium and oxygen.

c) Aluminium and oxygen.

Chemical energetics

Page 276

1. A reaction that releases heat energy to the surroundings.

2. A reaction that absorbs energy from the surroundings.

3. Polystyrene is a very good insulator and so very little energy is transferred to the surroundings.

Page 277

1. A high proportion of the energy released transfers to the surrounding air.

2. a) Weighing the spirit burner before and after burning the fuel.

b) Ethanol 29; Paraffin 33; Pentane 25; Octane 40.

i) Octane

ii) A 10 °C rise would be expected. The same amount of energy is transferred to double the volume of water.

3. The group of students using the metal should get more accurate results. The metal conducts the heat from the fuel to the water better than glass does.

Page 279

1. The reaction is endothermic.

2. The activation energy.

3. EXTENDED

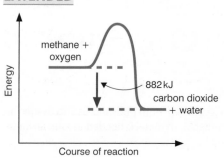

4. EXTENDED

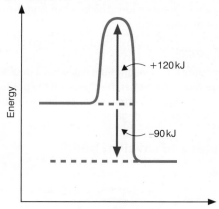

Reaction A

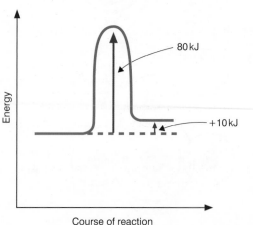

Reaction B

Page 282

1. The sign indicates whether the reaction is exothermic (negative sign) or endothermic (positive sign).

2. **EXTENDED** Energy is needed to break bonds.

3. **EXTENDED** In an endothermic reaction more energy is needed to break bonds than is recovered on forming bonds.

Rate of reaction

Page 288

1. In a physical change no new substances are made. In a chemical change at least one new substance is made.

2. The apparent change in mass is often because a gas has been either a reactant or a product (and is lost from the reaction vessel).

3. **EXTENDED** The particles must collide; there must be sufficient energy in the collision (to break bonds).

4. **EXTENDED** An effective collision is one which results in a chemical reaction between the colliding particles.

5. **EXTENDED** Student's diagram like Fig. 2.19. It is an energy barrier. Only collisions that have enough energy to overcome this barrier will lead to a reaction.

Page 291

1. A gas syringe will measure the volume of gas produced accurately.

2. **EXTENDED** No gas is being produced – the reaction hasn't started or it is finished.

3. **EXTENDED** The quicker reaction will have the steeper gradient.

Page 295

1. The units of concentration for solutions are mol/dm³.

2. **EXTENDED** The particles are closer together and there are more of them, so there will be more (effective) collisions per second.

3. **EXTENDED** Increasing the temperature means the particles have more (kinetic) energy. So more of the collisions will have energy greater than or equal to the activation energy and there will be more effective/successful collisions per second.

Page 296

1. A catalyst is a substance that changes the rate of a chemical reaction.

Redox reactions

Page 301

1. Reduction is the loss of oxygen or the gain of electrons.

2. **a)** +2

   b) +3

   c) +7

Acids, bases and salts

Page 305

1. Both solutions are alkalis. Solution A is a weakly alkaline whereas solution B is a strongly alkaline.

Page 310

1. A salt is formed when a replaceable hydrogen of an acid is replaced by a metal.

2. Sulfuric acid

3. Potassium chloride will be soluble in water (as are all potassium salts).

4. Calcium nitrate

5. Neutralisation is the reaction between and acid and an alkali or base to form a salt and water.

Identification of ions and gases

Page 317

1. **a)** A white precipitate, which does not dissolve in excess sodium hydroxide solution.

   b) A white precipitate, which does dissolve in excess soldium hydroxide solution.

2. Add sodium hydroxide solution. Fe^{2+} produces a green precipitate; Fe^{3+} produces a reddish-brown precipitate.

Page 320

1. Add dilute sodium hydroxide and heat. An alkaline gas (turns red litmus paper blue) indicates the presence of an ammonium compound.

2. **a)** Carbon dioxide

   b) Bubble the gas through limewater. A white precipitate forms.

3. The Fe^{3+} ion is present in solution X.

4. The Cl^- ion is present in solution Y.

5. **a)**

Name of cation	Colour of precipitate
Zinc/lead	white
Magnesium/calcium	white
Copper(II)	blue
Iron(II)	green/turns brown slowly
Iron(III)	rust brown/orange

b) HCl is added to remove any carbonate ions that may be present.

6. Plan needs to check for testing of both anion and cation for each sample and should include practical instructions.
   - Blue compound
   - Test for copper(II) – sodium hydroxide: result blue precipitate
     $Cu^{2+}(aq) + 2OH^-(aq) \rightarrow Cu(OH)_2(s)$
   - Test for sulfate – hydrochloric acid/barium chloride: result white precipitate
     $Ba^{2+}(aq) + SO_4^{2-}(aq) \rightarrow BaSO_4(s)$
   - White compound
   - Flame test for Na^+ – yellow
   - Test for carbonate – add dilute acid – effervescence/carbon dioxide evolved – turns limewater milky
     $CO_3^{2-}(s) + 2H^+(aq) \rightarrow H_2O(l) + CO_2(g)$

Page 321

1. Ammonia
2. Oxygen
3. Chlorine

SECTION 3 INORGANIC CHEMISTRY

The Periodic Table

Page 328

1. **a)** 20
   b) The proton number is the number of protons (which equals the number of electrons) in an atom of the element. Calcium atoms have 20 protons and 20 electrons.
   c) Group II
   d) Period 4
   e) Calcium is a metal.
2. The halogens
3. The halogens are non-metals.

Page 330

1. **EXTENDED** Aluminium has 3 electrons in the outer shell.
2. **EXTENDED** Oxygen will form an O^{2-} ion (the oxide ion) with a 2- charge.
3. **EXTENDED** Fluorine (F)
4. **EXTENDED** Barium (Ba)

Group I elements

Page 334

1. They react with water to form alkaline solutions.

2. They have one electron in the outer shell.
3. They are soft to cut (also have very low melting points).
4. **EXTENDED** The melting point of rubidium will be less than that of sodium. Melting points decrease down the group and rubidium is below sodium in the group.
5. **EXTENDED** The potassium atom is larger than the lithium atom so the outer electron is further from the attraction of the nucleus and can be more easily removed. Potassium is more reactive than lithium because of the distance of the outer electron from the nucleus.

Page 336

1. Sodium oxide is white.
2. Hydrogen. The solution formed is potassium hydroxide.
3. The compounds are soluble.
4. **a)** **EXTENDED** Rubidium will have a lower melting point.
5. **a)** **EXTENDED** A group is a vertical column of elements having similar chemical properties because of their outer shell electronic structure.
   b) **EXTENDED** lithium, sodium, potassium
   c) **EXTENDED** All the elements in Group I have one electron in the outer shell.
   d) **EXTENDED** The reactivity of these elements depends on the ease with which the outer electron is lost. One electron can easily be lost to form positive ions. The ease with which it can be lost increases down the group because the electron is less tightly held in the atom and therefore reactivity increases down the group.

Group VII elements

Page 343

1. Seven electrons in the outer shell.
2. The atoms need to gain only one electron to achieve a full outer shell.
3. Chlorine molecules are made up of two atoms combined/bonded together, Cl_2.
4. **EXTENDED** Solid. The trend down the group is gas, liquid, solid.
5. **EXTENDED** A displacement reaction involves one Group VII element being reduced (gaining electrons) and one being oxidised (losing electrons).

Pages 345

1. Chlorine kills any bacteria that might be present in the water.
2. Iodine

3. EXTENDED The non-stick surfaces on pans/ frying pans.

4. a) EXTENDED

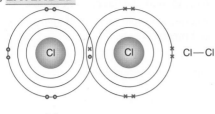

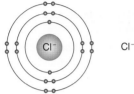

b) EXTENDED Chlorine is a more reactive halogen than bromine. Chlorine will displace bromine from a solution of a bromide ions. (Chlorine will oxidise bromide ions to bromine.)

Observations: chlorine water is pale green. When this is added to a colourless solution of potassium bromide the resulting solution will turn orange due to the presence of bromine.

5. a) EXTENDED The reactivity of fluorine is due to its electronic structure 2,7. Fluorine needs to gain only one electron to form a fluoride ion. This is very easy because of its small size and large attractive force of the nucleus.

b) EXTENDED Both chlorine and iodine are less reactive than fluorine. Fluorine could only be displaced from fluoride ions by a more reactive halogen. As there are no halogens that are more reactive than fluorine, fluorine will not be displaced from fluoride ions.

Transition metals and noble gases

Page 349

1. No. Copper is very unreactive – it is below hydrogen in reactivity series.

2. a) $FeSO_4(aq) + 2NaOH(aq) \rightarrow Fe(OH)_2(s) + Na_2SO_4(aq)$

b) Green

Metals

Page 357

1. A malleable metal can be hammered into shape.

2. An alloy is a mixture of a metal with one or more other elements.

3. Cupronickel is used for making coins.

4. The element added in the alloy disrupts the rows of aluminium atoms making them less likely to slide over each other when under strain.

Pages 361

1. No. Copper is below hydrogen in the reactivity series.

2. EXTENDED $2K(s) + 2H_2O(l) \rightarrow 2KOH(aq) + H_2(g)$

3. No. Carbon is below magnesium in the reactivity series.

4. EXTENDED $Mg(s) + PbO(s) \rightarrow MgO(s) + Pb(s)$

Pages 364

1. EXTENDED Iron ore (hematite), coke and limestone.

2. EXTENDED Iron(III) oxide

3. EXTENDED Carbon dioxide, carbon monoxide, nitrogen (from the air).

4. EXTENDED $2Fe_2O_3(s) + 3C(s) \rightarrow 4Fe(s) + 3CO_2(g)$

Air and water

Pages 372–373

1. Anhydrous means without water (water of crystallisation).

2. Cobalt(II) chloride will change from blue to pink.

3. a) The first filter is coarse gravel. The second filter is fine sand.

b) Chlorine is used to kill bacteria.

4. a) Nitrogen is 78%.

b) Carbon dioxide is 0.04%.

Pages 376

1. a) Methane is the major component of natural gas. It is also produced by decaying vegetable matter and by ruminant animals such as cows.

b) Carbon dioxide is another greenhouse gas.

Page 378

1. In a limited supply of air, carbon will form carbon monoxide.

2. a) copper(II) carbonate + sulfuric acid → copper(II) sulfate + carbon dioxide + water

b) EXTENDED $CuCO_3(s) + H_2SO_4(aq) \rightarrow CuSO_4(aq) + CO_2(g) + H_2O(l)$

3. a) calcium carbonate → calcium oxide + carbon dioxide

b) EXTENDED $CaCO_3(s) \rightarrow CaO(s) + CO_2(g)$

4. Rust is iron(III) oxide/hydrated iron(III) oxide.

5. Covering in grease, painting, plastic coating, coating with a metal will all stop air and water getting into contact with iron.

SECTION 4 ORGANIC CHEMISTRY

Fuels

Page 387

1. The supplies of petroleum are limited – it takes millions of years for crude oil to be formed.

2. Natural gas or methane. It is trapped in pockets above the oil.

3. Small chain of carbon atoms.

4. Long chain of carbon atoms.

5. These fractions readily form a vapour.

Page 389

1. Ethene is a member of the alkene homologous series.

2. The fractional distillation of crude oil produces a high proportion of long-chain hydrocarbons, which are not as useful as short-chain hydrocarbons. Cracking converts the long-chain hydrocarbons into more useful shorter chain hydrocarbons.

3. **EXTENDED** The conditions required for cracking oil fractions are a temperature of between 600 and 700 °C and a catalyst of silica or alumina.

Alkanes

Page 396–397

1. a) A compound that has no $C\!=\!C$ double bonds.

   b) A compound containing hydrogen atoms and carbon atoms only.

2. a) $C_{15}H_{32}$

   b) Carbon dioxide and water.

Page 397

1. Carbon dioxide and water are formed when propane burns completely.

2. To undergo complete combustion there must be a plentiful supply of oxygen (there must also be a source of ignition).

Alkenes

Page 402

1. It contains at least one $C\!=\!C$ double bond.

2. The manufacture of polymers (polyethene).

Answers: Physics

The answers given in this section have been written by the author and are not taken from examination mark schemes.

SECTION 1 MOTION

Motion

Pages 415

1. $10\,000/15 \times 60 = 10\,000/900 = 11.1\,\text{m/s}$.
2. $22.5\,\text{m}$
3. $3000\,\text{s} = 50\,\text{minutes}$

Page 416

1. The straight line will have a positive gradient for moving away, and a negative gradient for moving towards.
2. A graph will have axes which are labelled properly with a scale, plotted data points and the best line. A sketch graph is drawn to illustrate the main shape of the line. Only key values such as where the line changes direction, will be given on the axes.
3. Your graph is likely to be a curve, as shown in Fig. 1.8.

Page 417

1. **EXTENDED** $10\,\text{m/s}^2$
2. **EXTENDED** $-15\,\text{m/s}^2$

Page 418

1. a) Athlete: $8\,\text{m/s}$. Fun runner: $6.25\,\text{m/s}$.
   b) Athlete: horizontal line at $8\,\text{m/s}$, starting at $0\,\text{s}$ and finishing at $50\,\text{s}$. Fun runner: horizontal line at $6.25\,\text{m/s}$, starting at $0\,\text{s}$ and finishing at $64\,\text{s}$.

Page 419

1. **EXTENDED** The acceleration is the gradient of the graph.
2. **EXTENDED** The distance travelled is the area under the graph.
3. a) **EXTENDED** Athlete A: $8\,\text{m/s}$, athlete B: $6.25\,\text{m/s}$.
   b) **EXTENDED** Athlete A: horizontal line at $8\,\text{m/s}$, starting at $0\,\text{s}$ and finishing at $50\,\text{s}$. Athlete B: horizontal line at $6.25\,\text{m/s}$, starting at $0\,\text{s}$ and finishing at $64\,\text{s}$.
4. **EXTENDED** Area under speed-time graph is the same, i.e. $8\,\text{m/s} \times 50\,\text{s} = 6.25\,\text{m/s} \times 64\,\text{s}$.
5. **EXTENDED** $45\,\text{m}$

Mass and weight

Page 427

1. Mass is how much material is present in an object; weight is the force on the object due to gravity.
2. $600\,\text{N}$
3. $50\,\text{kg}$

Density

Page 432

1. a) Volume $= 2\,\text{cm} \times 4\,\text{cm} \times 5\,\text{cm} = 40\,\text{cm}^3$
   b) Density $=$ mass/volume $= 312\,\text{g}/40\,\text{cm}^3 = 7.8\,\text{g/cm}^3$
2. The bread contains more air spaces, making the overall density less.

Effects of forces

Page 438

1. a) A force can change the speed of an object, the shape of an object and the direction the object is moving in.
   b) Any three from: gravitational, electric, magnetic, (electromagnetic), strong nuclear force.
2. a) Gravity
   b) The mass of objects and the distance between their centres.
3. In a nucleus.
4. Electromagnetic

Page 445

1. **EXTENDED** $0.01\,\text{N}$
2. **EXTENDED** $0.2\,\text{N/cm}$ or $20\,\text{N/m}$
3. **EXTENDED** $20\,\text{cm}$

Page 447

1. No resultant force so stationary or constant speed.
2. Resultant force speed or direction of motion will change.
3. The gymnast's weight.

Page 448

1. Walking, driving – if there is no friction, you skid.
2. Where energy is transferred to thermal energy which is lost to the surroundings.

Pressure

Page 452

1. For the pin, the force is concentrated over a smaller area – there is a greater pressure.
2. Pressure = force/area = 100 N/0.2 m² = 500 Pa
3. 80 N
4. 1.28 m²

SECTION 2 WORK, ENERGY AND POWER

Work

Page 460

1. 250 J
2. 500 N
3. 80 J
4. 50 m
5. 400 N
6. $8.45 \times 10^{12}/8 \times 10^6 \,N = 1.06 \times 10^6 \,N$
7. a) The shuttle does a series of manoeuvres before landing to get rid of the excess energy.

   b) 1650 °C. The shuttle is covered with ceramic insulating materials designed to protect it from this heat. The materials include: reinforced carbon–carbon (RCC) on the wing surfaces and underside; high-temperature black surface insulation tiles on the upper forward fuselage and around the windows; white Nomex blankets on the upper payload bay doors, portions of the upper wing and mid/aft fuselage;
   low-temperature white surface tiles on the remaining areas.

Energy

Page 465

1. 100 J
2. 4 J

Page 467

1. Gravitational   kinetic → gravitational
2. Springs
3. Bonds between atoms

Page 468

1. Energy cannot be created or destroyed, only transferred from one form to another.
2. Electrical energy changes to light energy and heat.

3. a) Electrical

   b) Kinetic

   c) Sound, heat

Power

Page 475

1. The man has twice the weight, so the force is double, but the time is the same.
2. The machine transfers energy at a great rate.
3. The watt (W).
4. 240 W
5. a) 2940 J.

   b) 49 W.

Energy resources

Page 483

1. Light to electric.
2. Kinetic to electric.
3. Fuel is burned and steam is produced in a boiler. The steam turns a turbine. The turbine drives a generator. The generator produces electricity. The electricity is supplied to homes, industry, etc.
4. In certain parts of the world, water forms hot springs that can be used directly for heating. Water can also be pumped deep into the ground to be heated.

SECTION 3 THERMAL PHYSICS

Simple kinetic molecular model of matter

Page 491

1. It increases – either as faster and larger vibration in solids or as faster translational and vibrational motion in liquids and faster translational motion in gases.
2. Compressing a gas pushes the particles closer together; in a liquid they are already close to each other and will repel if pushed closer.
3. a) The particles are closely packed together in a regular arrangement in a solid.

   b) The particles are closely packed together in an irregular arrangement in a liquid.

   c) The particles are widely spaced in an irregular arrangement in a gas.
4. The volume of a gas depends on the shape of the container.

Page 492

1. The molecules are always moving about and spread out throughout the container.

2. Since volume is inversely proportional to pressure, increasing pressure means decreasing volume.

3. The mass of the gas must remain constant (that is, no particles move in or out of the system). The temperature must be measured using the Kelvin scale. The gas must be ideal (not liquefy or solidify).

Page 493

1. **EXTENDED** Higher temperature, increased flow of air across the surface, larger surface area.

2. **EXTENDED** Smaller surface area, so rate of cooling slower.

Matter and thermal properties

Page 500

1. Metals expand at different rates as their temperatures rise. So if strips of two metals are bound closely together, and are warmed, they bend as one metal expands more than the other.

2. The atoms vibrate more as the temperature goes up. So, even though they stay joined together, they move slightly further apart, and the solid expands a little in all directions.

3. Unlike other liquids, when its temperature falls below 4 °C, water begins to expand again, and becomes less dense.

4. First, we don't have to allow the gas to expand if it gets hotter; if we put it in a sealed container then we can just allow the pressure to increase instead. Second, if we do allow a gas to expand, then it will increase in volume much more than solids or liquids do as it gets hotter. Between 0 °C and 100 °C it will expand by a third, so 300 cm³ of gas will become 400 cm³.

Thermal processes

Page 506

1. They contain released electrons that can move freely and transfer energy.

2. There are no particles in outer space to transfer energy by colliding with each other.

3. Vibrations of the particles are passed on through the bonds between the particles.

4. Some wax is put on one end of a metal rod. The other end of the metal rod is heated until the wax on the other end melts.

5. **EXTENDED** Use petroleum jelly to attach drawing pins at regular distances along the copper strip. Heat one end of the strip and measure the time it takes for each drawing pin to fall off. Plot a graph of time until the drawing pin falls off against distance from the point that is being heated.

Page 507

1. The particles are free to move.

2. **EXTENDED** Warm air expands, which makes it less dense. Less dense air floats up above more dense (cooler) air.

3. Heat some potassium manganate(VII) crystals in water or show convection currents using smoke in air.

4. Fibres in the insulation create air pockets. This restricts the movement of the air and so convection currents cannot form.

Page 509

1. It can travel through a vacuum.

2. A hot object.

3. Two from: temperature, type of surface, area of surface.

4. **EXTENDED** The dull, black side.

5. **EXTENDED** B because it is at a higher temperature than A.

Page 511

1. Convection

2. The top of the room.

3. a) Vacuum has no particles to connect.

   b) Vacuum has no particles to form convection currents.

   c) Silver surfaces reflect the energy and do not absorb or emit it.

4. The flask reduces energy transfer in both directions, so hot drinks do not lose their energy and cold drinks are not warmed by energy entering from outside.

Page 514

1. More than half the energy wasted is through these two features.

2. Trapped air in the walls, roof, windows, etc.

3. Air is a bad conductor (good insulator) so keeping a layer near the body reduces heat loss by conduction. Keeping the layer trapped reduces heat loss by convection.

4. In hotter climates, you may want to lose heat, so loose clothing allows the air to move and transfer heat away from the body.

SECTION 4 PROPERTIES OF WAVES, INCLUDING LIGHT AND SOUND

General wave properties

Page 523

1. a) The waves travel by vibrations in the direction of travel of the wave.

b) The waves travel by vibrations at right angles to the direction of travel of the wave.

2. a) The distance between consecutive peaks or troughs of the wave.

b) The number of peaks/troughs that go past each second.

c) The size of the vibrations.

3. 15 m

4. Energy

Page 525

1. **EXTENDED** The density of each medium.

Light

Page 532

1. The light from the bottom of the pond is refracted away from the normal as it leaves the pond, so it enters your eye at a shallower angle. Your brain is fooled into thinking this shallower angle is the true angle of the bottom of the pond and so the pond looks shallower than it really is.

2. Being more dense than air, the carbon dioxide in the balloon will slow down the sound wave. This will tend to focus the sound together, like a lens. (A balloon filled with hydrogen will do the opposite.)

Electromagnetic spectrun

Page 540

1. 3×10^8 m/s

2. X-rays

3. Gamma, ultraviolet, visible, infrared, microwaves.

Page 543

1. Gamma rays, which as you see in Fig. 4.31, are the electromagnetic waves with the shortest wave-length, and therefore the greatest frequency and energy, are produced by radioactive nuclei; X-rays, which have the next highest energy radiation after gamma on the electromagnetic spectrum, are produced when high-energy electrons are fired at a metal target.

2. By exciting the atoms in a mercury vapour.

3. A photograph taken to show the infrared radiation given out from objects.

4. Water particles in food absorb the energy carried by microwaves. They vibrate more, making the food much hotter. Microwaves penetrate several centimetres into the food and so speed up the cooking process.

Sound

Page 550

1. Longitudinal

2. **EXTENDED** Small differences in air pressure.

3. 20–20 000 Hz

4. Sound with frequencies above the range for human hearing.

5. A sound source and two microphones are arranged in a straight line, with the sound source beyond the first microphone. The distance between the microphones (x), called microphone basis, is measured. The time of arrival between the signals (delay) reaching the different microphones (t) is measured. Then speed of sound = x/t.

SECTION 5 ELECTRICAL QUANTITIES

Electric charge

Page 560

1. You are the insulator. You will have rubbed electrons either onto or off yourself, perhaps by sliding your feet over a carpet. The metal handrail is a conductor – when you touch it, it allows the electrons to move to restore the balance and this is what you feel as a shock.

2. The key idea is friction – water droplets rubbing against each other, bits of dust, etc. Although water is a conductor, in a cloud it is isolated from the ground so the charge can build up.

Page 561

1. Electrons are extremely small and negatively charged. Electrons are also around the outside of atoms. These key ideas make it much easier for electrons to be moved about (than the positively charged parts of the atom) to account for all the electric effects we know.

2. The electron clouds in your feet repel the electron clouds in the floor. The force of repulsion is easily enough to stop you falling through.

3. a) Polythene rod should be negatively charged and cloth positively charged. Perspex rod should be positively charged and cloth negatively charged.

b) They will be equal and opposite.

Current, potential difference and electromotive force (e.m.f.)

Page 566

1. a) **EXTENDED** 15 C

b) **EXTENDED** 20 C

c) EXTENDED 92 C

d) EXTENDED 45 C

2. EXTENDED 0.5 A

3. EXTENDED 120 s

Resistance

Page 570

1. 10 V

2. 6 Ω

Page 575

1. The longer the wire, the further the electrons have to travel through the wire.

2. The thicker the wire, the more routes electrons have to travel through the wire.

3. **a)** EXTENDED 3R

   b) EXTENDED 1/3R, or R/3

SECTION 6 ELECTRIC CIRCUITS

Series and parallel circuits

Page 586

1. There is no current when the circuit is not complete.

2. In parallel each bulb takes the same energy from the battery, so the battery has to deliver twice as much energy as it would if the bulbs were in series, where its energy is shared.

3. Can be switched on and off independently.

4. **a)** In series dimmer because battery energy shared between two bulbs; **b)** in parallel full energy of battery given to each bulb.

Page 589

1. It is the same.

2. Each appliance can be designed to work with mains voltage supply; appliances can be switched on and off individually.

3. 122 Ω

4. EXTENDED 88.1 kΩ

Electrical energy

Page 593

1. EXTENDED 220 W

2. EXTENDED 12 V

3. EXTENDED 0.11 A

Page 594

1. EXTENDED No, because the voltage is not high enough from the mains.

2. EXTENDED 2160 J

3. EXTENDED 3.3 A

4. EXTENDED 218 s

5. EXTENDED 12 V

Dangers of electricity

Page 600

1. Brown wire: live, connected to fuse. Blue wire: neutral wire. Green and yellow wire: earth wire.

2. It melts if the current gets too high.

3. The student should not choose the 'nearest' fuse, but the 'nearest above'. If the appliance requires 6 A, the 5 A fuse will melt when the appliance is used.

4. The earth connection needs to be a low resistance path. This means that, in the event of a fault occurring, a high current will pass through the wire and this will melt the fuse (or trip the circuit breaker).

5. The casing cannot become live because it is not a conductor.

Index

Biology

Chemistry

Notes

Notes

Notes

Notes

Notes

Notes

Notes

Notes

Notes

Notes